T0281510

Lecture Notes in Computer Science   **14193**

Founding Editors

Gerhard Goos
Juris Hartmanis

The series Lecture Notes in Computer Science (LNCS), including its subseries Lecture Notes in Artificial Intelligence (LNAI) and Lecture Notes in Bioinformatics (LNBI), has established itself as a medium for the publication of new developments in computer science and information technology research, teaching, and education.

LNCS enjoys close cooperation with the computer science R & D community, the series counts many renowned academics among its volume editors and paper authors, and collaborates with prestigious societies. Its mission is to serve this international community by providing an invaluable service, mainly focused on the publication of conference and workshop proceedings and postproceedings. LNCS commenced publication in 1973.

Mickael Coustaty · Alicia Fornés
Editors

Document Analysis and Recognition – ICDAR 2023 Workshops

San José, CA, USA, August 24–26, 2023
Proceedings, Part I

 Springer

Editors
Mickael Coustaty
University of La Rochelle
La Rochelle, France

Alicia Fornés
Autonomous University of Barcelona
Bellaterra, Spain

ISSN 0302-9743   ISSN 1611-3349 (electronic)
Lecture Notes in Computer Science
ISBN 978-3-031-41497-8   ISBN 978-3-031-41498-5 (eBook)
https://doi.org/10.1007/978-3-031-41498-5

This Springer imprint is published by the registered company Springer Nature Switzerland AG
The registered company address is: Gewerbestrasse 11, 6330 Cham, Switzerland

Foreword from ICDAR 2023 General Chairs

We are delighted to welcome you to the proceedings of ICDAR 2023, the 17th IAPR International Conference on Document Analysis and Recognition, which was held in San Jose, in the heart of Silicon Valley in the United States. With the worst of the pandemic behind us, we hoped that ICDAR 2023 would be a fully in-person event. However, challenges such as difficulties in obtaining visas also necessitated the partial use of hybrid technologies for ICDAR 2023. The oral papers being presented remotely were synchronous to ensure that conference attendees interacted live with the presenters and the limited hybridization still resulted in an enjoyable conference with fruitful interactions.

ICDAR 2023 was the 17th edition of a longstanding conference series sponsored by the International Association of Pattern Recognition (IAPR). It is the premier international event for scientists and practitioners in document analysis and recognition. This field continues to play an important role in transitioning to digital documents. The IAPR-TC 10/11 technical committees endorse the conference. The very first ICDAR was held in St Malo, France in 1991, followed by Tsukuba, Japan (1993), Montreal, Canada (1995), Ulm, Germany (1997), Bangalore, India (1999), Seattle, USA (2001), Edinburgh, UK (2003), Seoul, South Korea (2005), Curitiba, Brazil (2007), Barcelona, Spain (2009), Beijing, China (2011), Washington, DC, USA (2013), Nancy, France (2015), Kyoto, Japan (2017), Sydney, Australia (2019) and Lausanne, Switzerland (2021).

Keeping with its tradition from past years, ICDAR 2023 featured a three-day main conference, including several competitions to challenge the field and a post-conference slate of workshops, tutorials, and a doctoral consortium. The conference was held at the San Jose Marriott on August 21–23, 2023, and the post-conference tracks at the Adobe World Headquarters in San Jose on August 24–26, 2023.

We thank our executive co-chairs, Venu Govindaraju and Tong Sun, for their support and valuable advice in organizing the conference. We are particularly grateful to Tong for her efforts in facilitating the organization of the post-conference in Adobe Headquarters and for Adobe's generous sponsorship.

The highlights of the conference include keynote talks by the recipient of the IAPR/ICDAR Outstanding Achievements Award, and distinguished speakers Marti Hearst, UC Berkeley School of Information; Vlad Morariu, Adobe Research; and Seiichi Uchida, Kyushu University, Japan.

A total of 316 papers were submitted to the main conference (plus 33 papers to the ICDAR-IJDAR journal track), with 53 papers accepted for oral presentation (plus 13 IJDAR track papers) and 101 for poster presentation. We would like to express our deepest gratitude to our Program Committee Chairs, featuring three distinguished researchers from academia, Gernot A. Fink, Koichi Kise, and Richard Zanibbi, and one from industry, Rajiv Jain, who did a phenomenal job in overseeing a comprehensive reviewing process and who worked tirelessly to put together a very thoughtful and interesting technical program for the main conference. We are also very grateful to the members of the Program Committee for their high-quality peer reviews. Thank you to our competition chairs, Kenny Davila, Chris Tensmeyer, and Dimosthenis Karatzas, for overseeing the competitions.

The post-conference featured 8 excellent workshops, four value-filled tutorials, and the doctoral consortium. We would like to thank Mickael Coustaty and Alicia Fornes, the workshop chairs, Elisa Barney-Smith and Laurence Likforman-Sulem, the tutorial chairs, and Jean-Christophe Burie and Andreas Fischer, the doctoral consortium chairs, for their efforts in putting together a wonderful post-conference program.

We would like to thank and acknowledge the hard work put in by our Publication Chairs, Anurag Bhardwaj and Utkarsh Porwal, who worked diligently to compile the camera-ready versions of all the papers and organize the conference proceedings with Springer. Many thanks are also due to our sponsorship, awards, industry, and publicity chairs for their support of the conference.

The organization of this conference was only possible with the tireless behind-the-scenes contributions of our webmaster and tech wizard, Edward Sobczak, and our secretariat, ably managed by Carol Doermann. We convey our heartfelt appreciation for their efforts.

Finally, we would like to thank for their support our many financial sponsors and the conference attendees and authors, for helping make this conference a success. We sincerely hope those who attended had an enjoyable conference, a wonderful stay in San Jose, and fruitful academic exchanges with colleagues.

David Doermann
Srirangaraj (Ranga) Setlur

Foreword from ICDAR 2023 Workshop Chairs

Our heartiest welcome to the proceedings of the ICDAR 2023 Workshops, which were organized as part of the 17th International Conference on Document Analysis and Recognition (ICDAR) held in San José, California, USA on August 21–26, 2023. The workshops were held after the main conference, from August 24–26, 2023, together with the Tutorials and the Doctoral Consortium. All the workshops were in-person events, as was the main conference.

The ICDAR conference included 8 workshops, covering diverse document image analysis and recognition topics, and also complementary topics such as Natural Language Processing, Computational Paleography and Digital Humanities. Overall the workshops received 60 papers and 43 of them were accepted (global acceptance rate 71.6%).

This volume collects the edited papers from 6 of these 8 workshops. We sincerely thank the ICDAR general chairs for trusting us with the responsibility for the workshops, and the publication chairs for assisting us with the publication of this volume. We also thank all the workshop organizers for their involvement in this event of primary importance in our field. Finally, we thank all the workshop presenters and authors.

August 2023

Mickael Coustaty
Alicia Fornés

Organization

General Chairs

David Doermann   University at Buffalo, The State University of New York, USA

Srirangaraj Setlur   University at Buffalo, The State University of New York, USA

Executive Co-chairs

Venu Govindaraju   University at Buffalo, The State University of New York, USA

Tong Sun   Adobe Research, USA

PC Chairs

Gernot A. Fink   Technische Universität Dortmund, Germany (Europe)

Rajiv Jain   Adobe Research, USA (Industry)

Koichi Kise   Osaka Metropolitan University, Japan (Asia)

Richard Zanibbi   Rochester Institute of Technology, USA (Americas)

Workshop Chairs

Mickael Coustaty   La Rochelle University, France

Alicia Fornes   Universitat Autònoma de Barcelona, Spain

Tutorial Chairs

Elisa Barney-Smith   Luleå University of Technology, Sweden

Laurence Likforman-Sulem   Télécom ParisTech, France

Competitions Chairs

Kenny Davila	Universidad Tecnológica Centroamericana, UNITEC, Honduras
Dimosthenis Karatzas	Universitat Autònoma de Barcelona, Spain
Chris Tensmeyer	Adobe Research, USA

Doctoral Consortium Chairs

Andreas Fischer	University of Applied Sciences and Arts Western Switzerland
Veronica Romero	University of Valencia, Spain

Publications Chairs

Anurag Bharadwaj	Northeastern University, USA
Utkarsh Porwal	Walmart, USA

Posters/Demo Chair

Palaiahnakote Shivakumara	University of Malaya, Malaysia

Awards Chair

Santanu Chaudhury	IIT Jodhpur, India

Sponsorship Chairs

Wael Abd-Almageed	Information Sciences Institute USC, USA
Cheng-Lin Liu	Chinese Academy of Sciences, China
Masaki Nakagawa	Tokyo University of Agriculture and Technology, Japan

Industry Chairs

Andreas Dengel	DFKI, Germany
Véronique Eglin	Institut National des Sciences Appliquées (INSA) de Lyon, France
Nandakishore Kambhatla	Adobe Research, India

Publicity Chairs

Sukalpa Chanda	Østfold University College, Norway
Simone Marinai	University of Florence, Italy
Safwan Wshah	University of Vermont, USA

Technical Chair

Edward Sobczak	University at Buffalo, The State University of New York, USA

Conference Secretariat

University at Buffalo, The State University of New York, USA

Program Committee

Senior Program Committee Members

Srirangaraj Setlur	Apostolos Antonacopoulos
Richard Zanibbi	Lianwen Jin
Koichi Kise	Nicholas Howe
Gernot Fink	Marc-Peter Schambach
David Doermann	Marcal Rossinyol
Rajiv Jain	Wataru Ohyama
Rolf Ingold	Nicole Vincent
Andreas Fischer	Faisal Shafait
Marcus Liwicki	Simone Marinai
Seiichi Uchida	Bertrand Couasnon
Daniel Lopresti	Masaki Nakagawa
Josep Llados	Anurag Bhardwaj
Elisa Barney Smith	Dimosthenis Karatzas
Umapada Pal	Masakazu Iwamura
Alicia Fornes	Tong Sun
Jean-Marc Ogier	Laurence Likforman-Sulem
C. V. Jawahar	Michael Blumenstein
Xiang Bai	Cheng-Lin Liu
Liangrui Peng	Luiz Oliveira
Jean-Christophe Burie	Robert Sabourin
Andreas Dengel	R. Manmatha
Robert Sablatnig	Angelo Marcelli
Basilis Gatos	Utkarsh Porwal

Program Committee Members

Harold Mouchere
Foteini Simistira Liwicki
Vernonique Eglin
Aurelie Lemaitre
Qiu-Feng Wang
Jorge Calvo-Zaragoza
Yuchen Zheng
Guangwei Zhang
Xu-Cheng Yin
Kengo Terasawa
Yasuhisa Fujii
Yu Zhou
Irina Rabaev
Anna Zhu
Soo-Hyung Kim
Liangcai Gao
Anders Hast
Minghui Liao
Guoqiang Zhong
Carlos Mello
Thierry Paquet
Mingkun Yang
Laurent Heutte
Antoine Doucet
Jean Hennebert
Cristina Carmona-Duarte
Fei Yin
Yue Lu
Maroua Mehri
Ryohei Tanaka
Adel M. M. Alimi
Heng Zhang
Gurpreet Lehal
Ergina Kavallieratou
Petra Gomez-Kramer
Anh Le Duc
Frederic Rayar
Muhammad Imran Malik
Vincent Christlein
Khurram Khurshid
Bart Lamiroy
Ernest Valveny
Antonio Parziale

Jean-Yves Ramel
Haikal El Abed
Alireza Alaei
Xiaoqing Lu
Sheng He
Abdel Belaid
Joan Puigcerver
Zhouhui Lian
Francesco Fontanella
Daniel Stoekl Ben Ezra
Byron Bezerra
Szilard Vajda
Irfan Ahmad
Imran Siddiqi
Nina S. T. Hirata
Momina Moetesum
Vassilis Katsouros
Fadoua Drira
Ekta Vats
Ruben Tolosana
Steven Simske
Christophe Rigaud
Claudio De Stefano
Henry A. Rowley
Pramod Kompalli
Siyang Qin
Alejandro Toselli
Slim Kanoun
Rafael Lins
Shinichiro Omachi
Kenny Davila
Qiang Huo
Da-Han Wang
Hung Tuan Nguyen
Ujjwal Bhattacharya
Jin Chen
Cuong Tuan Nguyen
Ruben Vera-Rodriguez
Yousri Kessentini
Salvatore Tabbone
Suresh Sundaram
Tonghua Su
Sukalpa Chanda

Mickael Coustaty
Donato Impedovo
Alceu Britto
Bidyut B. Chaudhuri
Swapan Kr. Parui
Eduardo Vellasques
Sounak Dey
Sheraz Ahmed
Julian Fierrez
Ioannis Pratikakis
Mehdi Hamdani
Florence Cloppet
Amina Serir
Mauricio Villegas
Joan Andreu Sanchez
Eric Anquetil
Majid Ziaratban
Baihua Xiao
Christopher Kermorvant
K. C. Santosh
Tomo Miyazaki
Florian Kleber
Carlos David Martinez Hinarejos
Muhammad Muzzamil Luqman
Badarinath T.
Christopher Tensmeyer
Musab Al-Ghadi
Ehtesham Hassan
Journet Nicholas
Romain Giot
Jonathan Fabrizio
Sriganesh Madhvanath
Volkmar Frinken
Akio Fujiyoshi
Srikar Appalaraju
Oriol Ramos-Terrades
Christian Viard-Gaudin
Chawki Djeddi
Nibal Nayef
Nam Ik Cho
Nicolas Sidere
Mohamed Cheriet
Mark Clement
Shivakumara Palaiahnakote
Shangxuan Tian

Ravi Kiran Sarvadevabhatla
Gaurav Harit
Iuliia Tkachenko
Christian Clausner
Vernonica Romero
Mathias Seuret
Vincent Poulain D'Andecy
Joseph Chazalon
Kaspar Riesen
Lambert Schomaker
Mounim El Yacoubi
Berrin Yanikoglu
Lluis Gomez
Brian Kenji Iwana
Ehsanollah Kabir
Najoua Essoukri Ben Amara
Volker Sorge
Clemens Neudecker
Praveen Krishnan
Abhisek Dey
Xiao Tu
Mohammad Tanvir Parvez
Sukhdeep Singh
Munish Kumar
Qi Zeng
Puneet Mathur
Clement Chatelain
Jihad El-Sana
Ayush Kumar Shah
Peter Staar
Stephen Rawls
David Etter
Ying Sheng
Jiuxiang Gu
Thomas Breuel
Antonio Jimeno
Karim Kalti
Enrique Vidal
Kazem Taghva
Evangelos Milios
Kaizhu Huang
Pierre Heroux
Guoxin Wang
Sandeep Tata
Youssouf Chherawala

Reeve Ingle
Aashi Jain
Carlos M. Travieso-Gonzales
Lesly Miculicich
Curtis Wigington
Andrea Gemelli
Martin Schall
Yanming Zhang
Dezhi Peng
Chongyu Liu
Huy Quang Ung
Marco Peer
Nam Tuan Ly
Jobin K. V.
Rina Buoy
Xiao-Hui Li
Maham Jahangir
Muhammad Naseer Bajwa

Oliver Tueselmann
Yang Xue
Kai Brandenbusch
Ajoy Mondal
Daichi Haraguchi
Junaid Younas
Ruddy Theodose
Rohit Saluja
Beat Wolf
Jean-Luc Bloechle
Anna Scius-Bertrand
Claudiu Musat
Linda Studer
Andrii Maksai
Oussama Zayene
Lars Voegtlin
Michael Jungo

Program Committee Sub Reviewers

Li Mingfeng
Houcemeddine Filali
Kai Hu
Yejing Xie
Tushar Karayil
Xu Chen
Benjamin Deguerre
Andrey Guzhov
Estanislau Lima
Hossein Naftchi
Giorgos Sfikas
Chandranath Adak
Yakn Li
Solenn Tual
Kai Labusch
Ahmed Cheikh Rouhou
Lingxiao Fei
Yunxue Shao
Yi Sun
Stephane Bres
Mohamed Mhiri
Zhengmi Tang
Fuxiang Yang
Saifullah Saifullah

Paolo Giglio
Wang Jiawei
Maksym Taranukhin
Menghan Wang
Nancy Girdhar
Xudong Xie
Ray Ding
Mélodie Boillet
Nabeel Khalid
Yan Shu
Moises Diaz
Biyi Fang
Adolfo Santoro
Glen Pouliquen
Ahmed Hamdi
Florian Kordon
Yan Zhang
Gerasimos Matidis
Khadiravana Belagavi
Xingbiao Zhao
Xiaotong Ji
Yan Zheng
M. Balakrishnan
Florian Kowarsch

Mohamed Ali Souibgui
Xuewen Wang
Djedjiga Belhadj
Omar Krichen
Agostino Accardo
Erika Griechisch
Vincenzo Gattulli
Thibault Lelore
Zacarias Curi
Xiaomeng Yang
Mariano Maisonnave
Xiaobo Jin
Corina Masanti
Panagiotis Kaddas
Karl Löwenmark
Jiahao Lv
Narayanan C. Krishnan
Simon Corbillé
Benjamin Fankhauser
Tiziana D'Alessandro
Francisco J. Castellanos
Souhail Bakkali
Caio Dias
Giuseppe De Gregorio
Hugo Romat
Alessandra Scotto di Freca
Christophe Gisler
Nicole Dalia Cilia
Aurélie Joseph
Gangyan Zeng
Elmokhtar Mohamed Moussa
Zhong Zhuoyao
Oluwatosin Adewumi
Sima Rezaei
Anuj Rai
Aristides Milios
Shreeganesh Ramanan
Wenbo Hu
Arthur Flor de Sousa Neto
Rayson Laroca

Sourour Ammar
Wenbo Hu
Gianfranco Semeraro
Andre Hochuli
Saddok Kebairi
Shoma Iwai
Cleber Zanchettin
Ansgar Bernardi
Vivek Venugopal
Abderrhamne Rahiche
Wenwen Yu
Abhishek Baghel
Mathias Fuchs
Yael Iseli
Xiaowei Zhou
Yuan Panli
Minghui Xia
Zening Lin
Konstantinos Palaiologos
Loann Giovannangeli
Yuanyuan Ren
Shreeganesh Ramanan
Shubhang Desai
Yann Soullard
Ling Fu
Juan Antonio Ramirez-Orta
Chixiang Ma
Truong Thanh-Nghia
Nathalie Girard
Kalyan Ram Ayyalasomayajula
Talles Viana
Francesco Castro
Anthony Gillioz
Yunxue Shao
Huawen Shen
Mathias Fuchs
Sanket Biswas
Haisong Ding
Solène Tarride

Contents – Part I

CBDAR

IWCP

Contents – Part II

WML

ADAPDA

ADAPDA 2023 Preface

Intelligent Document Analysis (DA) technologies are becoming increasingly pervasive in our daily life due to the digitalization of documents (both in the cultural and industrial domains) and the widespread use of paper tablets, pads, and smartphones to take notes and sign documents. In this respect, high-performing DA algorithms are needed that can deal with digitalized documents from different writers, in different languages (including ancient languages, modern slang terms, or writer-preferred abbreviations and symbols), and with different visual characteristics (due to the paper support and the writing tool), often very peculiar to the application domain. In this respect, domain adaptation and automatic personalization strategies are worth investigating to boost the performance of DA techniques in the scenarios mentioned above, which are of great cultural, practical, and economic interest.

The Automatically Domain-Adapted and Personalized Document Analysis (ADAPDA) workshop aims at gathering expertise and novel ideas for personalized DA tasks. These include but are not limited to: training and adaptation strategies of writer, language, and visual-specific models, new benchmarks, and data collection strategies to explore DA tasks in a personalized setting, as well as related works on the personalized DA topic.

ADAPDA is intended to be an occasion for the whole DA community to start the conversation on these aspects that open a new set of technical and scientific challenges but would also offer practical solutions to meet the current need for high-performing DA technologies in modern society where digitalization is pervasive.

To this end, the authors have been encouraged to submit either short or long papers presenting ongoing projects, datasets, final or preliminary results, as well as innovative methodologies and tools. From six submissions, we selected four long papers from authors in four different countries, both working in academia and industry. Each paper received two reviews, timely provided by members of the Program Committee, which we take the chance to thank for their dedication.

The workshop consisted of a positional talk in which the organizers presented their views on the topics of the workshop, an oral presentation of each accepted paper, and a discussion session based on these talks and the issues that they raised.

The participants were also asked for topics for the next workshop on the same subject, with the aim of building a fruitful conversation that we hope will continue in following ADAPDA editions, with which we plan to have an annual focus on the problems of adaptations and system evolutivity, alongside the large-scale systems that are going to become the trend.

June 2023

Rita Cucchiara
Eric Aquetil
Christopher Kermorvant
Silvia Cascianelli

Beyond Human Forgeries: An Investigation into Detecting Diffusion-Generated Handwriting

Guillaume Carrière[1,2]([✉]) [ID], Konstantina Nikolaidou[3] [ID], Florian Kordon[1] [ID], Martin Mayr[1] [ID], Mathias Seuret[1] [ID], and Vincent Christlein[1] [ID]

[1] Friedrich-Alexander-Universität Erlangen-Nürnberg, 91058 Erlangen, Germany
{guillaume.carriere,florian.kordon,martin.mayr,
mathias.seuret,vincent.christlein}@fau.de
[2] École Pour l'Informatique et les Techniques Avancées, Le Kremlin-Bicêtre, France
guillaume.carriere@epita.fr
[3] Luleå University of Technology, Luleå, Sweden
konstantina.nikolaidou@ltu.se

Abstract. Methods for detecting forged handwriting are usually based on the assumption that the forged handwriting is produced by humans. Authentic-looking handwriting, however, can also be produced synthetically. Diffusion-based generative models have recently gained popularity as they produce striking natural images and are also able to realistically mimic a person's handwriting. It is, therefore, reasonable to assume that these models will be used to forge handwriting in the near future, adding a new layer to handwriting forgery detection. We show for the first time that the identification of synthetic handwritten data is possible by a small Convolutional Neural Network (ResNet18) reaching accuracies of 90%. We further investigate the existence of distinct discriminative features in synthetic handwriting data produced by latent diffusion models that could be exploited to build stronger detection methods. Our experiments indicate that the strongest discriminative features do not come from generation artifacts, letter shapes, or the generative model's architecture, but instead originate from real-world artifacts in genuine handwriting that are not reproduced by generative methods.

Keywords: Forensics Analysis · Diffusion Models · Handwriting Generation · Synthetic Image Generation · Forgery Detection · Biometrics

1 Introduction

The analysis of handwriting is receiving attention due to its application in biometrics, where it can be used for identifying persons or detecting fraudulent documents. Especially in forensics, building strong systems capable of detecting forged handwriting is essential. Various methods have been developed in the

M. Coustaty and A. Fornés (Eds.): ICDAR 2023 Workshops, LNCS 14193, pp. 5–19, 2023.
https://doi.org/10.1007/978-3-031-41498-5_1

past for this purpose, varying from exploiting clear traces such as wrinkling of the ink [3] to more subtle spectral analysis [22] or the use of Gaussian mixture models [13]. More recently, deep learning emerged as an additional powerful candidate, shown to be effective on signatures [16] and general handwriting [27].

At the same time, the production of synthetic images using deep generative models has gained much popularity due to the impressive results brought by Generative Adversarial Networks (GAN) models and especially the more recent diffusion models [12]. The images obtained with these methods can achieve unparalleled realism and can be immensely useful to the general public, e.g., in the combination with text prompts using Latent Diffusion Models (LDMs) [33]. Recently, Nikolaidou *et al.* [28] showed that LDMs can be used to produce authentic-looking handwriting being indistinguishable to the naked eye to genuine handwriting. This poses a new threat to forgery detection, which focuses on handcrafted forgeries. To the best of our knowledge, there is no method that detects synthetic handwriting forgeries. Consequently, there might soon appear a need for strong methods able to detect forged handwriting produced by generative models.

This work investigates the characteristics of synthetic handwriting produced by generative models, specifically the most recent LDM [28]. We believe this type of data may possess inherent traces compared to the more generic images produced, for example, by text-to-image models trained on large datasets. Identifying these traces may prove essential for building strong forgery detection systems in the future.

We summarize our main contributions as follows:

- We show that synthetic handwriting produced by LDMs can be identified reliably with a small ResNet18 [18].
- We study our trained model to determine the most discriminative features of synthetic handwriting. In doing so, we show that these features do not depend on generation artifacts, letter shapes, or the actual generative model architecture used to produce the data.
- Instead, we observe strong evidence of artifacts in genuine handwriting that stem from real-world causes, which in turn get smoothened or even deleted by generative models.

This paper is organized as follows. First, we will briefly review related work in Sect. 2. This is followed by describing the studied synthetic data and how it was generated (Sect. 3). In Sect. 4, we present our experiments using this data and discuss their results before we draw conclusions in Sect. 5.

2   Related Work

The detection of synthetic data produced by LDMs is a relatively recent topic, and to the best of our knowledge, we are the first to tackle the identification of synthetic handwritten data. This section will briefly present these new generative models, as well as previous work on identifying discriminative traces on generated images.

2.1 Latent Diffusion Models (LDMs)

Denoising Diffusion Probabilistic Models (DDPMs) [19] are a class of generative models that include two processes: the forward diffusion and the backward denoising process. In the forward diffusion process, noise is injected into samples from a real data distribution across multiple time steps. In the backward denoising process, a neural network architecture is utilized to predict the injected noise and reconstruct the original input by gradually removing the predicted noise in backward steps. DDPMs have demonstrated impressive results in generating high-resolution samples and are most often used for text-to-image synthesis as seen in [32,35]. Their ability to create realistic images by conditioning on a text description is further facilitated by leveraging contrastive pre-training of image-caption pairs [31]. The conditions are typically inserted into the model using the cross-attention mechanism [36]. However, the use of this mechanism on whole high-resolution images, as well as the iterative denoising inherent to the diffusion process, both increase the sampling complexity. LDMs [33] aim to tackle this issue by projecting images in a lower dimensional space. This allows the generation of high-dimensional images with a lower computational cost than standard DDPMs.

2.2 Handwriting Generation

Several methods have been proposed in the context of handwriting generation. In GANwriting [21], the authors present a GAN-based method which uses conditions on both the style and the text to produce realistic word images. In concurrent works, whole text-lines are generated using the temporal component in the handwriting generation process [26] or space predictors and encoder networks [11]. GANwriting was later also extended to work on whole text lines [20]. The presence of small scale artifacts in word images generated by GANwriting is tackled in SmartPatch [25] by considering not only the whole image, but also smaller patches as individual entities. Other approaches experiment with transformer based architectures [2,30], or improve visual quality by focusing on the disentanglement of calligraphic styles and textual contents [14,15]. An LDM-based method has been recently introduced by Nikolaidou *et al.* [28] that also conditions on style and text similar to the other methods. They show that this approach is especially good in preserving the writer style so that typical writer identification methods [5–7] can easily be fooled. We use this method in our work to generate synthetic handwriting samples, further details on the data synthesis are provided in Sect. 3.

2.3 Artificial Traces

Generative models can leave specific signatures and anomalies in images that can be detected to reveal their origin. This has been shown on GAN architectures [4, 17,23,39], and has been used to build detectors that can identify synthetic images and even trace back the original architecture that was used to infer them [1,38].

More recent work shows evidence that DDPMs are affected as well [8,9,37]. Anomalies can often be observed as differences between genuine and synthetic data in the frequency domain or spectral distribution. These anomalies arise during the processing steps of the generative model, depending on its architecture and optimization policy. In the case of DDPMs, the artifacts typically appear when studying the power spectra and the autocorrelation of residual noise [8].

3   Handwriting Generation with LDMs

Samples produced by a diffusion-based method serve as an ideal candidate for forensics analysis, as they offer the possibility to study the characteristics that may be displayed by synthetic handwriting. Finding discriminative features in this type of artificial data is highly important, as diffusion models could be used in potential risk cases, such as handwriting forgery. To this end, we generate synthetic handwritten images using WordStylist [28]. WordStylist is an LDM inspired by [33] for styled handwritten image generation at the word-level. The main particularity of this model is its ability to reproduce characteristics of seen writing styles by conditioning the generation on a specific style and word prompt in a learned set.

To generate the synthetic samples, we use the same model structure and experimental setup as the original work [28]. The denoising network is based on the U-Net architecture [34] with reduced residual blocks [18] and intermediate multi-headed attention Transformer blocks [36] of dimension 320 and 4 heads. To encode and decode the images from pixel to latent space and vice versa, the same pre-trained Variational Autoencoder as in [28] is used. Regarding noise scheduling, noise is inserted linearly across $T = 1000$ timesteps from $\beta_1 = 10^{-4}$ to $\beta_T = 0.02$. To train WordStylist, a subset of IAM offline handwriting database at word-level [24] is used, similar to [21,25,28]. This training set includes images of 339 different writing styles and words that contain 2–7 characters. The images are resized to a fixed height of 64 pixels and a width of 256 pixels if the original width is larger than 256 pixels, or center-padded if the width is smaller than 256 pixels. Regarding training parameters, the AdamW optimizer with a learning rate of 10^{-4} is used with a batch size of 224 for 1K epochs.

The data produced by WordStylist can be divided into the following three categories: in-vocabulary (IV), "partial" in-vocabulary (pIV), and out-of-vocabulary (OOV). IV data corresponds to data produced by using pair-wise combinations of word and writer styles that appeared during the training process. "Partial" IV corresponds to data produced using an unseen combination of IV word and writing style, i.e., the word was seen during training but not with the specific writer style. Finally, OOV corresponds to data that was produced with OOV words. We choose to perform most experiments using only IV data, as it is the most challenging of the 3 categories to discriminate visually. As we are using the same combination of words and styles as in the genuine data, it also presents the advantage of having a perfect 50/50 ratio between genuine and synthetic data. We use the least possible amount of post-processing both on the

(a) Original   (b) Single-sided binarization

Fig. 1. (a) Example from the IAM dataset showing the clear difference in the background color between the bounding boxes and the rest of the image, (b) reducing the IAM artifacts through single-sided binarization.

genuine data that is extracted directly from the pre-processed training data of WordStylist, and the synthetic data produced by WordStylist.

4 Experiments

In this section, we describe our strategy to investigate the unique characteristics of handwriting data generated using WordStylist. We start by training a model to classify genuine and synthetic data accurately. Then, we analyze the classifier's features and evaluate whether they confirm our hypotheses and can serve as a discriminative criterion to distinguish between genuine and DDPM-based forgery. This is done through a combination of using additional data, augmentation techniques and explainability methods.

First, it should be noted that a minimum amount of pre-processing must be performed on our data due to the characteristics inherent to the IAM dataset. For single words, the bounding boxes of the connected components are separated from the rest of the image with a white background outside these boxes. This characteristic acts as an unnatural signature that is not reproduced perfectly by generative models, leading to a very easy classification if left unchecked. A visual example showing this signature can be seen in Fig. 1.

It is necessary to erase this bias in order to perform a meaningful analysis of the differences between synthetic and genuine data.

Therefore, when training our baseline, we applied "single-sided" binarization to the data. First, a binary mask of the image is computed using Otsu thresholding [29], separating the relevant text and the background area. Then, each image pixel corresponding to the mask's background is assigned a full-white color, whereas pixels in the foreground area retain their original values. A visual examples of this "single-sided" binarization is shown on Fig. 1.

The classifier model we use to create our baseline is a ResNet18 pre-trained on ImageNet. The model is fine-tuned on the aforementioned pre-processed data (totaling 88824 images) using an 80/20 train/test split where the training set is further split into a training and validation split (80/20). We train for 25 epochs using Adam optimizer, a learning rate of 10^{-6}, and a batch size of 4.

First, we will evaluate a baseline. In the subsequent experiments, we will investigate the following points:

Table 1. Baseline results of a ResNet18 model. R: recall, P: precision, NPV: negative predictive value, SP: specificity, ACC: accuracy. All results in percent.

	Predicted		
	Synthetic	Genuine	
Synthetic	45.15	4.70	R: 90.56
Genuine	5.72	44.42	SP: 88.58
	P: 88.75	NPV: 90.42	ACC: 89.57

- Measuring the importance of LDM artifacts for the baseline.
- Evaluating how much the baseline is agnostic in relation to the generative model.
- Evaluating the ability for the baseline to adapt to new unseen writer styles.
- Investigate the differences between OOV, IV and partially IV word images.

4.1   Baseline Results

This baseline reaches a fairly high accuracy of 89.57% in distinguishing forged from genuine samples and subsequently serves as the basis for our experiments. The full confusion matrix with additional numbers can be seen in Table 1.

4.2   Generation Artifacts

LDMs used in WordStylist [28] often leave artifacts in the Fourier domain and auto-correlation of residual noise. This means these fingerprints have a high chance of being present in our synthetic data. Although exploiting these artifacts can be an effective way of detecting synthetic images, they are not unique to the context of handwriting. Our objective is to discover characteristics inherent to synthetic handwritten words that could be valuable in the context of forensics analysis. Thus, we need to evaluate the importance of these generic artifacts with regards to our baseline's predictions.

These artifacts can also be generated by other means, such as interpolation resizing or JPEG compression. In fact, artifacts generated by JPEG compression can be so strong that they can hide or even completely destroy artifacts produced by generative models [8]. As a consequence, it is possible to remove diffusion artifacts from generated images by applying JPEG compression, even when using a high-quality compression factor. By pre-processing our data with JPEG compression before evaluation, we can check if the baseline model uses diffusion artifacts in its decision process. Visual examples of the effects of JPEG compression are shown in Fig. 2. This also allows us to show the performance of our classifier in a real-case scenario, as it is unlikely that fake data would be encountered without any other post-processing.

Results depicted in Fig. 3 show that the accuracy of the baseline stays relatively high as the JPEG compression quality is decreased. This implies that

(a) Original   (b) JPEG compression

Fig. 2. Visual examples of a word without/with JPEG compression with a quality factor of 20.

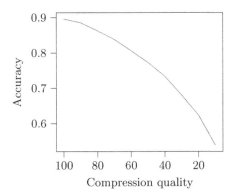

Fig. 3. Test accuracy depending on the quality factor of JPEG compression applied.

the baseline either does not exploit the diffusion artifacts, or at least not only them, as even a high JPEG quality factor such as 90 is already strong enough to hide most potential diffusion artifacts. The drop in accuracy when using lower quality factors is most likely due to the fact that JPEG compression is a destructive process, leading to some other useful features of the data potentially being erased. This experiment shows that there are indeed other characteristics that are easier to exploit than generic diffusion artifacts that can tell apart synthetic from genuine images.

4.3   Model-Agnostic Classifier

Through analysis of artifacts left by generative models, it is often possible to detect not only synthetic images but also the architecture of the original model that produced the data [10]. This implies that a classifier model using these artifacts as the basis of its decisions may only be able to detect synthetic data from one specific generative model architecture. More generally, even without exploiting the artifacts in the Fourier domain and auto-correlation that were investigated in the previous experiment, it is possible that a classifier model trained on data from only one generative model might have low performance when evaluating with data from another generative model.

Following this observation, we evaluate our baseline on synthetic data from different generative models. As for the previous experiment, this is also closer to

Table 2. Baseline accuracy on synthetic data only, using data from different generative models.

Generative Model	Accuracy [%]
WordStylist	90.56
GANwriting	78.02
Smartpatch	83.58
WordStylist (trained w. different seed)	90.62

a real-case scenario, where there would be no guarantee on the generative model used to produce fake data.

Two aspects were investigated in this experiment. First, the ability of our classifier model to be agnostic about the generative model's architecture, which we have evaluated using synthetic data from two different GAN models: GANwriting [21] and the improved version Smartpatch [25]. Second, the ability of our classifier model to be agnostic about the generative model's weights, evaluated by using data from another instance of WordStylist trained with a different seed in order to obtain different weights.

The results in Table 2 reveal a drop in performance when using data from a different architecture but show no difference when using the same architecture with different weights. Synthetic data from other architectures probably contain different traces, resulting in lower classification performance, but the classifier still keeps a high accuracy. This means that the classifier exploits traces not only from the synthetic data but also the genuine data to discriminate the images. It is also possible that telltale traces are shared by both diffusion models and GANs, but it is unlikely, considering the differences in their processing steps. This experiment also shows that these traces most likely depend on the design of the architecture as opposed to the actual weights of the generative model.

4.4   Letter Shape

In the following experiments, we further investigate whether meaningful and discriminatory features can be derived from the shape of the letters. Potential useful characteristics include the thickness and direction of strokes or the size of individual letter parts or the letter overall.

Writer Styles. One of the main particularities of the WordStylist model is its ability to condition the generation of images with a given author style. A qualitative evaluation from the original authors using writer identification models has shown that WordStylist is able to capture the style of specific writers accurately [28]. To complement this evaluation and further test our classifier in a real-case scenario, we evaluate the ability of our baseline to adapt to unseen writer styles. This is done by training another classifier, using the same architecture and pre-processing as our baseline, but splitting our data into training

Fig. 4. Visual examples of OOV/pIV images presenting clear inconsistencies.

and test sets by writer style. Doing this will give us an idea of the adaptability of our baseline on unseen writer styles.

The writer styles for these sets have been chosen randomly and made to match as closely as possible a ratio of 80/20 when splitting the corresponding images.

This new classifier achieves an accuracy of 88.73%, which is very close to the baseline's accuracy of 89.57%. These results suggest that little to no information inherent to the writer style can be exploited to discriminate between synthetic and genuine data. An alternative reason is that this information is retained for all writer styles during the image generation process.

Out-of-Vocabulary (OOV). The OOV and pIV categories in the data produced by WordStylist evidence some clear differences compared to the IV category. Generated images from these categories sometimes show inconsistencies in the shape of letters, resulting in unrealistic letter shapes or non-existent words that can be identified with the naked eye. This is most likely due to the fact that the combination of words and writer styles in these categories are not present in the genuine dataset. These inconsistencies are especially strong when sampling using a writer style with few corresponding images in the training dataset. Examples of these inconsistencies can be seen in Fig. 4.

We further investigate the presence of discriminative features originating from the shape of letters. For that purpose, we evaluate the baseline model where the IV synthetic data is replaced with data from the pIV and OOV categories. To thoroughly check if the aforementioned inconsistencies can aid the detection of synthetic images, we also train a secondary classifier on these categories using the same architecture and pre-processing as our baseline. It is important to mention that the amount of data from the OOV and pIV categories is lower than the IV category, leading to unbalanced training data.

The baseline achieves an accuracy of 89.40%, which is very similar to its accuracy of 89.57% on IV data. This indicates that the model does not benefit from the inconsistencies witnessed on OOV and pIV data, confirming that the discriminative features do not rely on letter shape. However, the results also show that features used by the baseline are still present on OOV and pIV data, as there is no degradation in performance either.

The classifier trained on both OOV and pIV splits achieves an accuracy of 89.49%, matching the baseline's accuracy. We argue that this is even stronger evidence that the letter shape does not allow for a meaningful feature to detect

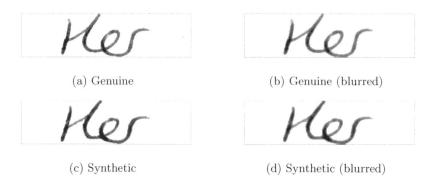

(a) Genuine

(b) Genuine (blurred)

(c) Synthetic

(d) Synthetic (blurred)

Fig. 5. Visual examples of a word with/without Gaussian blurring with a sigma value of 1.

synthetic images, even if the training set presents word-style pairings that are absent in the genuine data.

4.5 Effects of Gaussian Blurring

In addition to our initial experiments, we have also experimented with various transformations of the data. One of the transformations that strongly impacts the performance of our classifier is Gaussian blurring. Visual examples of the images used to evaluate this impact are shown in Fig. 5. Examples given in Fig. 6 show the effect on the baseline's accuracy depending on the intensity of the blurring. Note that these results were obtained by blurring before the "single-sided" binarization in the pre-processing steps. We can observe that Gaussian blurring results in a rapid performance drop once a certain intensity threshold is reached.

Following these observations, we trained another classifier with the same architecture as the baseline where we augmented the input images with a Gaussian blur with kernel size $k = 11$ and standard deviation $\sigma = 0.538$. The standard deviation was chosen to match the mean of the accuracies obtained when evaluating our baseline on the test set with various blurring intensities. The results of this new classifier model are also shown in Fig. 6.

All results indicate that applying Gaussian blurring erases discriminative features of genuine data or reproduces discriminative features of synthetic data. These features appear to be essential for synthetic data detection, as even the retrained model with augmented blurred images is still sensible to the intensity of blurring.

Our main hypothesis to explain this phenomenon is that the Gaussian blurring erases artifacts from the genuine data. Although we were unable to detect these visually, it is possible that artifacts could appear due to real-world causes like dust, pen malfunctions or interferences in the scanning process of the documents. These types of artifacts, which could be very subtle, e.g., an unusual spike in the image gradient, may not be reproduced by the LDM due to the latent

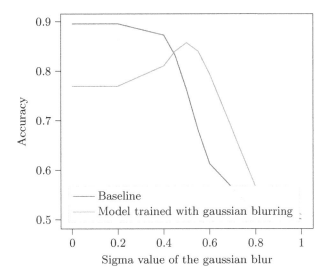

Fig. 6. Test accuracy when using a Gaussian blur of kernel size 11 at various sigma values.

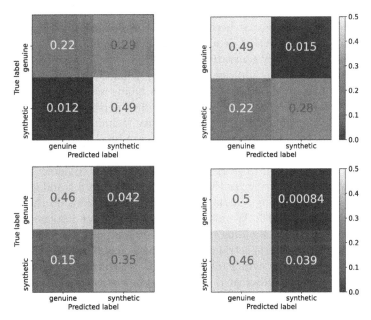

Fig. 7. Confusion matrices of classifier models evaluated on transformed images. Baseline model with a Gaussian blur of kernel size 11 and sigma 0.538 (top-left), blur augmented model without the Gaussian blur (top-right), baseline model with sharpening using a Gaussian blur of sigma 2 (bottom-left) and baseline model with salt and pepper applied to 1% of the image (bottom-right).

space operation or the denoising process. Gaussian blurring would then remove these artifacts, and the opposite action would lead the model to detect additional artificial artifacts of this type. In other terms, on handwriting, synthetic data produced by LDM may be too "smooth" compared to real data. Evidence towards this hypothesis is shown in the confusion matrices in Fig. 7 (top left vs. top right). These matrices show an increase in false synthetic predictions when adding blur and an increase in false genuine predictions when having a lighter blurring than during training. The presence of artifacts in genuine data is further confirmed by an increase in false genuine predictions when sharpening is applied (using unsharp masking) or when adding salt and pepper noise, which may simulate the genuine artifacts. If this hypothesis proves correct, these types of artifacts inherent in genuine handwriting would be an interesting clue for developing strong methods for detecting forged data.

5   Conclusion

The realism of synthetic images produced by generative models is increasing rapidly. Synthetic images obtained by diffusion models are already visually difficult to distinguish from real images. When using these new techniques solely on handwritten word images, it becomes almost impossible for a human viewer to tell whether the sample is of genuine or artificial origin.

Consequently, these new generative models are becoming powerful candidates for handwriting forgery. In addition, most forged handwriting detection methods assume the forgery is performed by human hands and exploit the resulting features that arise when trying to imitate another person's handwriting. This confirms the risk brought by these new generative techniques, as fake handwriting produced by specialized generative models might be able to fool both humans and machines.

Therefore, it is essential to start constructing methods to detect this type of artificially forged handwriting. In this work, we show that a standard Resnet18 model trained on classifying real/fake handwriting is able to do this, reaching performances of about 90%. We also investigate the most important features that could be exploited in building such detection methods. Our results suggest that neither the generative spectral artifacts, the shape of letters, nor the actual generative model plays a significant role in the recognition of synthetic data. Instead, in the case of handwriting, our results show strong evidence that some distinct artifacts are present in genuine data that most likely originate from real-world causes, like dust or interferences in the scanning process. These artifacts may prove essential in detecting forged synthetic handwriting, as generative approaches like DDPM do not seem to reproduce them.

References

1. Asnani, V., Yin, X., Hassner, T., Liu, X.: Reverse engineering of generative models: inferring model hyperparameters from generated images (2021)
2. Bhunia, A.K., Khan, S., Cholakkal, H., Anwer, R.M., Khan, F.S., Shah, M.: Handwriting transformers. In: 2021 IEEE/CVF International Conference on Computer Vision (ICCV), pp. 1066–1074 (2021)
3. Cha, S.H., Tappert, C.C.: Automatic detection of handwriting forgery. In: Proceedings Eighth International Workshop on Frontiers in Handwriting Recognition, pp. 264–267. IEEE (2002)
4. Chandrasegaran, K., Tran, N.T., Cheung, N.M.: A closer look at Fourier spectrum discrepancies for CNN-generated images detection. In: 2021 IEEE/CVF Conference on Computer Vision and Pattern Recognition (CVPR), pp. 7196–7205 (2021)
5. Christlein, V., Bernecker, D., Angelopoulou, E.: Writer identification using VLAD encoded contour-Zernike moments. In: 2015 13th International Conference on Document Analysis and Recognition (ICDAR), Nancy, pp. 906–910 (2015)
6. Christlein, V., Bernecker, D., Hönig, F., Maier, A., Angelopoulou, E.: Writer identification using GMM supervectors and exemplar-SVMs. Pattern Recogn. **63**, 258–267 (2017)
7. Christlein, V., Maier, A.: Encoding CNN activations for writer recognition. In: 13th IAPR International Workshop on Document Analysis Systems, Vienna, pp. 169–174 (2018)
8. Corvi, R., Cozzolino, D., Poggi, G., Nagano, K., Verdoliva, L.: Intriguing properties of synthetic images: from generative adversarial networks to diffusion models (2023). arXiv:2304.06408
9. Corvi, R., Cozzolino, D., Zingarini, G., Poggi, G., Nagano, K., Verdoliva, L.: On the detection of synthetic images generated by diffusion models. In: 2023 IEEE International Conference on Acoustics, Speech and Signal Processing (ICASSP), ICASSP 2023, pp. 1–5 (2023)
10. Cozzolino, D., Thies, J., Rössler, A., Riess, C., Nießner, M., Verdoliva, L.: ForensicTransfer: weakly-supervised domain adaptation for forgery detection (2019). arXiv:1812.02510
11. Davis, B.L., Morse, B.S., Price, B.L., Tensmeyer, C., Wigington, C., Jain, R.: Text and style conditioned GAN for the generation of offline-handwriting lines. In: 31st British Machine Vision Conference 2020, BMVC 2020, Virtual Event, UK, 7–10 September 2020. BMVA Press (2020)
12. Dhariwal, P., Nichol, A.: Diffusion models beat GANs on image synthesis. In: Ranzato, M., Beygelzimer, A., Dauphin, Y., Liang, P., Vaughan, J.W. (eds.) Advances in Neural Information Processing Systems, vol. 34, pp. 8780–8794. Curran Associates, Inc. (2021)
13. Fahn, C.S., Lee, C.P., Chen, H.I.: A text independent handwriting forgery detection system based on branchlet features and Gaussian mixture models. In: 2016 14th Annual Conference on Privacy, Security and Trust (PST), pp. 690–697 (2016)
14. Gan, J., Wang, W.: HiGAN: handwriting imitation conditioned on arbitrary-length texts and disentangled styles. In: Proceedings of the AAAI Conference on Artificial Intelligence, vol. 35, no. 9, pp. 7484–7492 (2021)
15. Gan, J., Wang, W., Leng, J., Gao, X.: HiGAN+: handwriting imitation GAN with disentangled representations. ACM Trans. Graph. **42**(1) (2022)
16. Gideon, S.J., Kandulna, A., Kujur, A.A., Diana, A., Raimond, K.: Handwritten signature forgery detection using convolutional neural networks. Procedia Comput.

Sci. **143**, 978–987 (2018). 8th International Conference on Advances in Computing & Communications (ICACC 2018)

17. Gragnaniello, D., Cozzolino, D., Marra, F., Poggi, G., Verdoliva, L.: Are GAN generated images easy to detect? A critical analysis of the state-of-the-art. In: 2021 IEEE International Conference on Multimedia and Expo (ICME), Los Alamitos, CA, USA, pp. 1–6. IEEE Computer Society (2021)

18. He, K., Zhang, X., Ren, S., Sun, J.: Deep residual learning for image recognition. In: 2016 IEEE Conference on Computer Vision and Pattern Recognition (CVPR), pp. 770–778 (2016)

19. Ho, J., Jain, A., Abbeel, P.: Denoising diffusion probabilistic models. In: Advances in Neural Information Processing Systems, vol. 33, pp. 6840–6851 (2020)

20. Kang, L., Riba, P., Rusiñol, M., Fornés, A., Villegas, M.: Content and style aware generation of text-line images for handwriting recognition. IEEE Trans. Pattern Anal. Mach. Intell. **44**, 8846–8860 (2021)

21. Kang, L., Riba, P., Wang, Y., Rusiñol, M., Fornés, A., Villegas, M.: GANwriting: content-conditioned generation of styled handwritten word images. In: Vedaldi, A., Bischof, H., Brox, T., Frahm, J.-M. (eds.) ECCV 2020. LNCS, vol. 12368, pp. 273–289. Springer, Cham (2020). https://doi.org/10.1007/978-3-030-58592-1_17

22. Kundu, S., Shivakumara, P., Grouver, A., Pal, U., Lu, T., Blumenstein, M.: A new forged handwriting detection method based on Fourier spectral density and variation. In: Palaiahnakote, S., Sanniti di Baja, G., Wang, L., Yan, W.Q. (eds.) ACPR 2019. LNCS, vol. 12046, pp. 136–150. Springer, Cham (2020). https://doi.org/10.1007/978-3-030-41404-7_10

23. Marra, F., Gragnaniello, D., Verdoliva, L., Poggi, G.: Do GANs leave artificial fingerprints? In: 2019 IEEE Conference on Multimedia Information Processing and Retrieval (MIPR), Los Alamitos, CA, USA, pp. 506–511. IEEE Computer Society (2019)

24. Marti, U.V., Bunke, H.: The IAM-database: an English sentence database for offline handwriting recognition. Int. J. Document Anal. Recogn. **5**, 39–46 (2002)

25. Mattick, A., Mayr, M., Seuret, M., Maier, A., Christlein, V.: SmartPatch: improving handwritten word imitation with patch discriminators. In: Lladós, J., Lopresti, D., Uchida, S. (eds.) ICDAR 2021. LNCS, vol. 12821, pp. 268–283. Springer, Cham (2021). https://doi.org/10.1007/978-3-030-86549-8_18

26. Mayr, M., Stumpf, M., Nicolaou, A., Seuret, M., Maier, A., Christlein, V.: Spatio-temporal handwriting imitation. In: Bartoli, A., Fusiello, A. (eds.) ECCV 2020. LNCS, vol. 12539, pp. 528–543. Springer, Cham (2020). https://doi.org/10.1007/978-3-030-68238-5_38

27. Nandanwar, L., Shivakumara, P., Kundu, S., Pal, U., Lu, T., Lopresti, D.: Chebyshev-harmonic-Fourier-moments and deep CNNs for detecting forged handwriting. In: 2020 25th International Conference on Pattern Recognition (ICPR), pp. 6562–6569. IEEE (2021)

28. Nikolaidou, K., et al.: WordStylist: styled verbatim handwritten text generation with latent diffusion models (2023). arXiv:2303.16576

29. Otsu, N.: A threshold selection method from gray-level histograms. IEEE Trans. Syst. Man Cybern. **9**(1), 62–66 (1979)

30. Pippi, V., Cascianelli, S., Cucchiara, R.: Handwritten text generation from visual archetypes. In: IEEE/CVF Conference on Computer Vision and Pattern Recognition (CVPR), pp. 22458–22467 (2023)

31. Radford, A., et al.: Learning transferable visual models from natural language supervision. In: International Conference on Machine Learning, pp. 8748–8763. PMLR (2021)

32. Ramesh, A., Dhariwal, P., Nichol, A., Chu, C., Chen, M.: Hierarchical text-conditional image generation with clip latents (2022). arXiv:2204.06125
33. Rombach, R., Blattmann, A., Lorenz, D., Esser, P., Ommer, B.: High-resolution image synthesis with latent diffusion models. In: 2022 IEEE/CVF Conference on Computer Vision and Pattern Recognition (CVPR), pp. 10674–10685 (2022)
34. Ronneberger, O., Fischer, P., Brox, T.: U-Net: convolutional networks for biomedical image segmentation. In: Navab, N., Hornegger, J., Wells, W.M., Frangi, A.F. (eds.) MICCAI 2015. LNCS, vol. 9351, pp. 234–241. Springer, Cham (2015). https://doi.org/10.1007/978-3-319-24574-4_28
35. Saharia, C., et al.: Photorealistic text-to-image diffusion models with deep language understanding. In: Advances in Neural Information Processing Systems, vol. 35, pp. 36479–36494 (2022)
36. Vaswani, A., et al.: Attention is all you need. In: Advances in Neural Information Processing Systems, vol. 30 (2017)
37. Yang, X., Zhou, D., Feng, J., Wang, X.: Diffusion probabilistic model made slim (2023)
38. Yu, N., Davis, L., Fritz, M.: Attributing fake images to GANs: learning and analyzing GAN fingerprints. In: 2019 IEEE/CVF International Conference on Computer Vision (ICCV), pp. 7555–7565 (2019)
39. Zhang, X., Karaman, S., Chang, S.F.: Detecting and simulating artifacts in GAN fake images. In: 2019 IEEE International Workshop on Information Forensics and Security (WIFS), pp. 1–6 (2019)

Leveraging Large Language Models for Topic Classification in the Domain of Public Affairs

Alejandro Peña[1]([⊠]) [ID], Aythami Morales[1] [ID], Julian Fierrez[1] [ID],
Ignacio Serna[1] [ID], Javier Ortega-Garcia[1] [ID], Íñigo Puente[2], Jorge Córdova[2],
and Gonzalo Córdova[2]

[1] BiDA - Lab, Universidad Autónoma de Madrid (UAM), 28049 Madrid, Spain
`alejandro.penna@uam.es`
[2] VINCES Consulting, 28010 Madrid, Spain

Abstract. The analysis of public affairs documents is crucial for citizens as it promotes transparency, accountability, and informed decision-making. It allows citizens to understand government policies, participate in public discourse, and hold representatives accountable. This is crucial, and sometimes a matter of life or death, for companies whose operation depend on certain regulations. Large Language Models (LLMs) have the potential to greatly enhance the analysis of public affairs documents by effectively processing and understanding the complex language used in such documents. In this work, we analyze the performance of LLMs in classifying public affairs documents. As a natural multi-label task, the classification of these documents presents important challenges. In this work, we use a regex-powered tool to collect a database of public affairs documents with more than 33K samples and 22.5M tokens. Our experiments assess the performance of 4 different Spanish LLMs to classify up to 30 different topics in the data in different configurations. The results shows that LLMs can be of great use to process domain-specific documents, such as those in the domain of public affairs.

Keywords: Domain Adaptation · Public Affairs · Topic Classification · Natural Language Processing · Document Understanding · LLM

1 Introduction

The introduction of the Transformer model [22] in early 2017 supposed a revolution in the Natural Language Domain. In that work, Vaswani *et al.* demonstrated that an Encoder-Decoder architecture combined with an Attention Mechanism can increase the performance of Language Models in several tasks, compared to recurrent models such as LSTM [8]. Over the past few years, there has been a significant development of transformer-based language model architectures, which are commonly known as Large Language Models (LLM). Its deployment sparked

M. Coustaty and A. Fornés (Eds.): ICDAR 2023 Workshops, LNCS 14193, pp. 20–33, 2023.
https://doi.org/10.1007/978-3-031-41498-5_2

a tremendous interest and exploration in numerous domains, including chat-bots (e.g., ChatGPT,[1] Bard,[2] or Claude[3]), content generation [2,16], virtual AI assistants (e.g., JARVIS [20], or GitHub's Copilot[4]), and other language-based tasks [9–11]. These models address scalability challenges while providing significant language understanding and generation abilities. That deployment of large language models has propelled advancements in conversational AI, automated content creation, and improved language understanding across various applications, shaping a new landscape of NLP research and development. There are even voices raising the possibility that most recent foundational models [1,12,13,21] may be a first step of an artificial general intelligence [3].

Large language models have the potential to greatly enhance the analysis of public affairs documents. These models can effectively process and understand the complex language used in such documents. By leveraging their vast knowledge and contextual understanding, large language models can help to extract key information, identify relevant topics, and perform sentiment analysis within these documents. They can assist in summarizing lengthy texts, categorizing them into specific themes or subject areas, and identifying relationships and patterns between different documents. Additionally, these models can aid in identifying influential stakeholders, tracking changes in public sentiment over time, and detecting emerging trends or issues within the domain of public affairs. By leveraging the power of large language models, organizations and policymakers can gain valuable insights from public affairs documents, enabling informed decision-making, policy formulation, and effective communication strategies. The analysis of public affairs documents is also important for citizens as it promotes transparency, accountability, and informed decision-making.

Public affairs documents often cover a wide range of topics, including policy issues, legislative updates, government initiatives, social programs, and public opinion. These documents can address various aspects of public administration, governance, and societal concerns. The automatic analysis of public affairs text can be considered a multi-label classification problem. Multi-label classification enables the categorization of these documents into multiple relevant topics, allowing for a more nuanced understanding of their content. By employing multi-label classification techniques, such as text categorization algorithms, public affairs documents can be accurately labeled with multiple attributes, facilitating efficient information retrieval, analysis, and decision-making processes in the field of public affairs.

This work focuses on NLP-related developments in an ongoing research project. The project aims to improve the automatic analysis of public affairs documents using recent advancements in Document Layout Analysis (DLA) and Language Technologies. The objective of the project is to develop new tools that allow citizens and businesses to quickly access regulatory changes that affect their

---

[1] https://openai.com/blog/chatgpt.
[2] https://blog.google/technology/ai/bard-google-ai-search-updates/
[3] https://www.anthropic.com/index/introducing-claude.
[4] https://github.com/features/preview/copilot-x.

present and future operations. With this objective in mind, a system is being developed to monitor the publication of new regulations by public organizations The block diagram of the system is depicted in Fig. 1. The system is composed of three main modules: *i)* Harvester module based on web scrappers; *ii)* a Document Layout Analysis (DLA) module; and *iii)* a Text Processing module. The Harvester monitors a set of pre-defined information sources, and automatically downloads new documents in them. Then, the DLA module conducts a layout extraction process, where text blocks are characterized and automatically classified, using Random Forest models, into different semantic categories. Finally, a Text Processing module process the text blocks using LLMs technology to perform multi-label topic classification, finally aggregating individual text predictions to infer the main topics of the document.

The full system proposed in Fig. 1 serves us to adapt LLMs to analyze documents in the domain of public affairs. This adaptation is based on the dataset used in our experiments, generated in collaboration with experts in public affairs regulation. They annotated over 92K texts using a semi-supervised process that included a regex-based tool. The database comprises texts related to more than 385 different public affairs topics defined by experts.

From all the analysis tool that can be envisioned in the general framework depicted in Fig. 1, in the present paper we focus in topic classification, with the necessary details of the Harverster needed to explain our datasets and interpret our topic classification results. Other modules such as the Layout Extractor are left for description elsewhere.

Specifically, the main contributions of this work are:

– Within the general document analysis system for analyzing public affairs documents depicted in Fig. 1, we propose, develop, and evaluate a novel functionality for multi-label topic classification.
– We present a new dataset of public affairs documents annotated by topic with more than 33K text samples and 22.5M tokens representing the main Spanish legislative activity between 2019 and 2022.
– We provide experimental evidence of the proposed multi-label topic classification functionality over that new dataset using four different LLMs (including RoBERTa [11] and GPT2 [16]) followed by multiple classifiers.

Our results shows that using a LLM backbone in combination with SVM classifiers suppose an useful strategy to conduct the multi-label topic classification task in the domain of public affairs with accuracies over 85%. The SVM classification improves accuracies consistently, even with classes that have a lower number of samples (e.g., less than 500 samples).

The rest of the paper is structured as follows: In Sect. 2 we describe the data collected for this work, including data preprocessing details. Section 3 describes the development of the proposed topic classification functionality. Section 4 presents the experiments and results of this work. Finally, Sect. 5 summarizes the main conclusions.

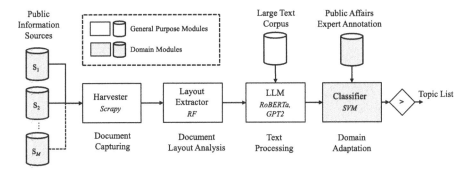

Fig. 1. Block diagram of an automatic public affairs document analysis system. The white blocks represent general-purpose modules, while the grey blocks represent domain-specific modules.

2 Data Collection and Analysis

The major decisions and events resulting from the legislative, judicial and administrative activity of public administrations are public data. Is a common practice, and even a legal requisite, for these administrations to publish this information in different formats, such as govermental websites or official gazettes[5]. Here, we use a regex-powered tool to follow up parliamentary initiatives from the Spanish Parliament, resulting in a legislative-activities text corpora in Spanish. Parliamentary initiatives involve a diverse variety of parliament interactions, such as questions to the government members, legislative proposals, etc.

Raw data were collected and processed with this tool, and comprise initiatives ranging from November 2019 to October 2022. The data is composed of short texts, which may be annotated with multiple labels. Each label includes, among others, topic annotations based on the content of the text. These annotations were generated using regex logic based on class-specific predefined keywords. Both topic classes and their corresponding keywords were defined by a group of experts in public affairs regulations. It is important to note that the same topic (e.g., "Health Policy") can be categorized differently depending on the user's perspective (e.g., citizens, companies, governmental agencies). We have simplified the annotation, adding a ID number depending on the perspective used (e.g., "Health Policy_1" or "Health Policy_2"). Our raw data is composed of 450K initiatives grouped in 155 weekly-duration sessions, with a total number of topic classes up to 385. Of these 450K samples, only 92.5K were labeled, which suppose roughly 20.5% of the samples. However, almost half of these are annotated with more than one label (i.e. 45.5K, 10.06% of samples), with a total number of labels of 240K. Figure 2 presents the distribution of the 30 most frequent topics in the data, where we can clearly observe the significant imbalance between classes. The most frequent topic in the raw data is "Healthcare Situation", appearing in more then 25K data samples. Other topics, such as

---

[5] https://op.europa.eu/en/web/forum.

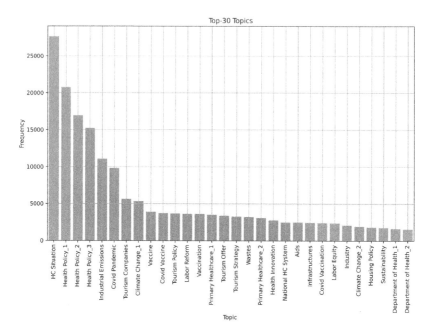

Fig. 2. Distribution of the top 30 most frequent topics in the raw data.

"*Health Policy*", have an important presence in the data as well. However, only 8 out of these 30 topics reach 5K samples, and only 5 of them are present in at least 10K. This imbalance, along with the bias towards health-related subjects in the most frequent topics, is inherent to the temporal framework of the database, as the Covid-19 pandemic situation has dominated significant public affairs over the past 3 years. Note that Fig. 2 depicts the thirty most frequent topics, whereas 385 topics are present in the data. To prevent the effects of major class imbalances, we will now focus on the 30 topics of Fig. 2.

2.1 Data Curation

We applied a data cleaning process to the raw corpora to generate a clean version of the labeled data. We started by removing duplicated texts, along with data samples with less than 100 characters. Some works addressing Spanish models applied a similar filtering strategy with a threshold of 200 characters [17,19,23] with the aim of obtaining a clean corpus to pre-train transformer models. Here we set the threshold to 100, as our problem here does not require us to be that strict (i.e., we do not want to train a transformer from scratch). Instead, we desired to remove extremely short text, which we qualitative assessed that were mainly half sentences, while retaining as much data as possible. In this sense, we filter text samples of any length starting with lowercase, to prevent half sentences to leak in. We also identified bad quality/noisy text samples to start with "CSV" or "núm", so we remove samples based on this rule. Finally, given the existence

Table 1. Summary of the parliamentary initiative database after the data cleaning process, which includes 33,147 data samples with multi-label annotations across 30 topics. We include a topic ID, the topic, and the number of samples annotated for each of them.

ID	Topic	#Samples	ID	Topic	#Samples
1	Healthcare Situation	13561	16	Primary Healthcare_1	1425
2	Health Policy_1	12029	17	Sustainability	1370
3	Health Policy_2	8229	18	Wastes	1294
4	Health Policy_3	8111	19	Aids	1216
5	Industrial Emissions	5101	20	Primary Healthcare_2	1189
6	Covid-19 Pandemic	3298	21	Tourism Offer	1181
7	Tourism Policy	2209	22	Labor Equity	1074
8	Tourism Companies	2033	23	Industry	1051
9	Climate Change_1	1930	24	Infrastructures	1029
10	Vaccination	1924	25	Covid-19 Vaccination	997
11	Vaccine	1751	26	National Healthcare System	964
12	Covid-19 Vaccine	1617	27	Climate Change_2	886
13	Tourism Strategy	1533	28	Housing Policy	744
14	Labor Reform	1529	29	Department of Health_1	541
15	Health Innovation	1469	30	Department of Health_2	518

of co-official languages different from Spanish in Spain (e.g., Basque, Galician or Catalan), which are used by a significant percentage of Spanish citizens, we filter data samples from these languages. Due to the lack of reliable language detectors in these co-official languages, and the use of some linguistic, domain-specific patterns in the parliamentary initiatives, we identified a set of words in these languages and use it to detect and filter out potential samples not written in Spanish. We applied this process several times to refine the set of words.

At data sample level, we clean texts by removing excessive white spaces and initiative identifiers in the samples. We then filter URLs and non-alphanumeric characters, retaining commonly used punctuation characters in Spanish written text (i.e., ()-.¿?¡!_;). After applying all the data curation process, we obtain a multi-label corpus of 33,147 data samples, with annotations on the 30 topics commented above. Table 1 presents the number of samples per topic category. Note that the number of samples of each topic has significantly decreased compared to the proportions observed in the raw data (see Fig. 2). The impact of the data curation process is different between topics, leading to some changes in the frequency-based order of the topics. The topic with most data samples in the curated corpus is still *"Healthcare Situation"*, but the number of samples annotated with this topic has been reduced by half. On the other hand, we have several topics with less than 1K samples, setting a lower limit of 518.

3   Methodology and Models

As we previously mentioned in Sect. 2, the samples in our dataset may present more than one topic label. Hence, the topic classification task on this dataset is a multi-label classification problem, where we have a significant number of classes that are highly imbalanced. This scenario (i.e., high number of classes, some of them with few data samples, with overlapped subjects between classes) leads us to discard a single classifier for this task. Instead of addressing the problem as a multi-label task, we break it into small, binary detection tasks, where an individual topic detector is trained for each of the 30 classes in a one vs all setup. This methodology, illustrated in Fig. 3, represents a big advantage, as it provides us a high degree of versatility to select the best model configuration for each topic to deploy a real system. During inference, new data samples can be classified by aggregating the predictions of the individual classifiers [5].

The architecture of the binary topic models is depicted in Fig. 3. We use a transformer-based model as backbone, followed by a Neural Network, Random Forest, or SVM classifier. In this work, we explore different transformer models, pretrained from scratch in Spanish by the Barcelona Supercomputing Center in the context of the MarIA project [7]. We included both encoder and decoder architectures. These model architectures are the following:

- **RoBERTa-base.** An encoder-based model architecture with 12 layers, 768 hidden size, 12 attention heads, and 125M parameters.
- **RoBERTa-large.** An encoder-based model architecture with 24 layers, 71,024 hidden size, 16 attention heads, and 334M parameters.
- **RoBERTalex.** A version [6] of RoBERTa-base, fine-tuned for the Spanish legal domain.
- **GPT2-base.** A decoder-based model architecture with 12 layers, 768 hidden size, 12 attention heads, and 117M parameters.

We listed above the configurations reported in [7] for the open-source models available in the HuggingFace repository of the models.[6] The RoBERTa models [11] are versions of BERT models [9], in which an optimized pre-training strategy and hyperparameter selection was applied, compared to the original BERT pre-training. The Spanish versions of these models were pre-trained following the original RoBERTa configuration, with a corpus of 570 GB of clean Spanish written text. The RoBERTalex model is a fine-tuned version of Spanish RoBERTa-base, trained with a corpus of 8.9 GB of legal text data. On the other hand, GPT2 [16] is a decoder-based model of the GPT family [2,12,13,15]. As such, the model is aimed to generative tasks (note that modern versions of GPT models, such as InstructGPT [13] or GPT4 [12] are fine-tuned to follow human instructions, so they cannot be considered generative models in the same way as earlier GPT models), different from the RoBERTa family, which is specialized in text understanding. The version used of GPT2 was trained using the same corpus as the RoBERTa models. All the models use byte-level BPE tokenizer [16]

---

[6] https://huggingface.co/PlanTL-GOB-ES.

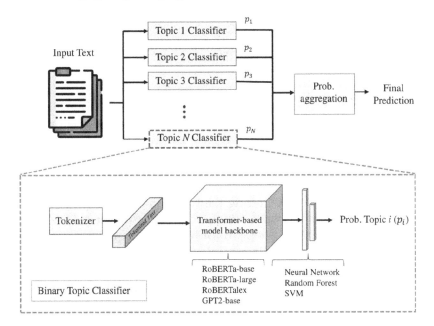

Fig. 3. Proposed multi-label topic classification system, in which an individual topic detector is applied to an input text before aggregating all the predictions, and the architecture of each binary topic classifier.

with vocab size of 50,265 tokens, and have the same length for the context windows, i.e. 512. While left padding is used in the RoBERTa models, right padding is advisable for the GPT2 model.

4   Experiments

As exposed in Sect. 3, due to the nature of the dataset collected for this work, we address multi-label topic classification by training a binary topic classifier for each class (one vs all), and then aggregating the individual predictions on a versatile way (e.g., providing rank statistics, topics over a fixed threshold, etc.). Hence, our experiments will focus on assessing the performance of different topic classifiers configurations, and the potential of the newly available Spanish language models in unconstrained scenarios (i.e., multi-label political data, with subjective annotations based on private-market interest). Section 4.1 will evaluate first the performance of different transformer-based models on our dataset, and then explore the combination of the best-performance model with SVM and Random Forest classifiers.

We conduct all the experiments using a K-fold cross validation setup with 5 folds, and report mean and average results between folds. We select True Positive Rate (TPR), and True Negative Rate (TNR) as our performance measures, due to the class imbalances in the parliamentary dataset. We use in our experiments the models available in the HuggingFace transformers library[7], along with several sklearn tools. Regarding the hardware, we conducted the experiments in a PC with 2 NVIDIA RTX 4090 (with 24 GB each), Intel Core i9, 32 GB RAM.

4.1   Topic Classification in the Domain of Public Affairs

Recalling from Fig. 3, our topic detector architecture is mainly composed of *i)* a transformer backbone, and *ii)* a classifier. We train the transformer models with a binary neural network classification output layer. For each topic, we train the detector using Weighted Cross Entropy Loss to address the class imbalance in a "One vs All" setup. Topic classifiers are trained for 5 epochs using a batch size of 32 samples, and freezing the transformer layers. Table 2 presents the results of the topics classifiers using the four transformer models explored in this work (i.e., RoBERTa-base [7], RoBERTa-large [7], RoBERTalex [6], and GPT2-base [7]). We can observe a general behavior across the RoBERTa models. The classifiers trained for the topics with more samples obtain higher TPR means, close to the TNR mean values. In these cases, the classifiers are able to distinguish reasonably well text samples in which the trained topic is present. These results are, in general, consistent across folds, exhibiting moderate deviation values. This behavior degrades from Topic 9 onwards, where the low number of samples (i.e., less than 2K) leads to an increase of the TNR to values over 90% with a decay of TPR. However, we can observe some exceptions in the classifiers using RoBERTa-base as backbone (topics 11, 12, 24), where TNR scales to values close to 100% while preserving TPR performances over 80%. Furthermore, RoBERTa-base classifiers exhibit better results than the RoBERTa-large classifiers (probably due to the constrained number of samples), and even than the RoBERTalex models. Remember that both RoBERTa-base and RoBERTalex are the same models, the latter being the RoBERTa-base model with a fine-tuning to the legal domain that, a priori, should make it more appropriate for the problem at hand. Regarding GPT2-based classifiers, we observe similar trends to those of the RoBERTa models, but exhibiting lower performances. This is not surprising, as the GPT model was trained for generative purposes, rather than text understanding like RoBERTa.

It's worth noting here the case of Topic 1, which obtains the lowest TNR mean value in all models, with deviation values over 0.15, despite being the topic with more data samples (i.e. a third of the data). We hypothesize that the low performances when detecting negative samples is mostly due to the overlap with the rest of the topics, as this topic focuses on general healthcare-related aspects (remember from Table 1 that half of the topics are related with healthcare).

From the results presented in Table 2, we can conclude that RoBERTa-base is the best model backbone for our task. Now, we want to assess if a specialized classifier, such as Support Vector Machines (SVM) or Random Forests (RF), can be used to fine tune the performance to the specific domain. For these classifiers, we used RoBERTa-base as feature extractor to compute 768-dimensional text

---

[7] https://huggingface.co/docs/transformers/index.

Table 2. Results of the binary classification for each topic (one vs all), using different transformer models with a Neural Network classifier. We report True Positive Rate (TPR) and True Negative Rate (TNR) as mean$_{std}$ (in parts per unit), computed after a K-fold cross validation (5 folds).

ID	RoBERTa-b [7]		RoBERTa-l [7]		GPT2-b [7]		RoBERTalex [6]	
	TPR	TNR	TPR	TNR	TPR	TNR	TPR	TNR
1	$.80_{07}$	$.75_{19}$	$.78_{08}$	$.76_{0.19}$	$.58_{14}$	$.60_{15}$	$.79_{10}$	$.70_{18}$
2	$.87_{09}$	$.88_{04}$	$.84_{11}$	$.86_{05}$	$.61_{25}$	$.82_{05}$	$.83_{10}$	$.82_{06}$
3	$.83_{08}$	$.87_{04}$	$.81_{09}$	$.87_{04}$	$.65_{18}$	$.79_{07}$	$.79_{09}$	$.84_{05}$
4	$.86_{07}$	$.89_{03}$	$.83_{10}$	$.88_{03}$	$.69_{17}$	$.79_{07}$	$.80_{10}$	$.86_{04}$
5	$.76_{05}$	$.81_{06}$	$.72_{07}$	$.81_{07}$	$.63_{06}$	$.74_{09}$	$.67_{08}$	$.80_{06}$
6	$.82_{05}$	$.87_{02}$	$.83_{05}$	$.87_{03}$	$.67_{04}$	$.63_{08}$	$.68_{06}$	$.83_{04}$
7	$.85_{04}$	$.93_{03}$	$.83_{06}$	$.91_{05}$	$.64_{08}$	$.78_{08}$	$.75_{07}$	$.94_{03}$
8	$.82_{02}$	$.89_{05}$	$.81_{03}$	$.88_{06}$	$.63_{04}$	$.78_{07}$	$.69_{02}$	$.91_{04}$
9	$.79_{10}$	$.90_{04}$	$.77_{11}$	$.89_{06}$	$.58_{08}$	$.76_{08}$	$.68_{07}$	$.91_{03}$
10	$.76_{26}$	$.96_{03}$	$.67_{31}$	$.95_{03}$	$.49_{42}$	$.91_{10}$	$.62_{34}$	$.95_{04}$
11	$.89_{11}$	$.98_{02}$	$.72_{31}$	$.98_{01}$	$.55_{44}$	$.93_{09}$	$.70_{32}$	$.97_{03}$
12	$.88_{12}$	$.98_{02}$	$.73_{30}$	$.98_{01}$	$.57_{41}$	$.94_{09}$	$.72_{30}$	$.97_{02}$
13	$.76_{09}$	$.89_{06}$	$.75_{08}$	$.86_{07}$	$.33_{09}$	$.79_{09}$	$.58_{14}$	$.91_{05}$
14	$.76_{12}$	$.93_{03}$	$.72_{12}$	$.93_{03}$	$.39_{13}$	$.81_{06}$	$.65_{13}$	$.94_{02}$
15	$.61_{09}$	$.85_{04}$	$.58_{10}$	$.86_{03}$	$.53_{06}$	$.82_{05}$	$.54_{08}$	$.90_{02}$
16	$.75_{03}$	$.90_{03}$	$.71_{05}$	$.88_{04}$	$.43_{04}$	$.77_{03}$	$.64_{05}$	$.91_{03}$
17	$.71_{25}$	$.94_{06}$	$.64_{32}$	$.96_{05}$	$.59_{31}$	$.93_{05}$	$.65_{31}$	$.92_{05}$
18	$.62_{08}$	$.90_{03}$	$.54_{10}$	$.85_{05}$	$.36_{07}$	$.79_{07}$	$.51_{05}$	$.91_{02}$
19	$.69_{10}$	$.92_{02}$	$.69_{09}$	$.91_{03}$	$.45_{11}$	$.86_{05}$	$.49_{12}$	$.95_{01}$
20	$.73_{05}$	$.93_{02}$	$.73_{06}$	$.90_{03}$	$.32_{04}$	$.86_{03}$	$.58_{05}$	$.94_{02}$
21	$.67_{04}$	$.89_{05}$	$.67_{06}$	$.86_{06}$	$.48_{05}$	$.84_{05}$	$.45_{03}$	$.93_{03}$
22	$.71_{05}$	$.95_{02}$	$.66_{03}$	$.94_{02}$	$.40_{04}$	$.89_{04}$	$.51_{03}$	$.97_{01}$
23	$.70_{08}$	$.96_{02}$	$.57_{17}$	$.96_{02}$	$.24_{17}$	$.96_{01}$	$.43_{17}$	$.98_{01}$
24	$.83_{08}$	$.97_{04}$	$.69_{11}$	$.98_{01}$	$.20_{24}$	$.98_{01}$	$.55_{18}$	$.98_{01}$
25	$.80_{16}$	$.97_{04}$	$.54_{36}$	$.97_{04}$	$.44_{40}$	$.98_{03}$	$.57_{34}$	$.97_{04}$
26	$.52_{10}$	$.95_{01}$	$.48_{13}$	$.96_{01}$	$.17_{03}$	$.96_{02}$	$.40_{10}$	$.98_{01}$
27	$.72_{08}$	$.97_{02}$	$.62_{08}$	$.97_{02}$	$.25_{07}$	$.97_{01}$	$.56_{05}$	$.98_{01}$
28	$.44_{05}$	$.97_{02}$	$.32_{15}$	$.96_{03}$	0_{0}	1_{0}	$.20_{08}$	$.99_{01}$
29	$.46_{06}$	$.98_{01}$	$.17_{.04}$	$.99_{0}$	0_{0}	1_{0}	$.18_{04}$	$.99_{0}$
30	$.43_{06}$	$.98_{01}$	$.15_{03}$	$.99_{0}$	0_{0}	1_{0}	$.15_{03}$	$.99_{0}$

embeddings from each of the text samples. We explored two approaches for these embeddings: *i)* using the embedding computed for the [CLS] token, and *ii)* averaging all the token embeddings (i.e., mean pooling). In the original BERT model [9], and hence the RoBERTa model, the [CLS] is a special token appended at the start of the input, which the model uses during training for the Next

Table 3. Results of the binary classification for each topic (one vs all), using RoBERTa-base [7] in combination with SVM and Random Forest classifiers. We report True Positive Rate (TPR) and True Negative Rate (TNR) as mean$_{std}$ (in parts per unit), computed after a K-fold cross validation (5 folds).

ID	RoBERTa-b [7] + SVM		RoBERTa-b [7] + RF	
	TPR	TNR	TPR	TNR
1	$.80_{.07}$	$.76_{.20}$	$.70_{.11}$	$.81_{.21}$
2	$.87_{.09}$	$.88_{.04}$	$.74_{.18}$	$.94_{.03}$
3	$.83_{.07}$	$.88_{.04}$	$.64_{.18}$	$.97_{.02}$
4	$.86_{.07}$	$.90_{.02}$	$.67_{.18}$	$.98_{.02}$
5	$.80_{.05}$	$.80_{.06}$	$.12_{.06}$	$.99_{.01}$
6	$.85_{.05}$	$.85_{.03}$	$.23_{.04}$	$.99_{0}$
7	$.90_{.02}$	$.90_{.05}$	$.49_{.06}$	1_{0}
8	$.89_{.01}$	$.86_{.07}$	$.33_{.02}$	1_{0}
9	$.88_{.07}$	$.87_{.04}$	$.24_{.05}$	1_{0}
10	$.84_{.18}$	$.94_{.03}$	$.51_{.41}$	1_{0}
11	$.92_{.08}$	$.97_{.02}$	$.57_{.38}$	1_{0}
12	$.93_{.07}$	$.98_{.02}$	$.56_{.36}$	1_{0}
13	$.87_{.04}$	$.85_{.08}$	$.08_{.02}$	1_{0}
14	$.87_{.07}$	$.88_{.04}$	$.14_{.04}$	1_{0}
15	$.70_{.08}$	$.80_{.06}$	$.06_{.02}$	1_{0}
16	$.89_{.03}$	$.88_{.04}$	$.13_{.05}$	1_{0}
17	$.79_{.18}$	$.92_{.06}$	$.59_{31}$	1_{0}
18	$.78_{.06}$	$.81_{.05}$	$.09_{.03}$	1_{0}
19	$.87_{.03}$	$.85_{.03}$	$.14_{.10}$	1_{0}
20	$.89_{.03}$	$.90_{.03}$	$.14_{.06}$	1_{0}
21	$.88_{.02}$	$.79_{.08}$	$.09_{.02}$	1_{0}
22	$.90_{.03}$	$.88_{.03}$	$.16_{.03}$	1_{0}
23	$.89_{.04}$	$.89_{.05}$	$.27_{.15}$	1_{0}
24	$.90_{.05}$	$.95_{.02}$	$.37_{.23}$	1_{0}
25	$.90_{.07}$	$.95_{.04}$	$.41_{.31}$	$.99_{.01}$
26	$.83_{.06}$	$.89_{.04}$	$.17_{.11}$	1_{0}
27	$.91_{.04}$	$.90_{.04}$	$.33_{.04}$	1_{0}
28	$.87_{.04}$	$.86_{.06}$	$.06_{.01}$	1_{0}
29	$.84_{.06}$	$.89_{.03}$	$.10_{.03}$	1_{0}
30	$.85_{.05}$	$.89_{.03}$	$.08_{.03}$	1_{0}

Sentence Prediction objective. Thus, the output for this embedding is used for classification purposes, serving the [CLS] embedding as a text representation. We repeated the experiment using both types of representations, and end up selecting the first approach after exhibiting better results. Table 3 presents the results of the topic models using RoBERTa-base text embeddings together with a SVM and Random Forest classifier. In all cases, we use a complexity parameter of 1 and RBF kernel for the SVM, and a max depth of 1,000 for the Random Forest. We note that these parameters can be tuned for each topic to improve the results. The first thing we notice in Table 3 is the poor performance of the RF-based classifiers, which are the worst among all the configurations. Almost for all the topics under 2K samples, the TNR saturates to 1, and the TPR tends to extremely low values. From this, we can interpret that the classifier is not learning, and just predicting the negative, overrepresented class. However, the performance on the topics over 2K samples is far from the one observed for the RoBERTa models of Table 2. This could be expected, as the RF classifier is not the best approach to work with input data representing a structured vector subspace with semantic meaning, such as text/word embedding subspaces, specially when the number of data samples is low. On the other hand, the SVM performance clearly surpass all previous configurations in terms of TPR. While the results are comparable with those of RoBERTa-base with NN for the first 5 topics, this behavior is maintained for all topics, regardless of the number of data samples. Almost all classifiers achieve a TPR over 80%, except for topics 15, 17 and 18. Nevertheless, the results in these topics increase with the SVM (e.g., for topic 15, where RoBERTa-base with the NN classifier achieved a TPR mean of 61%, here we obtain a 70%). TNR values are, in general, slightly lower, but this could be caused because in previous configurations, topic classifiers tend to exhibit bias towards the negative class as the number of samples falls (i.e., similar to the behavior of the RF classifier). Interestingly, the high deviation observed in the Topic 1 TNR appears too in both SVM and RF classifiers, which could support our previous hypothesis. As we commented before, we suspect that an hyperparameter tuning could improve even more the SVM results on our data.

5 Conclusions

This work applies and evaluates Large Language Models (LLMs) for topic classification in public affairs documents. These documents are of special relevance for both citizens and companies, as they contain the basis of all legislative updates, social programs, public announcements, etc. Thus, enhancing the analysis of public documents using the recent advances of the NLP community is desirable.

To this aim, we collected a Spanish text corpora of public affairs documents, using a regex-powered tool to process and annotate legislative initiatives from the Spanish Parliament during a capture period over 2 years. The raw text corpora is composed of more than 450K initiatives, with 92K of them being annotated in a multi-label scenario with up to 385 different topics. Topic classes were defined by experts in public affairs regulations. We preprocess this corpus and generate a clean version of more than 33K multi-label texts, including annotations for the 30 most frequent topics in the data.

We use this dataset to assess the performance of recent Spanish LLMs [6,7] to perform multi-label topic classification in the domain of public affairs. Our experiments include text understanding models (three different RoBERTa-based models [11]) and generative models [16], in combination with three different classifiers (i.e., Neural Networks, Random Forests, and SVMs). The results show how text understanding models with SVM classifiers supposes an effective strategy for the topic classification task in this domain, even in situations where the number of data samples is limited.

As future work, we plan to study in more depth biases and imbalances [4] like the ones mentioned before presenting Fig. 2, and compensating them with imbalance-aware machine learning procedures [18]. More recent LLMs can be also tested for this task, including multilingual and instruction-based models, which have shown great capacities in multiple NLP tasks, even in zero-shot scenarios. We will also continue our research by exploring the incorporation of other NLP tasks (e.g. text summarization, named entity recognition) and multimodal methods [14] to our framework, with the objective of enhancing automatic analysis of public affairs documents.

Acknowledgments. This work was supported by VINCES Consulting under the project VINCESAI-ARGOS and BBforTAI (PID2021-127641OB-I00 MICINN/ FEDER). The work of A. Peña is supported by a FPU Fellowship (FPU21/00535) by the Spanish MIU. Also, I. Serna is supported by a FPI Fellowship from the UAM.

References

1. Anil, R., Dai, A.M., Firat, O., Johnson, M., et al.: PaLM 2 technical report. arXiv:2305.10403 (2023)
2. Brown, T., Mann, B., Ryder, N., Subbiah, M., et al.: Language models are few-shot learners. In: NIPS, vol. 33, pp. 1877–1901 (2020)
3. Bubeck, S., Chandrasekaran, V., Eldan, R., Gehrke, J., et al.: Sparks of artificial general intelligence: early experiments with GPT-4. arXiv:2303.12712 (2023)
4. DeAlcala, D., Serna, I., Morales, A., Fierrez, J., et al.: Measuring bias in AI models: an statistical approach introducing N-Sigma. In: COMPSAC (2023)
5. Fierrez, J., Morales, A., Vera-Rodriguez, R., Camacho, D.: Multiple classifiers in biometrics. Part 1: Fundamentals and review. Inf. Fusion **44**, 57–64 (2018)
6. Gutiérrez-Fandiño, A., Armengol-Estapé, J., Gonzalez-Agirre, A., Villegas, M.: Spanish legalese language model and corpora. arXiv:2110.12201 (2021)
7. Gutiérrez-Fandiño, A., Armengol-Estapé, J., Pàmies, M., Llop, J., et al.: MarIA: Spanish language models. Procesamiento del Lenguaje Nat. **68** (2022)
8. Hochreiter, S., Schmidhuber, J.: Long short-term memory. Neural Comput. **9**(8), 1735–1780 (1997)
9. Kenton, J., Chang, M., Lee, K., Toutanova, K.: BERT: pre-training of deep bidirectional transformers for language understanding. In: NAACL, pp. 4171–4186 (2019)
10. Lewis, M., Liu, Y., Goyal, N., Ghazvininejad, M., et al.: BART: denoising sequence-to-sequence pre-training for natural language generation, translation, and comprehension. In: ACL, pp. 7871–7880 (2020)
11. Liu, Y., Ott, M., Goyal, N., Du, J., et al.: RoBERTa: a robustly optimized BERT pretraining approach. arXiv:1907.11692 (2019)
12. OpenAI: GPT-4 technical report. Technical report (2023)
13. Ouyang, L., Wu, J., Jiang, X., Almeida, D., et al.: Training language models to follow instructions with human feedback. In: NIPS, vol. 35, pp. 27730–27744 (2022)

14. Peña, A., Serna, I., Morales, A., Fierrez, J., et al.: Human-centric multimodal machine learning: recent advances and testbed on AI-based recruitment. SN Comput. Sci. **4**, 434 (2023). https://doi.org/10.1007/s42979-023-01733-0
15. Radford, A., Narasimhan, K., Salimans, T., Sutskever, I., et al.: Improving language understanding by generative pre-training. Technical report (2018)
16. Radford, A., Wu, J., Child, R., Luan, D., et al.: Language models are unsupervised multitask learners. OpenAI Blog **1**(8), 9 (2019)
17. Raffel, C., Shazeer, N., Roberts, A., Lee, K., et al.: Exploring the limits of transfer learning with a unified text-to-text transformer. J. Mach. Learn. Res. **21**(1), 5485–5551 (2020)
18. Serna, I., Morales, A., Fierrez, J., Obradovich, N.: Sensitive loss: improving accuracy and fairness of face representations with discrimination-aware deep learning. Artif. Intell. **305**, 103682 (2022)
19. Serrano, A., Subies, G., Zamorano, H., Garcia, N., et al.: RigoBERTa: a state-of-the-art language model for Spanish. arXiv:2205.10233 (2022)
20. Shen, Y., Song, K., Tan, X., Li, D., et al.: HuggingGPT: solving AI tasks with ChatGPT and its friends in HuggingFace. arXiv:2303.17580 (2023)
21. Touvron, H., Lavril, T., Izacard, G., Martinet, X., et al.: LLaMA: open and efficient foundation language models. arXiv:2302.13971 (2023)
22. Vaswani, A., Shazeer, N., Parmar, N., Uszkoreit, J., et al.: Attention is all you need. In: Advances in Neural Information Processing Systems, vol. 30 (2017)
23. Xue, L., Constant, N., Roberts, A., Kale, M., et al.: mT5: a massively multilingual pre-trained text-to-text transformer. In: NAACL, pp. 483–498 (2021)

The Adaptability of a Transformer-Based OCR Model for Historical Documents

Phillip Benjamin Ströbel[1(✉)] [iD], Tobias Hodel[2] [iD], Walter Boente[3] [iD], and Martin Volk[1] [iD]

[1] Department of Computational Linguistics, University of Zurich, Zurich, Switzerland
{pstroebel,volk}@cl.uzh.ch
[2] Walter Benjamin Kolleg, University of Bern, Bern, Switzerland
tobias.hodel@unibe.ch
[3] Faculty of Law, University of Zurich, Zurich, Switzerland
walter.boente@rwi.uzh.ch

Abstract. We tested the capabilities of Transformer-based text recognition technology when dealing with (multilingual) real-world datasets. This is a crucial aspect for libraries and archives that must digitise various sources. The digitisation process cannot rely solely on manual transcription due to the complexity and diversity of historical materials. Therefore, text recognition models must be able to adapt to various printed texts and manuscripts, especially regarding different handwriting styles. Our findings demonstrate that Transformer-based models can recognise text from printed and handwritten documents, even in multilingual environments. These models require minimal training data and are a suitable solution for digitising libraries and archives. However, it is essential to note that the quality of the recognised text can be affected by the handwriting style.

Keywords: OCR · Handwritten Text Recognition · Transformers · Digital Humanities · Cultural Heritage

1 Introduction

The emergence of the Transformer architecture in 2017 [25] and the subsequent introduction of Bi-directional Encoder Representations from Transformers (BERT) [4] in 2018 have revolutionised the approach to natural language processing. Instead of training fully-fledged systems from scratch, large pre-trained models are used and fine-tuned for tasks such as question-answering systems, summarization, and machine translation. BERT-style models are excellent transfer learners [16], meaning they can adapt to various tasks. This adaptability has led to their use in image processing as well [5,24], resulting in the development of large pre-trained Transformers for images [2].

When it comes to automatic text recognition (ATR), a common method involves using a convolutional neural network (CNN) [6,11] to extract features

© The Author(s), under exclusive license to Springer Nature Switzerland AG 2023
M. Coustaty and A. Fornés (Eds.): ICDAR 2023 Workshops, LNCS 14193, pp. 34–48, 2023.
https://doi.org/10.1007/978-3-031-41498-5_3

from word or line images and then using a recurrent neural network (RNN) [9,17] to translate those features into a sequence of characters (see [8,27]). Some recent advancements in Transformers for language and image processing suggest combining these architectures could replace the CNN backbone and use a more powerful language model instead of the RNN. TrOCR, a Transformer-based OCR model introduced by [12], follows this approach. It uses an image Transformer as an encoder and a language Transformer as a decoder, trained on a large dataset of synthetically generated line images (687.3M printed and 17.9M handwritten) and their corresponding English transcriptions.

We aim to assess TrOCR's adaptability to real-world and multilingual datasets since it has been primarily trained on synthetic English-language data. This is especially important for libraries and archives that face challenges in digitising their material for easy access and availability. E. g., the Swiss platform *e-rara*[1] has 42 transcriptions out of 7,282 digitised 17th-century prints. Similarly, *e-manuscripta*[2], another Swiss platform, has 529 transcriptions despite having a collection of 52,034 manuscripts[3]. Our investigation is necessary to ensure that TrOCR can effectively handle the challenges of digitising historical materials since manual transcription, although attempted (see [3]), would not scale [23].

The variability of printed texts and manuscripts, particularly in handwriting, means that ATR models require constant re-training (or training from scratch). This, in turn, necessitates the creation of more ground truth data, which is an expensive process. Therefore, it would be beneficial to have an architecture that can easily adapt to new styles of print or handwriting with minimal input. This paper evaluates the potential transferability and adaptability of TrOCR, a relatively new model. Our contributions include the introduction of three new datasets and testing TrOCR's transferability on printed data (newspapers) and correspondence data by adding two more datasets. Furthermore, our experiments provide useful insights into the "best practices" when working with Transformer-based ATR models.

2   Related Work

Recent advancements in neural OCR techniques have led to significant improvements in recognizing text from images of historical documents. A study by [20] showed that *OCRopy*[4] achieved exceptional character accuracy, consistently over 94%, on a diachronic book corpus dating from 1487 to 1914, all of which were in blackletter[5] fonts. The models generalised and performed equally well on various books. The study found that a training set of 100 to 200 lines produced comparable results to models with a considerably larger training set, as recommended by [19].

---

[1] https://www.e-rara.ch/.

[2] https://www.e-manuscripta.ch/.

[3] These statistics are as of November 2022.

[4] https://github.com/tmbdev/ocropy.

[5] Also called Gothic font.

The *READ*[6] project developed the *Transkribus*[7] framework [14] to recognise handwriting. This tool helps transcribe manuscripts by analysing a page's layout and dividing it into text regions, lines, baselines, and words. Researchers can train Handwritten Text Recognition (HTR) models using transcriptions produced with or imported into Transkribus. Previously, users could train *HTR+* models [27] (now replaced by *PyLaia* models). HTR+ and PyLaia models are neural network architectures that recognise handwriting with limited training data. Even for challenging datasets, [10] have shown that CERs between 1% and 10% are possible. Additionally, HTR models can successfully recognise printed documents, resulting in low character error rates (CERs) [22].

[13] found that ten pages of ground truth data can achieve a 1.4% CER for German blackletter newspapers using a combination of CNNs and RNNs, image binarisation, padding, and synthetic data generation. They used a simpler architecture than [27] proposed.

In a study by [28], OCRopy and *Calamari*[8] were compared for transcribing medieval printed books in German and Latin. The amount of training data used was found to significantly impact the systems' performance, with both systems showing improved CER as the number of lines in the training set increased, starting from 60 lines. E. g., when Calamari was trained with 3,000 lines instead of 60, the CER lowered from 4.9% to 0.43% for data from 1488. However, the study did not identify an optimal number of lines for the training set.

Researchers in Digital Humanities use various OCR models to extract text from historical documents. However, understanding the distinct processing steps can be challenging since no standardised workflow exists. To simplify the process of training OCR models for collections, *OCR-D*[9] [15] creates standards for applying and testing different OCR software instead of restricting users to a single system. This makes it easier to compare different systems.

3   Data

3.1   Newspapers – The NZZ

To include a representative for printed data, we worked with the German-language *Neue Zürcher Zeitung* (NZZ) dataset [22][10]. It comprises 167 pages in blackletter. This amounts to 43,151 lines equalling 304,286 words. For reference, we used the same train, validation, and test split as published on GitHub.

TrOCR has neither seen blackletter nor German during training. We check whether and how quickly it adapts to this data. Furthermore, because blackletter is close to Gothic handwriting and TrOCR has both a pre-trained printed and handwritten model, we test which is more advantageous for blackletter fonts.

---

[6] https://readcoop.eu/.

[7] https://readcoop.eu/transkribus/?sc=Transkribus.

[8] https://github.com/Calamari-OCR/calamari.

[9] https://ocr-d.de/.

[10] Available on GitHub: https://github.com/impresso/NZZ-black-letter-ground-truth.

3.2 ICFHR 2018 Competition Dataset

The data used in this analysis is from the 2018 International Conference on Frontiers in Handwriting Recognition competition [21]. The challenge was to minimise CERs using a pre-trained model trained on general data. It also required adapting to smaller training sets that included handwriting styles the pre-trained model had not seen. We chose this dataset because it represents the state-of-the-art in traditional ATR architectures and has been utilised in subsequent studies (e. g., [1]).

The dataset is divided into two segments: 1) a *pre-training* set of 17 documents, each comprising roughly 25 pages. This equals 11,925 lines containing a total of 75,145 words. 2) a *fine-tuning* set of five documents, each with a distinct script. The second part contains 2,878 lines and 16,399 words. For each document in the fine-tuning part, there is one document to form a

Table 1. ICFHR2018 fine-tuning data set sizes. The columns correspond to the number of lines in one, four, or 16 pages.

	# of lines			
# of pages:	1	4	16	test set
Konzilprotokolle	29	116	469	358
Schiller	21	84	328	244
Ricordi	19	88	383	280
Patzig	38	156	641	576
Schwerin	68	264	1,057	718

test set of similar size. The whole dataset statistics are available in [21]. The competition design included size ablation experiments to determine how many lines of a new document a pre-trained model needs to see to produce satisfactory results. Therefore, we can further divide the fine-tuning data into different sizes, as displayed in Table 1 from [1].

Upon manual inspection, we found that the data set is highly diverse concerning languages (English, (Old) German, Latin, Italian, Swedish, and Norwegian), making it multilingual. However, the fine-tuning part only contains German and Italian. This has implications for pre-training and specific fine-tuning, which we will address in the experiments outlined in Sect. 4.

We use this dataset to conduct controlled experiments on the impact of pre-fine-tuning and further fine-tuning on TrOCR. These experiments will guide us in how to work with other handwritten data that we will introduce later.

3.3 Bullinger Dataset

The Bullinger dataset contains scanned images and text lines from the correspondence of Heinrich Bullinger (1504–1575). It is one of the largest letter networks of the 16th century, comprising nearly 12,000 letters preserved and digitised by the Zurich States Archive and the Central Library of Zurich. There are 3,141 edited transcriptions available [7] plus an estimated 5,400 "working transcriptions" produced by scholars or students.

We aligned scanned documents with available transcriptions using Transkribus and its *Text2Image* tool developed by CITlab Rostock[11]. The

---

[11] https://github.com/CITlabRostock/CITlabModule.

Text2Image process involves an initial layout analysis to identify text lines in the scan, followed by a text recognition step that compares the recognised text to passages in the available transcriptions. It is an iterative, semi-supervised approach where the base HTR model is continually trained with the aligned material to improve subsequent alignments. However, this process is prone to errors: misalignments occur frequently at the beginning and end of lines.

Table 2. Summary of the Bullinger correspondence data aligned via Text2Image.

	# lines	%	# words
LA	134,236	81.02	1,073,106
ENHG	31,437	18.98	253,372
Total	165,673		1,326,478

Table 2 summarises the training data generated with Text2Image and can be accessed on GitHub[12]. We separated the data into Latin and Early New High German (ENHG) using the language identifier developed by [26].

After obtaining layout analysis from Transkribus, we had multiple options for image pre-processing, illustrated in Fig. 1. The layout analysis identifies the lines in an image and outlines a polygon around the handwriting within the line. The different pre-processing techniques include a) cropping the line image to follow the polygon outline and placing it on a white background, b) rectangular cropping based on the highest and lowest points of the polygon, and c) cropping with the average colour of the line as the background. These techniques may affect the training process, which we will examine further in Sect. 4.

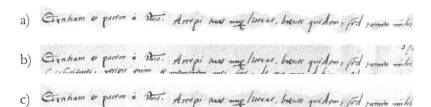

Fig. 1. Different pre-processing strategies.

With this dataset, we assess TrOCR's simultaneous adaptability to diverse hands, for some of which have only ten to twenty lines. Additionally, we will develop a model specifically for Bullinger's handwriting, as he is the primary contributor to the dataset with 33,675 lines.

3.4 Gwalther Dataset

We gathered a dataset of excerpts from Rudolf Gwalther's *Lateinische Gedichte*[13] using scans and partial transcriptions by Prof. em. Dr. Peter Stotz

---

[12] https://github.com/pstroe/bullinger-htr.

[13] *Gwalther*, Rudolf: [Lateinische Gedichte]. [Zürich], 1540-1580. Zentralbibliothek Zürich, Ms D 152.

(1942–2020) from e-manuscripta. As of May 2022, we have transcripts of 142 pages, resulting in 4,037 lines and 26,088 words. We aligned the transcriptions with the lines using Transkribus and made minor corrections for consistent capitalisation and punctuation. The resulting dataset is available on Zenodo[14].

This dataset will prove helpful because the Bullinger dataset contains approximately 75 letters penned by Gwalther to Bullinger. Figure 2 highlights the contrast between Gwalther's handwriting in (a) his manuscript *Lateinische Gedichte* and (b) a letter addressed to Bullinger. Hence, we can determine whether the quality of the recognised text is affected by the domain (correspondence or book volume).

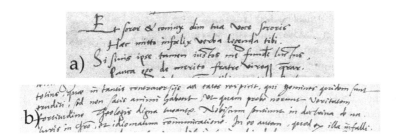

Fig. 2. Gwalther's handwriting from (a) a manuscript, and (b) from a letter.

3.5   Huber Dataset

We have a dataset of letters by Eugen Huber (1849–1923), a Swiss lawyer and politician who wrote almost daily for over eight years to his deceased wife, containing details about his daily life and work. We aligned existing transcriptions and manually transcribed 3,757 pages from the letter collection using Transkribus. The Huber lines were pre-processed in the same manner as the ones from the Bullinger dataset. Our efforts have produced a ground truth of 88,154 lines comprising 679,533 words. This dataset is approximately 22 times larger than the Gwalther dataset regarding the number of lines, but only 2.5× the size of the Bullinger's account in the Bullinger dataset. The Huber dataset is available on GitHub[15] (Fig. 3).

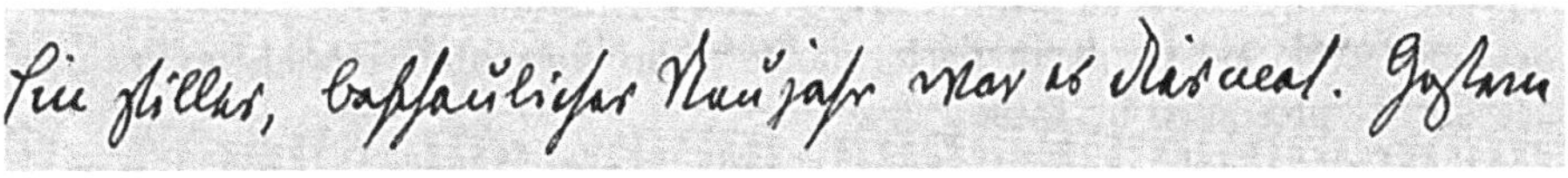

Fig. 3. Example line from a preprocessed line from the Huber dataset.

---

[14] https://zenodo.org/record/4780947.
[15] https://github.com/pstroe/huber-htr.

The handwriting in this dataset is from a later period than the ICFHR2018, Bullinger, and Gwalther datasets. With the Huber dataset, we analyse how the period of a source document can impact models. Moreover, since we will have trained models on the single-author Gwalther dataset, we can make assumptions about single-author models. By additionally training on Bullinger's handwriting only we provide further insights into how TrOCR adapts to a single hand.

4   Experiments and Results

4.1   TrOCR Performance on Blackletter

We utilised the data mentioned in Sect. 3.1 and split it into 80/10/10 for training, validation, and testing. After fine-tuning both the printed and handwritten pre-trained TrOCR$_{LARGE}$ models with the data[16], we computed the CER. The results of fine-tuning the TrOCR model pre-trained on printed data are remarkable. With just one epoch of fine-tuning, the CER improved from 88.92% to 0.99%. This is an impressive 8,981% improvement and demonstrates that TrOCR competes with traditional OCR methods. For instance, an HTR+ model in Transkribus achieves a CER of 0.67%, while a Transkribus PyLaia model achieves 0.6% (Table 3).

Table 3. CERs for different numbers of training epochs for TrOCR

epochs	model type	CER
zero-shot		88.92
1		0.99
5	printed	0.89
10		0.97
15		1.22
zero-shot		33.41
10	handwritten	1.21
15		1.3

One unique aspect of the results is that the handwritten TrOCR model had a better zero shot performance, but it quickly lost its advantage over the printed model.

4.2   TrOCR Performance on Handwritten Data

ICFHR2018. To test the effectiveness of pre-finetuning and specialised fine-tuning for different dataset sizes, we used the ICFHR2018 dataset. We pre-fine-tuned the TrOCR$_{LARGE}$ model for 1, 3, 5, 10, and 15 epochs and used the resulting models as base models to further fine-tune on 1, 4, or 16 documents for each fine-tuning set, as shown in Table 1. The CERs obtained by averaging over document sizes and zero-shot performance of pre-fine-tuned models on specific datasets are presented in Table 4.

As we fine-tune the model for more epochs, its performance steadily improves. However, we observed only a slight increase in performance from 5 to 15 epochs. Therefore, the best-performing model is fine-tuned for around 5 to 10 epochs, considering a dataset of approximately 11k line images. TrOCR's performance is slightly lower than the state-of-the-art, averaging 16.54% as reported in [1].

---

[16] See Sect. 4.2 why we only fine-tuned LARGE models.

Table 4. CERs on the test sets of the different fine-tuning sets. (Fine-tuning) Set size 0 means zero-shot evaluation of the pre-fine-tuned model.

FT set	set size	epochs				
		1	3	5	10	15
Konzilpr.	0	15.81	12.09	12.23	12.31	12.65
	1	13.86	13.40	10.19	12.28	11.61
	4	11.99	9.43	7.99	8.62	8.62
	16	7.84	4.89	**4.62**	5.84	6.44
Schiller	0	23.50	20.24	19.84	20.16	19.15
	1	30.85	20.43	20.55	20.58	20.46
	4	23.19	19.79	18.18	18.71	18.21
	16	18.08	13.70	12.54	**12.07**	13.46
Ricordi	0	35.24	37.78	34.38	37.66	38.52
	1	33.46	32.01	30.30	30.84	31.33
	4	28.68	22.86	23.55	23.45	25.79
	16	17.69	14.11	**13.53**	13.75	15.72
Patzig	0	29.85	27.94	28.11	27.74	26.99
	1	27.80	22.89	22.15	22.55	23.52
	4	21.80	16.88	15.72	16.50	16.71
	16	15.62	**10.65**	10.96	11.00	11.21
Schwerin	0	28.49	28.89	28.04	27.58	27.31
	1	19.61	14.89	13.71	14.02	14.21
	4	12.10	8.30	7.57	7.68	8.53
	16	5.34	4.41	4.20	**3.99**	4.32
Average		21.04	17.78	**16.92**	17.37	17.74

We used the random warp grid distortion (RWGD) technique [29] for data augmentation to create ten modified versions of each line, as it has been shown to improve performance and reduce the risk of overfitting in the field [1,21]. RWGD involves perturbing the intersections between horizontal and vertical lines in a grid placed over the line image, using a normal distribution to determine the direction and distance. We followed the procedure outlined in [1] and scaled the grid size and standard deviation to our line images. This technique resulted in a tenfold increase in the number of lines for pre-fine-tuning (i. e., 114,681). An example of RWGD applied to a line image is shown in Fig. 4.

Then, we conducted pre-fine-tuning with the augmented data from the pre-training training set while specific fine-tuning was restricted to the original dataset. During our testing, we observed a gradual increase in CER from 19.94 to 27.1 for epochs one to fifteen. Our best result (obtained on one epoch) still fell 3.02% points short of the top pre-fine-tuned models that did not use augmented data. The CERs suggest suboptimal learning rates rather than overfitting because we should have seen a decrease in CER.

To improve model performance, a warmup[17] period was added with an Adafactor learning rate schedule [18]. The models pre-fine-tuned on one and three epochs were re-trained with and without data augmentation. 10% of the overall optimisation steps were used as a warmup period. The 3-epoch model without data augmentation had a CER of 17.08%, while the 1-epoch model with

---

[17] Warmup describes a procedure by which a very small learning rate is used at the beginning for a certain number of steps to adapt the model to new data slowly.

data augmentation had a CER of 16.31%, beating the best model from earlier and the state-of-the-art in [1]. This performance gain is only due to the warmup, showing that slowly adapting an already powerful model like TrOCR to new data is beneficial. Especially for relatively small datasets, data augmentation in combination with warmup needs to be considered.

For a small dataset, fine-tuning fewer epochs with warmup periods, particularly with data augmentation, are beneficial. The composition of languages in the pre-fine-tuning data appears to affect performance. Specifically, the Ricordi dataset is the only fine-tuning set containing Italian lines, while the pre-fine-tuning data only contains Latin lines. The zero-shot performances of models pre-fine-tuned for one epoch show that the data-augmented model per-

Fig. 4. Example of RWGD applied to a line image (original on top) from the *Konzilprotokolle* dataset to produce different versions (here five)

forms 4.64% points better than the one without. This is true for most fine-tuning sets, possibly due to the number of Latin lines in the pre-fine-tuning data.

Bullinger. We utilised the Bullinger dataset featured in Sect. 3.3 for our experiments. Specifically, we worked with the pre-processed version c) displayed in Fig. 1. Our objective is to analyse TrOCR's behaviour concerning (1) the required number of epochs to train, (2) the influence of pre-processing, (3) the influence of data augmentation, and (4) the influence of multilingualism.

We split the data into training, validation, and test sets at an 80/10/10 ratio. To investigate TrOCR's bias towards frequent writers, we evaluated the models on two additional test sets: one with 1,235 lines from writers who had written at least ten letters to Bullinger and another with 1,013 lines from infrequent writers. Since the writers' distribution is almost Zipfian, with few writing frequently and many writing only once or rarely, we expect all fine-tuned models to be biased towards frequent writers. We also separated the frequent and infrequent writer test sets' training data into Latin and Early New High German (ENHG) using the tool from Sect. 3.3.

Table 5 shows the results concerning issues (1) and (4). Based on the test set's performance, we can conclude that the results stabilise after five epochs. Fine-tuning TrOCR for ten epochs resulted in a CER of 9.7%, which leads us to expect no more decrease in CER after six epochs. We also conducted experiments by fine-tuning a TrOCR model on the Latin and ENHG data for five epochs. Overall, we observed that the performance on frequent-writer data is better across all models, which confirms our hypothesis that there is a bias towards frequent writers. However, the gap between the performance is relatively small. TrOCR performs well in recognising 16th century handwriting.

Table 5. CERs of different TrOCR models trained on different language splits and evaluated on different datasets.

			CER					
			frequent writers			infrequent writers		
training data	# epochs	test set	multi	Latin	ENHG	multi	Latin	ENHG
multi	1	9.38	9.53	9.49	9.70	11.58	11.07	12.79
	2	8.25	8.38	8.51	7.71	10.46	10.18	11.11
	3	7.53	7.81	8.07	6.50	9.95	9.61	10.74
	4	7.20	**7.21**	**7.61**	5.27	9.68	**9.31**	10.55
	5	7.07	7.31	7.85	5.25	9.69	9.54	**10.25**
	6	**7.06**	7.41	7.89	**5.09**	**9.61**	9.39	10.11
Latin	5	7.02	11.09	**7.99**	26.21	16.10	**9.69**	31.21
ENHG	5	9.97	28.41	33.02	**6.06**	27.89	34.83	**11.43**

Regarding language, we have observed that the multilingual model surpasses the performance of fine-tuned models for a specific language. For instance, the multilingual model demonstrates a lower CER for the ENHG test splits in both the frequent and infrequent writers test sets.

To test the impact of issue (3) (data augmentation) on TrOCR fine-tuning, we created additional line versions using the same method as in Sect. 4.2. However, due to the ample amount of data present in the Bullinger dataset, we only produced up to six different versions to avoid an excessive amount of training data. The results showed that the more data-augmented lines we use, the better the performance. TrOCR delivered the best results with six augmented versions of a line and two fine-tuning epochs, outperforming the model fine-tuned for six epochs by 0.4% points. This demonstrates that data augmentation is a valuable technique to enhance performance, even for larger datasets, albeit the gains are smaller.

To find answers to issue (2), we tested whether using exact polygon cut-outs on a white background would negatively affect TrOCR's performance by fine-tuning TrOCR on such images for one to three epochs. The resulting CERs were 20.51%, 8.38%, and 7.61%, respectively. The high CER of 20.51% suggests that TrOCR is influenced by being presented with clean cut-outs of line images. However, continuous fine-tuning helped TrOCR focus on the line images' relevant areas, as evidenced by the considerably lower CER after two and three epochs. This could be promising for applying TrOCR models, as they can learn the location of actual handwriting in a line they need to detect.

Bullinger Model. The author distribution shows a heavily skewed image. This requires exploring different training strategies. One could be to train specific models for the authors for which we have the most training material. Starting with Bullinger himself, who contributes roughly 20% to the overall Bullinger dataset, we fine-tuned a model just to Bullinger's handwriting. We achieved the best CER performance of 9.2% while training for ten epochs. The model trained on one epoch achieved a CER of 11.43%. We note that there were no considerable performance gains between fine-tuning one to ten epochs.

Gwalther. To compare with Transkribus directly, we divided our training and testing data into a 90/10 split[18]. Our experiment's baseline was an HTR+ model in Transkribus that was trained for 50 epochs using a base model[19]. We used the same approach and fine-tuned both the

Table 6. CER of HTR+ versus TrOCR while TrOCR has been trained on several epochs and with two configurations.

Model	fine-tuning epochs					epochs
	3	5	10	15	20	50
HTR+	-	-	-	-	-	2.74
TrOCR$_{BASE}$	3.84	3.72	**3.18**	3.31	3.62	-
TrOCR$_{LARGE}$	2.94	2.72	2.58	**2.55**	2.62	-

BASE and LARGE instances of the TrOCR model with the Gwalther training set. The results of our experiment are displayed in Table 6.

First, the baseline set by the HTR+ with a CER of 2.74% is quite low. As concerns TrOCR, the BASE model is incapable of beating the baseline. However, the LARGE model beats the baseline by 0.19% points after only 15 epochs of fine-tuning. For the Gwalther data, the CER decreases with an increasing number of fine-tuning epochs, although after a certain number of epochs, the CER increases again. The differences in CER between models fine-tuned for different numbers of epochs are sometimes small. Judging from these results, there is a slight preference for the LARGE model.

It is worth noting that the HTR+ sets a very low baseline with a CER of only 2.74%. Regarding TrOCR, the BASE model falls short of beating the baseline. However, after just 15 epochs of fine-tuning, the LARGE model outperforms the baseline by 0.19% points. As for the Gwalther data, the CER tends to decrease with more fine-tuning epochs, although it increases again after a certain point. While the differences in CER between models fine-tuned for different numbers of epochs may be minor, the LARGE model appears to have a slight edge based on these results.

Surprisingly, our fine-tuned TrOCR model performed well in the experiments despite not having encountered any Latin data before. During fine-tuning on the

---

[18] In Transkribus, only a training and validation set can be defined.

[19] The base model, Acta_17 HTR+, was compiled by Alvermann et al. from the University of Greifswald, containing approximately 600k words in Latin and Low German written by 1k different authors from the period of 1580–1705. It was trained over 1k epochs.

Gwalther dataset, the model was only exposed to 23k Latin words. Despite the training and test set having a vocabulary overlap of only 68.9%, TrOCR outperformed the baseline.

We get an entirely different picture when we apply the Gwalther model trained on the book volume to a letter containing 42 lines. The best model is the LARGE model trained for three epochs with a CER of 39.3%. Although the writer is the same and the handwriting is similar, we do not achieve CERs near the ones presented above. This means that although TrOCR is a versatile model, it has its limits.

Huber. We split the Huber dataset into training, validation and test sets using an 80/10/10 split. We fine-tuned the handwritten $TrOCR_{LARGE}$ model on the training set for one to three epochs[20]. We found CERs on the test set of 14.42%, 4.94%, and 4.36% respectively. With the best results, the Huber model is second-best compared to the other single-hand models trained on Gwalther's (2.55%) and Bullinger's (9.2%) hands. The fact of a worse performance than Gwalther despite having been trained on more data shows that quantity is not the only decisive factor (see Sect. 5).

5   Conclusion and Outlook

TrOCR has proven to be a flexible model based on our experiments and their results. It can accurately recognise printed material and decode various handwriting styles, including those in multilingual settings. Moreover, TrOCR requires minimal training data to adapt to new handwriting styles and languages. As a result, pre-trained Transformer-based ATR models are excellent candidates for use in the digitalisation process of libraries and archives.

It is worth noting that handwriting style impacts the overall text quality, specifically regarding CER. While we achieved good results for Gwalther and Huber, the Bullinger model performed significantly worse by 6.65% points compared to Gwalther and 4.84% points compared to Huber. Bullinger's handwriting is notoriously challenging to decipher, even for humans. This also explains why the Huber model performs worse than the Gwalther model despite being fine-tuned on considerably more data. Comparing the two handwriting styles shows that Huber's is more difficult to read than Gwalther's. In this context, fine-tuning on more data does not always improve the results. Therefore, when applying and improving ATR models, it is essential to consider the circumstances under which letters were written and the intended audience. However, fine-tuning models like TrOCR for individual authors can still be beneficial.

One possible avenue for future research is to design training procedures that can handle diverse sets of handwriting samples and to identify the ones that pose greater challenges than others. Additionally, it remains unclear whether

---

[20] We gathered information from previous experiments, i. e., we trained with warmup and line images with a random background.

the errors made by TrOCR are similar to those made by other OCR techniques. A deeper analysis of these errors could help us decide whether to improve the image encoding or the decoding side of Transformer-based ATR models.

Acknowledgements. This work has been supported by the Hasler Foundation, Switzerland. We would also like to thank Anna Janka, Raphael Müller, Peter Rechsteiner, Dr. Patricia Scheurer, David Selim Schoch, PD Dr. Raphael Schwitter, Christian Sieber, Martin Spoto, Jonas Widmer, and Dr. Beat Wolf for their contributions to the dataset and ground truth creation.

References

1. Aradillas, J.C., Murillo-Fuentes, J.J., Olmos, P.M.: Boosting offline handwritten text recognition in historical documents with few labeled lines. IEEE Access **9**, 76674–76688 (2021). https://ieeexplore.ieee.org/document/9438636
2. Bao, H., Dong, L., Wei, F.: BEiT: BERT pre-training of image transformers. In: International Conference on Learning Representations (2021). https://openreview.net/pdf?id=p-BhZSz59o4
3. Causer, T., Grint, K., Sichani, A.M., Terras, M.: 'Making such bargain': transcribe Bentham and the quality and cost-effectiveness of crowdsourced transcription. Digit. Sch. Hum. **33**(3), 467–487 (2018). https://doi-org.ezproxy.uzh.ch/10.1093/llc/fqx064
4. Devlin, J., Chang, M.W., Lee, K., Toutanova, K.: BERT: pre-training of deep bidirectional transformers for language understanding. In: Proceedings of NAACL-HLT, pp. 4171–4186 (2019). https://doi.org/10.48550/arXiv.1810.04805
5. Dosovitskiy, A., et al.: An image is worth 16×16 words: transformers for image recognition at scale. In: International Conference on Learning Representations (2021). https://openreview.net/forum?id=YicbFdNTTy
6. Fukushima, K.: Neocognitron: a self-organizing neural network model for a mechanism of pattern recognition unaffected by shift in position. Biol. Cybern. **36**(4), 193–202 (1980). https://doi.org/10.1007/BF00344251
7. Gäbler, U., et al. (eds.): Heinrich Bullinger Briefwechsel. Heinrich Bullinger Werke, Theologischer Verlag Zürich (1974–2020)
8. Granell, E., Chammas, E., Likforman-Sulem, L., Martínez-Hinarejos, C.D., Mokbel, C., Cirstea, B.I.: Transcription of Spanish historical handwritten documents with deep neural networks. J. Imaging **4**, 15 (2018). https://doi.org/10.3390/jimaging4010015
9. Hochreiter, S., Schmidhuber, J.: Long short-term memory. Neural Comput. **9**(8), 1735–1780 (1997). https://doi.org/10.1007/978-3-642-24797-2_4
10. Hodel, T.M., Schoch, D.S., Schneider, C., Purcell, J.: General models for handwritten text recognition: feasibility and state-of-the art. German Kurrent as an example. J. Open Hum. Data **7**(13), 1–10 (2021). https://doi.org/10.5334/johd.46
11. LeCun, Y., et al.: Backpropagation applied to handwritten zip code recognition. Neural Comput. **1**(4), 541–551 (1989). https://doi.org/10.1162/neco.1989.1.4.541
12. Li, M., et al.: TrOCR: transformer-based optical character recognition with pre-trained models. arXiv (2021). https://doi.org/10.48550/arXiv.2109.10282
13. Martínek, J., Lenc, L., Král, P.: Training strategies for OCR systems for historical documents. In: MacIntyre, J., Maglogiannis, I., Iliadis, L., Pimenidis, E. (eds.) AIAI 2019. IAICT, vol. 559, pp. 362–373. Springer, Cham (2019). https://doi.org/10.1007/978-3-030-19823-7_30

14. Mühlberger, G., Seaward, L., Terras, M., et al.: Transforming scholarship in the archives through handwritten text recognition: Transkribus as a case study. J. Documentation **75**(5), 954–976 (2019). https://www.emerald.com/insight/content/doi/10.1108/JD-07-2018-0114/full/html

15. Neudecker, C., et al.: OCR-D: an end-to-end open source OCR framework for historical printed documents. In: Proceedings of the 3rd International Conference on Digital Access to Textual Cultural Heritage, pp. 53–58 (2019). https://doi.org/10.1145/3322905.3322917

16. Ruder, S., Peters, M.E., Swayamdipta, S., Wolf, T.: Transfer learning in natural language processing. In: Proceedings of the 2019 Conference of the North American Chapter of the Association for Computational Linguistics: Tutorials, pp. 15–18 (2019). https://aclanthology.org/N19-5004/

17. Schuster, M., Paliwal, K.K.: Bidirectional recurrent neural networks. IEEE Trans. Signal Process. **45**(11), 2673–2681 (1997). https://doi.org/10.1109/78.650093

18. Shazeer, N., Stern, M.: Adafactor: Adaptive learning rates with sublinear memory cost. In: International Conference on Machine Learning, pp. 4596–4604 (2018). http://proceedings.mlr.press/v80/shazeer18a.html

19. Springmann, U., Fink, F., Schulz, K.U.: Automatic quality evaluation and (semi-) automatic improvement of OCR models for historical printings. arXiv (2016). https://doi.org/10.48550/arXiv.1606.05157

20. Springmann, U., Lüdeling, A.: OCR of historical printings with an application to building diachronic corpora: A case study using the RIDGES herbal corpus. Digit. Hum. Q. **11**(2) (2017). https://arxiv.org/abs/1608.02153

21. Strauß, T., Leifert, G., Labahn, R., Hodel, T., Mühlberger, G.: ICFHR2018 competition on automated text recognition on a READ dataset. In: 16th International Conference on Frontiers in Handwriting Recognition, pp. 477–482 (2018). https://ieeexplore.ieee.org/document/8583807

22. Ströbel, P.B., Clematide, S.: Improving OCR of black letter in historical newspapers: the unreasonable effectiveness of HTR models on low-resolution images. In: Proceedings of the Digital Humanities 2019 (2019). https://www.zora.uzh.ch/id/eprint/177164/

23. Terras, M.: The role of the library when computers can read. In: The Rise of AI: Implications and Applications of Artificial Intelligence in Academic Libraries, pp. 137–148. ALAstore (2022). https://www.pure.ed.ac.uk/ws/portalfiles/portal/255303209/Rise_of_AI_Chapter_11.pdf

24. Touvron, H., Cord, M., Douze, M., Massa, F., Sablayrolles, A., Jégou, H.: Training data-efficient image transformers & distillation through attention. In: International Conference on Machine Learning, pp. 10347–10357 (2021). https://proceedings.mlr.press/v139/touvron21a

25. Vaswani, A., et al.: Attention is all you need. In: Advances in Neural Information Processing Systems, pp. 5998–6008 (2017). https://dl.acm.org/doi/10.5555/3295222.3295349

26. Volk, M., et al.: Nunc profana tractemus. Detecting code-switching in a large corpus of 16th century letters. In: Proceedings of the 13th Language Resources and Evaluation Conference, pp. 2901–2908 (2022). https://aclanthology.org/2022.lrec-1.311/

27. Weidemann, M., Michael, J., Grüning, T., Labahn, R.: HTR engine based on NNs P2 building deep architectures with TensorFlow. Technical report (2018). https://read.transkribus.eu/wp-content/uploads/2018/12/D7.9_HTR_NN_final.pdf

28. Wick, C., Reul, C., Puppe, F.: Comparison of OCR accuracy on early printed books using the open source engines Calamari and OCRopus. Spec. Issue Autom. Text Layout Recogn. **33**(1), 79–96 (2018). https://doi.org/10.21248/jlcl.33.2018. 219
29. Wigington, C., Stewart, S., Davis, B., Barrett, B., Price, B., Cohen, S.: Data augmentation for recognition of handwritten words and lines using a CNN-LSTM network. In: 14th IAPR International Conference on Document Analysis and Recognition, pp. 639–645 (2017). https://ieeexplore.ieee.org/document/8270041

Using GANs for Domain Adaptive High Resolution Synthetic Document Generation

Tahani Fennir[1,3], Bart Lamiroy[2,3(✉)] [iD], and Jean-Charles Lamirel[4,5] [iD]

[1] Vasa, 4, Rue Gustave Eiffel, Rosières-près-Troyes, France
`tahani.fennir@vasa.fr`
[2] Université Reims Champagne-Ardenne, CReSTIC, EA 3804, Reims, France
`bart.lamiroy@univ-reims.fr`
[3] Université de Lorraine, LORIA, UMR 7503, Metz, France
[4] Université de Strasbourg, LORIA, UMR 7503, Strasbourg, France
[5] Dalian University, Dalian, China

Abstract. In this paper we are addressing one specific problem of Context-Adaptive Document Analysis: the generation of specific learning data. While many Document Analysis solutions exist for well described cases, it is often difficult to adapt them to other contexts. We present an ongoing research methodology for generating synthetic documents that are not only language and layout-specific but also compatible with user-defined application contexts. In the contemporary digital landscape, corporations face a considerable challenge: the scarcity of high-resolution, accurately labeled data, specifically in the context of French language documents, which often exhibit stylistic nuances distinct from English counterparts. This paper addresses the need for high-resolution synthetic documents, establishing it as the primary impetus behind our research. While the DocSynth model [2] (on which our work is based) represented a preliminary step towards the generation of realistic document images, it was impeded by its incapacity to produce high-resolution outputs. Our research aims to overcome this restriction, enhancing the DocSynth model's capacity to generate high-resolution document images. Additionally, we strive to equip it with the ability to recognize and generate text, effectively addressing a crucial demand within the industry.

1 Introduction

The field of Document Analysis and Recognition continues to grapple with the challenge of automatic document understanding. In the evolving digital landscape, both scanned paper documents and digitally-born documents coexist, particularly within corporate scenarios. These documents, ranging from forms, invoices, to contracts, exhibit significant variability. The escalating adoption of Robotic Process Automation tools within paperless offices underscores the need for an automated approach to managing information in document workflows. This approach necessitates a comprehensive understanding and interpretation of these documents.

As a tool for document understanding, the syntactical description of a its layout offers a comprehensive perspective of the document, providing context to individual components such as named entities, graphical symbols, and key-value associations. Consequently, when driven by the document layout, the performance of information extraction is significantly enhanced. The Office Document Architecture [4] proposes standardized guidelines for document representation, emphasizing two crucial aspects: one perspective views a document as an image intended for display or printing, while the other dissects its textual and graphical elements to interpret the document's layout and logical structure. The layout structure primarily comprises various layout objects, such as text blocks, images, tables, lines, words, characters, among others. In contrast, the logical structure encapsulates the semantic relationships between conceptual elements like the company logo, signature, title, body, paragraph regions, *etc.*

The recognition of a document's layout, a challenge that has persisted for several decades, remains an essential precursor for information extraction. Moreover, in the domain of business intelligence, the large-scale extraction of information from document content is paramount for effective decision-making processes. While layout detection seems to achieve quite robust and good results on generic documents [1], it remains an issue to fine-tune and adapt these algorithms to specific collections having particular properties. This is even more the case when the documents are written in low-resource languages.

The advent of deep learning, while providing new insights into layout understanding, has also highlighted the need for adequately labeled data to supervise learning tasks. Many companies, particularly those operating in low-resource language regions, confront the issue of a shortage of such data. This scarcity hampers the effective training of machine learning models that aim to accurately understand and interpret a diverse range of documents. Given these challenges, it is obvious that there is a need for increased efforts in data collection and labeling, coupled with data augmentation strategies such as the synthetic generation of realistic images. These efforts will contribute to enhancing the performance and reliability of automatic document understanding systems.

This paper describes an empirical approach aiming at designing a tool for the generation of convincing replicas of printed documents. These replicas contain credible layout components, as evidenced in Fig. 1. This model, termed the automated document image synthesizer, has been architected to facilitate the management of an array of document types, spanning paper-based to electronic formats, via a singular, user-centric platform. The consequential practical application of this development is to improve the efficiency of visual search engines and information retrieval systems. Under the customary *modus operandi* in document retrieval tasks, a user typically seeks to index within a repository of genuine documents. Alternatively, our approach promotes the generation of query sample variations, consequently enhancing the performance metrics of the retrieval model.

In an attempt to build upon the preliminary work executed using the Doc-Synth model [2], we have targeted our efforts towards the realization of two goals. Primarily, we want to elevate the image resolution, thereby yielding a significant improvement in the quality of the synthesized document images. Furthermore, we aim to expand the language proficiencies of the model, specifically embedding support for French. The projected outcome is the evolution of the DocSynth model into a multilingual image document generator, capable of delivering high-resolution image documents, while simultaneously retaining the beneficial attributes inherent to the original model.

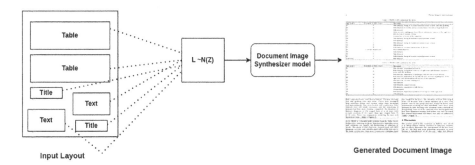

Fig. 1. Schematic Representation of the Automated Document Image Synthesizer Producing Synthetic Document Replicas: Provided with an input document layout consisting of object bounding boxes and categories assembled within an image lattice, our model draws from a normal distribution for the semantic and spatial attributes of each layout object. Consequently, it can generate a multitude of plausible document images as per the user's requirement.

The approach developped in this paper builds upon recent advances in neural rendering technology. We use deep generative models, such as Generative Adversarial Networks (GANs) and Variational Autoencoders (VAEs), that have the capacity to learn complex transformations from captured images and transform them into new synthetic images. These models successfully integrate knowledge projection and geometry, with learned components, thereby creating algorithms for controllable image generation.

Consequently, the adoption of GANs for this task is warranted given their demonstrated prowess in generating high-quality synthetic images, effectively addressing the challenge of reconstruction and rendering complex document layouts.

The rest of this paper is structured as follows:

- Section 2 presents a review of the existing literature .
- In Sect. 3, we articulate the primary methodological contributions derived from our research investigation.
- Section 4 is dedicated to delineating the specific processes and parameters related to the experimental training phase of our study.

- Section 5 delves into Experimental Training.
- Section 6 enumerates the results acquired from our experiment.
- Section 7 presents the Evaluation
- Section 8 is reserved for concluding remarks, summarizing the pivotal findings of our research, and the potential implications for the field.

2   Literature Review

The examination of structural and spatial correlations within intricate document layouts still represents a considerable challenge for Document Analysis and Recognition. The identification of physical and logical layouts in documents is fundamental to operations such as Optical Character Recognition and document image transcription, document categorization, information extraction *etc.* For an in-depth exploration of the latest advancements in the field of document layout analysis, please refer to [1].

In alignment with the objective outlined in Sect. 1, the primary goal of this research is the development of a generative neural model. This model is designed to create visually plausible document images from a provided reference layout, emphasizing particularly on high-resolution images. The underpinning concept of this approach finds its roots in the DocSynth model [2], which advocates for data augmentation and the generation of corresponding ground truth using automatically synthesized images – a strategy that has incited considerable interest within the Computer Vision community. Post the introduction of Generative Adversarial Networks (GANs) by Goodfellow in 2014 [5], there has been a proliferation of variants that have displayed remarkable proficiency in generating realistic images, ranging from handwriting digits and faces to natural scenes [11]. Yet, the precise control over image generation, specifically in relation to the composition and arrangement of objects, remains an intriguing scientific challenge in this domain.

Recent research has started to make headway into addressing this issue. Notably, Lake et al. [7] proposed a hierarchical generative model capable of assembling complete objects from individual parts. The model showcased its effectiveness by generating Omniglot characters composed of separate strokes, paving the way for more controlled image synthesis. Another significant contribution comes from Zhao et al. [13], who devised a model that can generate a collection of realistic images with objects situated in specific locations. The model utilizes a reference spatial distribution of bounding boxes and object labels as inputs, thus offering a degree of control over the image generation process. Our research is significantly influenced by these advances, with the aim to further the progress of controlled synthetic image generation.

The fidelity of layout representations has significant relevance across diverse contexts within the graphical design arena, with a particular focus on intricate, highly structured, and content-rich entities. Addressing this facet necessitates methodologies that can navigate the complex interplay of these elements. Recent research has demonstrated substantial progress in this regard. A notable exemplar is the work of Li et al. [9] introducing LayoutGAN. Their work is based on the implementation of Generative Adversarial Networks (GANs), amalgamated with a wireframe rendering layer, with a specific orientation towards the generation of realistic document layouts. This approach combines the subtleties of genuine design representation and the robust capabilities inherent to artificial design techniques. Consequently, LayoutGAN stands as a prominent advancement in harmonizing artificial design methods with actual layout representations. Contrastingly, Zheng et al.'s [14] research provides a GAN-based methodology that diverges from the aforementioned study, as it centralizes on generating document layouts that are inherently content-aware. Their method capitalizes predominantly on the content of the document as an auxiliary prior, offering a layout generation process that is substantially more sensitive to context. This approach, thus, presents a compelling alternative that further enriches the field of layout representation studies.

In conclusion, despite the significant strides made by modern generative models in generating diverse synthetic image documents, they remain encumbered by limitations when tasked with generating high-resolution images. This acknowledged limitation within the field serves as the impetus for our research pursuit. Our primary objective is to rectify this deficit, thereby augmenting the capabilities of generative models to produce high-resolution synthetic document images.

3   Proposed Approach and Implementation

In this section, we will begin by outlining the notation conventions used in this paper. Subsequently, we will delve into an explanation of the proposed methodology, provide an overview of the network architecture, and conclude with a comprehensive discussion on the specifics of the implementation.

3.1   Notations

Let us start with a formal definition of the problem. Define Ω as an image lattice with dimensions of 256×256 pixels, and $\mathcal{I}$ as a document image delineated on this lattice. The layout, denoted as $L(\mathcal{I})$, consists of n discrete object instances O_i, each furnished with an assigned class category l_i and bounding box $bbox_i$. Each instance within this layout is defined as $O_i = (l_i, bbox_i)$, for i in the range of 1 to n. The total number of document object categories (for instance, title, text block, image, margin note, etc.) is represented by $|O|$. As such, the layout of the image $\mathcal{I}$ can be represented as

$$L(\mathcal{I}) = \{O_i = (l_i, bbox_i)\}_{i=1...n}$$

Let Z_{obj} denote the cumulative sampled latent vector encompassing each object instance O_i within the layout $L(\mathcal{I})$. This can be expressed as $Z_{obj} = \{z_{obji}\}$, where i spans from 1 to n. These latent vectors have been sampled randomly for the objects from a standard Normal distribution $\mathcal{N}(0,1)$, under the assumption of independent and identically distributed (i.i.d) variables.

The taks of layout-guided document image synthesis can be conceptualized as the acquisition of a generative function, denoted as G. This function is designed to establish a mapping from a prescribed document layout input, denoted as $\mathcal{I}$, to a synthetically generated output image, represented as $\hat{I}$. This process is succinctly represented by the equation:

$$\hat{I} = G(\mathcal{I}; \Phi_G)$$

where Φ_G designates the parameters of the generative function G, which the model tries to learn. The generative model $G(\cdot)$ is designed to apprehend the underlying conditional data distribution, represented as $p(\hat{I}|\mathcal{I}; \Phi_G)$, present in a multidimensional space. This distribution is defined to be equivariant with respect to the spatial locations of document layout objects, denoted as $bbox_i$.

4   Generative Network Architecture

Our proposed architecture for generating synthetic document images is fundamentally composed of three pivotal constituents. Firstly, two object predictors, labelled as A and A', serve as the groundwork for the model. These predictors are supplemented by a conditioned image generator symbolized as B. A global layout encoder, represented by F, carries out the task of encoding the layout, bringing global contextual understanding to the architecture. Additionally, an image decoder, indicated by J, reconstructs the image based on encoded information. Finally, the model is equipped with a pair of discriminators, denoted by D_{ob} and D_{im}, which perform the role of discriminating between the objects and images respectively, thus fine-tuning the generative process.

4.1   Object and Layout Encoding

The initial step in the model is object encoding, where object latent estimates, denoted as Z_{crop}^{obj}, are sampled from the ground-truth image, I, through the object predictor A. This process serves to model the variability in object appearances and plays a key role in generating the reconstructed image $\tilde{I}$. Specifically, the object predictor A predicts the mean and variance of the posterior distribution for every cropped object O_i derived from the input image. To enhance the consistency between the generated output image I' and its object estimates, the model incorporates an additional predictor, A'. This predictor infers the mean and variances for the generated objects O' that are cropped from I'. It's noteworthy that predictors A and A' are constituted of several convolutional layers, culminating in two dense fully-connected layers.

Post the sampling of object latent estimation $z_i \in \mathbb{R}^n$ from either the posterior or the prior distribution, the model proceeds to construct a layout encoding, denoted by F_i. This construction utilizes the input layout information, denoted as $L = (l_i, bbox_i)_{i=1}^n$, for every individual object O_i present in the image I. Each feature map F_i is crafted to contain the spatial and semantic information disentangled to correspond with layout L and the appearance of the objects O_i, as interpolated by the latent estimation z_i. In this step, the object category label l_i is transformed into a label embedding $e_i \in \mathbb{R}^n$ and then concatenated with the latent vector z_i. Subsequently, the resultant feature map F_i for each object is filled with the corresponding bounding box information $bbox_i$, leading to a tuple represented by $<l_i, z_i, bbox_i>$. This layout information encoded in these feature maps is then processed by a global layout encoder network F. This network, composed of multiple convolutional layers, generates downsampled feature maps.

4.2   Integration of Conv-LSTM into Spatial Reasoning for Image Synthesis

In our quest to generate plausible synthetic document images using the encoded input layout information, an essential element is the subsequent step involving the conditioned generator B. This generator is responsible for producing a robust hidden feature map, denoted as h. The feature map h must meet several criteria:

– It should encode global features, effectively associating an object representation with its neighboring ones within the document layout.
– It should encode local features imbued with spatial information relevant to each object.
– It should invoke spatial reasoning, evaluating the plausibility of the generated document relative to its contained objects.

To fulfill these requirements, we propose the integration of a spatial reasoning module embedded within a Convolutional Long-Short-Term Memory (Conv-LSTM) network architecture. This differs from traditional LSTM networks as Conv-LSTMs by replacing the hidden state vectors with feature maps. Additionally, the various gates within this network are implemented by convolutional layers, which helps in preserving spatial information with higher precision.

4.3   Image Reconstruction and Generation

Building on the basis set by the hidden layout feature map (h), an artifact created by our spatial reasoning module, we carefully progress towards our primary research objective. We utilize the capabilities of an image decoder (J), designed with a series of deconvolutional layers, to decode the feature map (h) into two distinct, yet related visual representations: I' and $\widetilde{I}$. Image I', a detailed reconstruction of the original input image (I), is crafted using latent estimation $(Z_{\mathrm{crop}}^{obj})$, a process supported by its affiliated objects (O). Concurrently, the image $\widetilde{I}$ appears as an independently generated visual entity, its creation facilitated by Z_{obj} sampled directly from a pure standard normal distribution $N(0,1)$.

4.4 Image and Object Discriminators

In pursuit of rendering synthetic document images with a degree of realism, and to ensure the saliency of their constituting objects, we employ a pair of discriminators - D_{im} and D_{ob}. These discriminators serve the dual purpose of classifying an input image either as genuine or synthetic. This is achieved by the maximization of the Generative Adversarial Network (GAN) objective, as detailed in the associated equation.

$$\mathcal{L}_{GAN} = \mathbb{E}_{x \sim Prcal} logD\left(x\right) + \mathbb{E}_{x \sim Pfake} log(1 - D\left(y\right))$$

5 Experimental Training

The model designed to operate within an adversarial training paradigm. This model intricately incorporates a generative mechanism paired with dual discriminators, all in a bid to streamline the optimization of multifaceted learning objectives throughout the training process.

The model introduced two distinct adversarial loss mechanisms - the Image Adversarial Loss ($\mathcal{L}_{GAN}^{im}$), and the Object Adversarial Loss ($\mathcal{L}_{GAN}^{ob}$). These adversarial losses catalyze the reciprocal rivalry between the generative and discriminative elements of the network, fortifying the overall process of synthetic document creation.

To further enhance the robustness of the model, four auxiliary loss mechanisms integrated into the system: the Kullback-Leibler (KL) Divergence Loss $\mathcal{L}_{kl}$, the Image Reconstruction Loss ($\mathcal{L}_1^{im}$), the Object Reconstruction Loss ($\mathcal{L}_{ob}$), and the Auxiliary Classification Loss ($\mathcal{L}_{AC}^{ob}$). These mechanisms work collaboratively to augment the fidelity and realism of the generated synthetic documents, whilst ensuring the overall network's performance optimization.

Our comprehensive loss function amalgamates these diverse loss components, and its formulation is detailed in the subsequent equation. This holistic approach ensures a robust and adaptive training regime, thereby enhancing the efficacy of synthetic document generation.

$$\mathcal{L}_{GAN} = \lambda_1 \mathcal{L}_{GAN}^{im} + \lambda_2 \mathcal{L}_{GAN}^{ob} + \lambda_3 \mathcal{L}_{AC}^{ob} + \lambda_4 \mathcal{L}_{KL} + \lambda_5 \mathcal{L}_1^{im} + \lambda_6 \mathcal{L}_1^{ob} \quad (1)$$

To stabilize the training process for our generative network, the Spectral-Normalization [10] Generative Adversarial Network (SNGAN) is incorporated as the foundational architecture of the model. Also to enhance the normalization of the object feature maps, Conditional Batch Normalization (CBN) [3] has been applied within the object predictors.

The model proposed herein, dubbed as 'DocSynth', was originally designed to cater specifically to image sizes of 64×64 and 128×128 pixels. Our intent was to extend this capacity and enable the model to generate images of higher resolutions, specifically of 256×256 pixels or more. Consequently, to actualize this goal, we amended the decoder component within the generator module, thus empowering the model to generate images at these increased resolutions.

Our generator decoder represents a series of convolutional and upsampling operations, with ReLU activations and batch normalization between layers. We denote the input to the Decoder as $h \in R(CxHxW)$ where C represents the number of channels, H the height, and W the width of the feature map.

The Decoder functionality in this context is illustrated as a sequence of computational operations:

- Initially, a convolution operation Conv(h) is performed on the input tensor h using a 2D convolution kernel.
- Subsequently, a batch normalization $BN(h')$ is implemented on the output h' from the convolution operation, yielding h''. represented as $h'' = BN(h')$
- Finally, a rectified linear unit (ReLU) activation function ReLU(h'') is applied to the batch-normalized tensor h'', resulting in h'''. This is expressed as: $h''' = ReLU(h'')$.

5.1 Datasets

Our framework was evaluated using the PubLayNet dataset [18]. The dataset, a product of the PubMed Central library, predominantly features images culled from the repository of scientific literature. It categorizes document objects into five distinct classifications: text, title, lists, tables, and figures, thus providing a diverse basis for evaluation. The dataset is extensive, comprising 35,703 images for training and an additional subset of 11,245 images dedicated to validation.

6 Results

As an illustration of our model's generation capacity we provide a visual depiction in Fig. 2, showcasing a variety of synthetically generated data samples varying in image dimensions. The synthetic samples with resolutions of 64 and 128×128 pixels were generated via the DocSynth model [2], while our model created the 256×256 pixel image.

Upon examination of the results, it becomes apparent that our model is capable of generating and identifying the document layout. However, a discernible degradation in image quality becomes evident upon up-scaling the resolution to 512×512 pixels, implying sub-optimal functionality. This observation illuminates the resolution as a potential point of frailty within the model's operational framework, which has a subsequent adverse effect on the quality of output layouts. Thus, it necessitates an architectural modification within our model.

In 512×512 image sample, we observe that the model's performance deteriorates when the generator decoder undergoes up-sampling. It consequently fails to accurately capture the inherent features of the image. This observation signifies that merely up-sampling the generator decoder may not rectify the issue; an additional intervention is requisite.

Fig. 2. Comparative Visualization of Synthetic Document Images Generated at Varying Resolutions

We suggest the incorporation of a high-resolution loss to the generator as a potential solution. This loss function will be primarily responsible for the content, thereby assisting in the accurate augmentation of pertinent information to the generator. This approach will help circumvent bias, particularly since the output at a 512×512 resolution exhibited a marked predisposition. Hence, by employing a high-resolution loss, we can mitigate this problem and improve the reliability and accuracy of our model's outputs.

Furthermore, our examination of the results elucidated that our model does not perform recognition of the text it generates. Instead, it renders text in the form of image batches, visually mimicking the appearance of text. Therefore, the output text is not genuinely text, but rather an image that bears a striking resemblance to it. This insight further indicates the necessity for improvement in the model's design and functionality.

7   Evaluation

The FID metric [6] is a standard GAN performance metric to compute distances between the feature vectors of real images and the feature vectors of synthetically generated ones. A lower FID score denotes a better quality of generated samples and more similar to the real ones. In this work, the Inception-v3 [12] pretrained model were used to extract the feature vectors of the generated samples of document images.

From the results in Table 1 the FID score of our model is relatively high, suggesting that there is a significant difference between the distributions of the generated images and the real images. This score implies that the quality of the synthetically generated images could be relatively low, and the images are not closely resembling the real ones.

Table 1. FID for synthetic document generation

Method	FID
Document Image Synthesizer model (512 × 512)	368.97
Document Image Synthesizer model (256 × 256)	248.54
DocSynth model (128 × 128)	33.71
DocSynth model (64 × 64)	28.37

8 Conclusion

We have shown that we can adapt the existing DocSynth method to generate higher resolution images. However, this work is still in progress and we have discovered that the upscaling cannot yet achieve full high resolution, especially if we want it to incorporate reasonably realistic text sections and not simple textures resembling text.

We have a series of improvements and further research directions we are going to explore to solve these shortcommings. They are described below.

One viable approach to augment the output quality of our model involves leveraging Super-Resolution Generative Adversarial Networks (SRGANs) [8]. This technique merges an adversarial network with a deep learning network, aiming to produce higher resolution images. During the training phase, a high-resolution (HR) image is scaled down to a low-resolution (LR) image. The GAN generator then upscales these LR images to generate super-resolution (SR) images. A discriminator is deployed to differentiate the HR images. The training of both the discriminator and the generator is facilitated by back-propagating the GAN loss.

The core constituents of the network architecture for both the generator and discriminator include convolution layers, batch normalization, and Parameterized Rectified Linear Unit (PReLU). The generator also employs skip connections, close to the structure of the Residual Network (ResNet).

The loss function for the generator is formulated by the content loss (also known as reconstruction loss) and the adversarial loss. Content loss can be determined by comparing the Mean Square Errors (MSE) between the HR and SR images on a pixel-by-pixel basis. However, this approach computes distance in a purely mathematical context. Therefore, SRGANs apply a perceptual loss to calculate the MSE of features that are extracted by the VGG-19 network, aiming for their features to align for a specific layer within VGG-19 (i.e., achieving the minimum MSE for these features).

In the context of content loss, the MSE between the higher-resolution and super-resolution images is perceived as perceptual mapping. This method, if utilized, could potentially be employed to upscale the resolution. Our generator model exhibits excellent quality in generating 64 × 64 images. However, a compromise in quality is observed when the image resolution is increased. Therefore, the implementation of reconstruction loss could potentially serve as a counter-measure to this challenge.

In order to address the challenge of text generation within our model, we propose the integration of a language model into our generator. Our aim is to facilitate the translation of text into images through the utilization of Trans-former models. We are currently conducting experiments that involve training our text model with GPT-2.

In this phase of the research, we have extracted all the text present in the PubLayNet images. This text has been preprocessed to create a dataset, which has been subsequently divided into training, testing, and validation sets. The training process is currently underway.

Acknowledgement. The main author acknowledges support for this work during her MSc. thesis at the Université de Lorraine, LORIA, UMR 7503, France. The extension currently under investigation will be integrated by Vasa (Actisolutions SAS). Upon successful completion of this research and achievement of the desired results, this will have a significant impact on Vasa Company's research and development. It will notably enhance the company's database and catalyze the improvement of results for their document analysis models.

References

1. Binmakhashen, G.M., Mahmoud, S.A.: Document layout analysis: a comprehensive survey. ACM Comput. Surv. **52**(6), 1–36 (2019)
2. Biswas, S., Riba, P., Lladós, J., Pal, U.: DocSynth: a layout guided approach for controllable document image synthesis. In: Lladós, J., Lopresti, D., Uchida, S. (eds.) ICDAR 2021. LNCS, vol. 12823, pp. 555–568. Springer, Cham (2021). https://doi.org/10.1007/978-3-030-86334-0_36
3. De Vries, H., Strub, F., Mary, J., Larochelle, H, Pietquin, O., Courville, A.C.: Modulating early visual processing by language. In: Advances in Neural Information Processing Systems, vol. 30 (2017)
4. Furuta, R., Scofield, J., Shaw, A.: Document formatting systems: survey, concepts, and issues. ACM Comput. Surv. (CSUR) **14**(3), 417–472 (1982)
5. Goodfellow, I., et al.: Generative adversarial networks. Commun. ACM **63**(11), 139–144 (2020)
6. Heusel, M., Ramsauer, H., Unterthiner, T., Nessler, B., Hochreiter, S.: GANs trained by a two time-scale update rule converge to a local nash equilibrium. In: Advances in Neural Information Processing Systems, vol. 30 (2017)
7. Lake, B.M., Salakhutdinov, R., Tenenbaum, J.B.: Human-level concept learning through probabilistic program induction. Science **350**(6266), 1332–1338 (2015)
8. Ledig, C., et al.: Photo-realistic single image super-resolution using a generative adversarial network. In: Proceedings of the IEEE Conference on Computer Vision and Pattern Recognition, pp. 4681–4690 (2017)
9. Li, J., Yang, J., Hertzmann, A., Zhang, J., Xu, T.: LayoutGAN: generating graphic layouts with wireframe discriminators. arXiv preprint arXiv:1901.06767 (2019)
10. Miyato, T., Kataoka, T., Koyama, M., Yoshida, Y.: Spectral normalization for generative adversarial networks. arXiv preprint arXiv:1802.05957 (2018)

11. Radford, A., Metz, L., Chintala, S.: Unsupervised representation learning with deep convolutional generative adversarial networks. arXiv preprint arXiv:1511.06434 (2015)
12. Szegedy, C., Vanhoucke, V., Ioffe, S., Shlens, J., Wojna, Z.: Rethinking the inception architecture for computer vision. In: Proceedings of the IEEE Conference on Computer Vision and Pattern Recognition, pp. 2818–2826 (2016)
13. Zhang, J., et al.: Controllable person image synthesis with spatially-adaptive warped normalization. arXiv preprint arXiv:2105.14739 (2021)
14. Zheng, X., Qiao, X., Cao, Y., Lau, R.W.H.: Content-aware generative modeling of graphic design layouts. ACM Trans. Graph. (TOG) **38**(4), 1–15 (2019)

GREC

GREC 2023 Preface

Graphics recognition is the subfield of document recognition dealing with graphic entities such as tables, charts, illustrations in figures, and notations (e.g., for music and mathematics). Graphics often help describe complex ideas much more effectively than text alone. As a result, recognizing graphics is useful for understanding the information content in documents, the intentions of document authors, and for identifying the domain of discourse in a document. Since the 1980s, researchers in the graphics recognition community have addressed the analysis and interpretations of graphical documents (e.g., electrical circuit diagrams, engineering drawings, etc.), handwritten and printed graphical elements/cues (e.g., logos, stamps, annotations, etc.), graphics-based information retrieval (comics, music scores, etc.) and sketches, to name just a few of the challenging topics in this area.

The GREC workshops provide an excellent opportunity for researchers and practitioners at all levels of experience to meet and share new ideas and knowledge about graphics recognition methods. The workshops enjoy strong participation from researchers in both industry and academia.

The aim of this workshop is to maintain a very high level of interaction and creative discussions between participants, maintaining a *workshop* spirit, and not being tempted by a 'mini-conference' model.

The 15th edition of the International Workshop on Graphic Recognition (GREC 2023) built on the success of the fourteen previous editions held at Penn State University (USA, 1995), Nancy (France, 1997), Jaipur (India, 1999), Kingston (Canada, 2001), Barcelona (Spain, 2003), Hong Kong (China, 2005), Curitiba (Brazil, 2007), La Rochelle (France, 2009), Seoul (Korea, 2011), Lehigh (USA, 2013), Nancy (France, 2015), Kyoto (Japan, 2017), Sydney (Australia, 2019) and Lausanne (Switzerland, 2021).

Traditionally, for each paper session at GREC, an invited presentation describes the state-of-the-art and open questions for the session's topic, followed by short presentations of each paper. Each session concludes with a panel discussion moderated by the invited speaker, in which the authors and attendees discuss the papers presented along with the larger issues identified in the invited presentation.

For this 15th edition of GREC, the authors had the opportunity to submit short or long papers depending on the maturity of their research. From 15 submissions, we selected 11 papers from 6 different countries, comprised of 9 long papers and 3 short papers. Each submission was reviewed by at least two expert reviewers. We would like to take this opportunity to thank the program committee members for their meticulous reviewing efforts.

August 2023

Nathalie Girard
Samit Biswas
Jorge Calvo-Zaragoza
Jean-Christophe Burie

Organization

General Chair

Jean-Christophe Burie   La Rochelle University, France

Program Co-chairs

Nathalie Girard   IRISA, Université de Rennes, France
Jorge Calvo-Zaragoza   University of Alicante, Spain
Samit Biswas   Indian Institute of Engineering Science and Technology, Shibpur, India

Steering Committee

Alicia Fornés   Universitat Autònoma de Barcelona, Spain
Bart Lamiroy   Université de Lorraine, France
Rafael Lins   Federal University of Pernambuco, Brazil
Josep Lladós   Universitat Autònoma de Barcelona, Spain
Jean-Marc Ogier   Université de la Rochelle, France

Program Committee

Eric Anquetil   IRISA, Rennes, France
Bertrand Couasnon   IRISA/INSA Rennes, France
Mickael Coustaty   La Rochelle University, France
Alicia Fornes   Universitat Autonoma de Barcelona - CVC, Spain
Alexander Gribov   ESRI, USA
Nina Hirata   Sao Paulo University, Brazil
Motoi Iwata   Osaka Metropolitan University, Japan
Bart Lamiroy   Université de Lorraine - LORIA, Nancy, France
Josep Llados   Universitat Autonoma de Barcelona - CVC, Spain
Sekhar Mandal   Indian Institute of Engineering Science and Technology, India
Wataru Ohyama   Tokyo Denki University, Tokyo, Japan
Alexander Pacha   TU Wien, Austria

Oriol Ramos Terrades   Universitat Autonoma de Barcelona -
                                       CVC, Spain
Christophe Rigaud   La Rochelle University, France
Richard Zanibbi   Rochester Institute of Technology, USA

A Survey and Approach to Chart Classification

Anurag Dhote[1,3] , Mohammed Javed[1,3(✉)] , and David S. Doermann[2]

[1] Indian Institute of Information Technology Allahabad, Prayagraj, India
{mit2021082,javed}@iiita.ac.in
[2] Department of CSE, University at Buffalo, Buffalo, NY, USA
doermann@buffalo.edu
[3] Computer Vision & Biometrics Lab, Dept. of IT, IIIT Allahabad, Prayagraj, India

Abstract. Charts represent an essential source of visual information in documents and facilitate a deep understanding and interpretation of information typically conveyed numerically. In the scientific literature, there are many charts, each with its stylistic differences. Recently the document understanding community has begun to address the problem of automatic chart understanding, which begins with chart classification. In this paper, we present a survey of the current state-of-the-art techniques for chart classification and discuss the available datasets and their supported chart types. We broadly classify these contributions as traditional approaches based on ML, CNN, and Transformers.

Furthermore, we carry out an extensive comparative performance analysis of CNN-based and transformer-based approaches on the recently published CHARTINFO UB-UNITECH PMC dataset for the CHART-Infographics competition at ICPR 2022. The data set includes 15 different chart categories, including 22,923 training images and 13,260 test images. We have implemented a vision-based transformer model that produces state-of-the-art results in chart classification.

Keywords: Chart Classification · Deep Learning · Chart Mining

1 Introduction

Charts provide a compact summary of important information or research findings in technical documents and are a powerful visualization tool widely used by the scientific and business communities. In the recent literature, the problem of chart mining has attracted increased attention due to numerous advantages, as suggested in the comprehensive survey published by Davila et al. in 2019 [11]. The term Chart mining refers to the process of extracting information represented by charts. Another motivating factor in the increased attention paid to this problem is a series of competitions held in conjunction with significant conferences to address the critical challenges in the chart mining pipeline [10,12,13].

Since a variety of charts are possible, chart classification is often the first step in chart mining. The task of chart image classification can be formalized as, given

M. Coustaty and A. Fornés (Eds.): ICDAR 2023 Workshops, LNCS 14193, pp. 67–82, 2023.
https://doi.org/10.1007/978-3-031-41498-5_5

a chart image extracted from a document, classifying the image into one of N defined categories. The wide variety of chart types in the literature adds to the complexity of the task [6,11,34]. Some additional problems include interclass similarity, noise in authentic chart images, and more state-of-the-art datasets that cover multiple chart types and incorporate 2.5 or 3D charts and noise into the training samples [34]. The rise of robust deep learning models has contributed significantly to the success of chart classification. Deep learning approaches have outperformed traditional machine learning approaches regarding robustness and performance. Yet there need to be more state-of-the-art solutions that can provide stable results and are robust enough to address noise in some data sets. In this paper, we provide a performance comparison of several deep learning models that are state-of-the-art in the ImageNet [28] classification task. In addition, we report the performances of several popular vision transformers, which, to the best of our knowledge, have yet to be used for chart classification, except for the recent ICPR 2022 CHART-Infographics competition [13].

This paper is organized as follows. Section 2 summarizes the existing chart classification literature covering traditional and deep learning-based methods, including a brief discussion on transformer-based chart classification. Section 3 reports and summarizes publicly available datasets. Section 4 briefly highlights the popular ImageNet pre-trained deep learning-based models that will be used for our comparative study. Section 5 describes the latest edition of the UB PMC dataset, the training and testing protocols, and a discussion on their performance for chart classification. Section 6 provides information on possible improvements and suggestions for future research. Finally, Sect. 7 concludes with a summary of the paper.

2   Chart Classification Techniques

Based on the type of approaches used to implement the chart classification task in the literature, they can be grouped into traditional ML, CNN-based deep learning, and Transformer-based deep learning. Each type of approach is described briefly below.

2.1   Traditional ML Approaches

Traditional approaches rely on feature extraction methods that are often manual and general-purpose. Features are extracted and then represented in mathematical form for direct processing by machine learning classifiers. Savva et al. [29] present a system that automatically reformats visualizations to increase visual comprehension. The authors use low-level image features for classification in conjunction with text-level features. The system uses a multiclass SVM classifier trained on a corpus containing 2601 chart images labeled with ten categories, following Gao et al.'s manual extraction approach. In [14], researchers propose VIEW, a system that automatically extracts information from raster-format charts. The authors used an SVM to separate the textual and graphical components and classify the chart images based on the graphic elements extracted from the visual components. The text is typically found in three chart categories - bar charts, pie charts, and line graphs, with 100 images for each category collected from various real-world digital resources.

Instead of taking an image as input, Karthikeyani and Nagarajan [19] present a system to recognize chart images from PDF documents using eleven texture features that are part of a Gray Level Co-Occurrence Matrix. A chart image is located in the PDF Document database, and the features are extracted and fed to the learning model. SVM, KNN, and MLP are the classifiers used for classification. Cheng et al. [7] employ a multimodal approach that uses text and image features. These features are provided as input to an MLP. The output is characterized as a fuzzy set to get the final result. The corpus contains 1707 charts with three categories and a 96.1% classification result.

2.2   CNN-Based Deep Learning Approaches

Liu et al. [22] used a combination of Convolutional Neural Networks (CNNs) and Deep Belief networks (DBNs) to capture high-level information present in deep hidden layers. Fully Connected Layers of Deep CNN are used to extract deeply hidden features. A DBN is then used to predict the image class using the deep hidden features. The authors use transfer learning and perform fine-tuning to prevent overfitting. They use a data set that includes more than 5,000 images of charts, including pie, scatter, line, bar, and flow classes. Deep features are useful over primitive features to provide better stability and scalability to the proposed framework. The proposed method achieves an average accuracy of 75.4%, which is 2.8% more than the method that uses only deep ConvNets.

Given the results of CNN in the classification of natural images, Siegel et al. [30] used two CNN-based architectures for chart classification. They evaluated AlexNet and ResNet-50, which are pre-trained on the ImageNet data set and then fine-tuned for chart classification. This transfer learning approach is prevalent in subsequent works addressing this particular problem. The proposed frameworks outperformed the state-of-the-art model at the time, such as ReVision, by a significant margin. ResNet-50 achieved the best classification accuracy of 86% on a data set that contained more than 60000 images spread over seven categories.

Amara et al. [1] proposed a CNN-based on LeNet to classify images from their corpus of 3377 images into 11 categories. The model comprises eight layers, one input layer, five hidden layers, one fully connected layer, and one output layer. The fully connected layer is used as a classifier, while the hidden layers are convolution and pooling layers designed to extract features automatically. A fully connected layer employs softmax activation to classify images into defined classes. For evaluation of the model's performance, an 80-20 split is performed on the data set for training and assessment. The proposed model performs better than the LeNet and pretrained LeNet architectures with an accuracy of 89.5%.

Jung et al. [18] present a classification method using the deep learning framework Caffe and evaluate its efficacy by comparing it with ReVision [29]. The authors use GoogLeNet [32] for classification and compare its results with shallower networks like LeNet-1 and AlexNet [20]. GoogLeNet outperforms LeNet-1 and AlexNet with an accuracy of 91.3%. Five-fold cross-validation is used for calculating the accuracy on an image corpus with 737–901 images for each chart type. The test concludes that ChartSense provides higher classification accuracy for all chart types than ReVision.

With studies adapting the deep learning approach for chart image classification, a comparative study of traditional vs. CNN architectures was required. Chagas et al. [6] provide a comparative analysis of conventional vs. CNN techniques. Authors evaluated CNN architectures (VGG19 [31], Resnet-50 [15], and Inception-V3 [33]) for chart image classification for ten classes of charts. The performance is compared with conventional machine learning classifiers, Naive Bayes, HOG features combined with KNN, Support Vector Machines, and Random Forests. Pre-trained CNN models with fine-tuned last convolutional layers were used. The authors concluded that CNN models surpass traditional methods with an accuracy of 77.76% (Resnet-50) and 76.77% (Inception-V3) compared to 45.03% (HOG + SVM).

Dia et al. [9] employ four deep learning models on a corpus of 11,174 chart images of five categories. Of AlexNet [20], VGG16 [31], GoogLeNet [32] and ResNet [15], the authors get the best accuracy of 99.55% for VGG16 model. VGG16 outperforms the models used in ChartSense paper by a large margin.

Significant roadblocks to chart mining research are caused by the fact that current chart data sets must be larger and contain sufficient diversity to support deep learning. To address this problem, Jobin et al. [21] presented DocFigure, a chart classification data set with 33,000 charts in 28 different classes. To classify charts, the author's proposed techniques utilize deep features, deep texture features, and a combination of both. Among these baseline classification techniques, the authors observed that combining deep features and deep texture features classifies images more efficiently than individual features. The average classification accuracy improved by 3.94% and 2.10% by concatenating FC-CNN and FV-CNN over individual use of FC-CNN and FV-CNN, respectively. The overall accuracy of the combined feature methods turned out to be 92.90%.

Luo et al. proposed a unified method to handle various chart styles [26], where they show that generalization can be obtained in deep learning frameworks with rule-based methods. The experiments were performed on three different datasets of over 300,000 images with three chart categories. In addition to the framework, an evaluation metric for the bar, line, and pie charts is also introduced. The authors concluded that the proposed framework performs better than traditional rules-based and pure deep learning methods.

Araújo et al. [2] implemented four classic CNN models that performed well on computer vision tasks, including Xception [8], VGG19 [31], ResNet152 [15] and MobileNet [16]. The weights of these models were pre-trained on the ImageNet dataset, and the authors further performed hyperparameter tuning to obtain a stable learning rate and weight decay. These models were employed on a self-aggregated chart image corpus of 21,099 images with 13 different chart categories. Xception outperforms the other models by hitting an accuracy of 95%.

The problem of small datasets has been prevalent since the problem of chart mining was first introduced. Most work tries to increase the size of the dataset. However, Bajic and Job [4] use a Siamese CNN network to work with smaller datasets. The authors show that an accuracy of 100% can be achieved with 50 images per class, which is significantly better than using a vanilla CNN.

With the increase in datasets for chart images and the rise of deep learning models being employed on said datasets, an empirical study of these deep learning models was due. Thiyam et al. [35] compared 15 different deep-learning models on a self-aggregated dataset of 110,182 images spfeatures24 different chart categories. In addition, the authors tested the performance of these models on several preexisting test sets. They concluded that Xception (90.25%) and DenseNet121(90.12%) provide the most consistent and stable performance of all the deep learning models. The authors arrived at this decision by employing a five-fold cross-validation technique and calculating the standard deviation for each model across all datasets.

Davila et al. [10] summarized the work of different participants in the competition's first edition by harvesting raw tables from Infographics that provided data and tools for the chart recognition community. Two data sets were provided for the classification task. One was a synthetically generated AdobeSynth dataset, and the other UB-PMC data set was gathered from the PubMedCentral open-access library. The highest average F1-measure achieved for the synthetic data set was 99.81% and the highest F1-measure achieved for the PMC data set was 88.29%. In the second edition of the competition, the PMC set was improved and included in the training phase. An ensemble of ResNet152 and DenseNet121 achieved the highest F1-score of 92.8%. The third edition of the competition was recently held at ICPR 2022. The corpus of real chart images was made up of 36,183 chart images. The winning team achieved an F1 score of 91% with a base Swin transformer model with a progressive resizing technique. We summarize the competition details in Table 1.

Table 1. Competition on Harvesting Raw Tables from Infographics (CHART-Infographics)

Competition	Dataset	#Classes	Train Size	Test Size	Top performing Model	F1-measure
ICDAR 2019 [10]	AdobeSynth	10	198,010	4540	ResNet-101	99.81%
	PMC	7		4242		88.29%
ICPR 2020 [12]	Adobe Synth	12	14,400	2,999	DenseNet-121 +	100%
	UB PMC	15	15,636	7,287	ResNet-152	92.8%
ICPR 2022 [13]	UB PMC	15	22,923	13,620	Swin Transformer	91%

Table 2. Published Literature on Chart Classification

Authors	Dataset	Model	Metric	Performance
Savva et al. [29]	Self-acquired	SVM	Accuracy	96.00%
Gao et al. [14]	Self-acquired	SVM	Accuracy	97.00%
Kartikeyani and Nagarajan [19]	Self-acquired	MLP	Accuracy	69.68%
		KNN		78.06%
		SVM		76.77%
Cheng et al. [7]	Self-acquired	MLP	Accuracy	96.10%
Liu et al. [22]	DeepChart	CNN + DBN	Accuracy	75.40%
Siegel et al. [30]	ChartSeer	AlexNet	Accuracy	84.00%
		ResNet-50		86.00%
Amara et al. [1]	Self-acquired	CNN	Accuracy	89.50%
Jung et al. [18]	Chart-Sense	GoogleNet	Accuracy	91.30%
Balaji et al. [5]	Self-acquired	CNN	Accuracy	99.72%
Chagas et al. [6]	Chart-Vega	ResNet-50	Accuracy	76.76%
		Inception-V3		76.77%
Dai et al. [9]	Self-acquired	ResNet	Accuracy	98.89%
		GoogLeNet		99.07%
		AlexNet		99.48%
		VGG-16		99.55%
Liu et al. [23]	Self-acquired	VGG-16	Accuracy	96.35%
Davila et al. [10]	Synthetic	ResNet-101	F1-measure	99.81%
	UB-PMC	ResNet-101		88.29%
Jobin et al. [21]	DocFigure	FC-CNN + FV-CNN	Accuracy	91.30%
Bajic et al. [3]	Self-acquired	VGG-16	Accuracy	89.00%
Araujo et al. [2]	Self-acquired	Xception	Accuracy	95.00%
Luo et al. [26]	Chart-OCR	CNN	Custom(Bar)	91.90%
			Custom(Pie)	91.80%
			Custom(Line)	96.20%
Davila et al. [12]	UB-PMC	DenseNet-121 + ResNet-152	F1-measure	92.80%
Bajic and Job [4]	Self-acquired	Siamese CNN	Accuracy	100%
Thiyam et al. [35]	Self-acquired	Xception	Accuracy	90.25%
		DenseNet121		90.12%
		DenseNet201		90.53%
Davila et al. [13]	UB-PMC	Swin Transformer	F1-measure	91.00%

2.3   Transformer-Based Deep Learning Approaches

Since the inception of Vision Transformer, there has been a lot of development in various computer vision tasks such as image classification, object detection, and image segmentation. Vision transformer has outperformed CNN-based models in these tasks on the ImageNet dataset. However, there has not been widespread application of vision transformers to chart image classification. To our knowledge, only the Swin transformer [24] has been used for chart classification as reported in [13], which won the CHART-Infographics challenge ICPR2022. The authors applied a Swin Transformer Base Model with a progressive resizing technique. The models were initially trained on a scale (input size) of 224 followed by 384 [13].

The existing models in the literature are summarised in Table 2.

3   Chart Classification Datasets

There has been a significant increase in the size of datasets both in terms of the number of samples and the number of chart types. The Revision dataset [29] had only 2,601 images and 10 chart types. The recent publicly available dataset [13] comprises around 33,000 chart images of 15 different categories. The details of several publicly available datasets are discussed in this section.

Table 3. Chart Classification Datasets

Dataset	Year	#Samples	#Category	Public (Y/N)
ReVision [29]	2011	2601	10	Y
View [14]	2012	300	3	N
Self [19]	2012	155	8	N
Self [7]	2014	1707	3	N
DeepChart [22]	2015	5000	5	Y
ChartSeer [30]	2016	60000	7	N
Self [1]	2017	3377	11	N
Chart-Sense [18]	2017	6997	10	Y
Chart-Text [5]	2018	6000	2	N
Chart-Vega [6]	2018	14471	10	Y
Chart decoder [9]	2018	11,174	5	N
Self [23]	2019	2500	2	N
Synthetic [10]	2019	202550	10	Y
UB-PMC [10]		4242	7	Y
DocFigure [21]	2019	33000	28	Y
Self [3]	2020	2702	10	N
Self [2]	2020	21099	13	N
Chart-OCR [26]	2021	386966	3	N
UB-PMC [12]	2021	22924	15	Y
Self [4]	2021	3002	10	N
Self [35]	2021	110182	24	N
UB-PMC [13]	2022	33186	15	Y

ChartSense [18]*:* The ChartSense dataset was put together using the ReVision dataset, and the authors manually added some additional charts. The corpus has 5659 chart images that cover ten chart categories.

ChartVega [6]*:* This dataset has ten chart types and was created due to a need for a benchmark dataset for chart image classification [6]. The dataset contains both synthetic and real chart images. The set contains 14,471 chart images, of which 12059 are for training and 2412 are for testing. In addition, a validation set of 2683 real chart images is provided. No separate annotations are provided, as chart images are separated according to their types.

DocFigure [21]: This corpus consists of 28 categories of annotated figure images. There are 33,000 images that include non-chart categories like natural images, tables, 3D objects, and medical images. The train set consists of 19,797 images, and the test set contains 13173 images. The labels are provided in a text document.

ChartOCR [26]: The dataset contains 386,966 chart images created by the authors by crawling public excel sheets online. The dataset contains only three classes of chart images. The dataset is divided into the train, validation, and test sets. The training corpus contains 363,078 images, the validation set contains 11,932 images, and the test set contains 11,965 images. The annotations for the chart images are provided in JSON format.

UB-PMC CHART-Infographics: This dataset was introduced in the first edition of Competition on Harvesting Raw Tables from Infographics (ICPR 2019 CHART Infographics) [10]. This dataset has synthetic images created using matplotlib. For the testing, a large set of synthetic data and a small set of real chart images harvested from PubMedCentral[1] were used. The training set has 198,010 images, whereas the synthetic test set has 4540 images, and the real test set has 4242 images. The dataset has ten different chart categories.

The second edition of the competition [12] provided a dataset containing 22923 real chart images of 15 different chart categories in both training and testing sets. The training set has 15636 images, while the test set has 7287 images. The annotations for the chart image samples are provided in both JSON and XML formats. The dataset presented as a part of the third and most recent competition comprises 36183 images of 15 different chart categories. The training set contains 22,923 images, while the test set contains 13,260 images. Similar to the previous edition, the annotations are provided in JSON and XML formats. To the best of our knowledge, this is the largest publicly available dataset for chart image classification.

The existing classification data sets for charts are summarized in Table 3, and the composition of the publicly available datasets is reported in Table 4.

4   Deep Learning Models for Comparative Analysis

In this section, we briefly discuss prominent deep-learning models that have been used to study the performance of chart classification. We have selected two categories of deep learning models - CNN-based and Transformer-based for the comparative study. For CNN-based models, we have considered the proven state-of-the-art models for image classification on the large-scale benchmark dataset ImageNet [28] over the years. For vision transformer models, we have chosen the models that have been proven to outperform CNN-based models in computer vision tasks.

---

[1] https://www.ncbi.nlm.nih.gov/pmc/.

Table 4. Composition of publicly available datasets

Chart Type	UB-PMC [13]	DocFigure [21]	Chart-Sense [18]	Chart-OCR [26]	Chart-Vega [6]
Arc	–	–	–	–	1440
Area	308	318	509	–	1440
Block	–	1024	–	–	–
Bubble	–	339	–	–	–
Flowchart	–	1074	–	–	–
Heatmap	377	1073	–	–	–
Horizontal Bar	1421	–	–	–	–
Horizontal Interval	586	–	–	–	–
Line	13956	9022	619	122890	1440
Manhattan	256	–	–	–	–
Map	906	1078	567	–	–
Parallel Coordinate	–	–	–	–	1339
Pareto	–	311	391	–	–
Pie	433	440	568	76922	1440
Polar	–	338	–	–	–
Radar	–	309	465	–	–
Re-orderable Matrix	–	–	–	–	1440
Scatter	2597	1138	696		1640
Scatter-Line	3446	–	–	–	–
Sunburst	–	–	–	–	1440
Surface	283	395	–	–	–
Table	–	1899	594	–	–
Treemap	–	–	–	–	1440
Venn	206	889	693	–	–
Vertical Bar	9199	1196	557	187154	1512
Vertical Box	1538	605	–	–	–
Vertical Interval	671	–	–	–	–
Total	36183	33071	5659	386966	14471

4.1   ResNet [15]

The Deep Residual Network was introduced in 2015 and was significantly deeper than the previous deep learning networks. The motivation behind the model was to address the degradation problem: Degrading training accuracy with increasing depth of the model. The authors added shortcut connections, also known as skip connections, that perform the proposed identity mapping and are significantly easier to optimize than unreferenced mappings. Despite being deeper than the previous models, ResNet still needed to be simplified. It achieved the top-5 error of 3.57% and claimed the top position in the 2015 ILSVRC classification competition [28]. We use a 152-layer version of this Deep Residual Network called ResNet-152 for our classification problem.

4.2   Xception [8]

Xception is a re-interpretation of the inception module. The said inception module is replaced with depth-wise separable convolutions. The number of parameters in both Inception V3 and Xception is the same, so the slight performance improvement is due to the more efficient use of parameters. Xception shows a better performance improvement than Inception V3 on the JFT dataset on the ImageNet dataset. It achieves the top five accuracy of 94.5%. Xception also shows promising results in the chart classification literature, as reported by [2] and [35].

4.3   DenseNet [17]

The Dense Convolutional Network, introduced in 2018, connects each layer in the network architecture to all other layers. This allows for the exchange of feature maps at every level and considers the same input as input gathered from all the previous layers rather than just one preceding layer. The difference between DenseNet and Resnet lies in the way that they combine features. ResNet combines features through summation, whereas DenseNet combines them through concatenation. DenseNet is easier to train due to the improved flow of gradients and other information through the network. The vanilla DenseNet has fewer parameters than the vanilla ResNet network. We used DenseNet-121 for our classification task as it was one of the best models for the chart image dataset as reported in [35].

4.4   ConvNeXt [25]

ConvNeXt model was introduced as a response to hierarchical transformers outperforming convnets in image classification tasks. Starting with a standard ResNet architecture, the model is carefully modified to adapt the specific characteristics of a typical hierarchical transformer. This resulted in a CNN-based model that matches the transformers in robustness and scalability across all benchmarks. ConvNeXt achieves a top-1 accuracy of 87.8% on ImageNet.

4.5   DeIT Transformer [36]

The authors proposed the Data Efficient Image Transformer(DeIT) with 86M parameters to make the existing vision transformer more adoptable. This convolution-free approach achieves competitive results against the existing state-of-the-art models on ImageNet. The proposed vision transformer achieved a top-1 accuracy of 85.2% on the ImageNet classification task. We use the base Base DeIT transformer for the chart classification task.

4.6   Swin Transformer [24]

A hierarchical transformer that employs shifting windows to obtain representations for vision tasks. The authors note that the hierarchical architecture provides linear computational complexity and scalability concerning image size. The limitation of self-attention calculation concerning noncoincident local windows due to the shifting windows allows for better cross-window connectivity. The qualities above contribute to the Swin transformer's excellent performance across computer vision tasks. It achieves 87.3% top-1 accuracy on the ImageNet-1k dataset. We perform experiments with all the 13 available Swin Transformer models and report their performance in Table 5. Furthermore we refer to the best performing Swin Transformer model as Swin-Chart in Table 6.

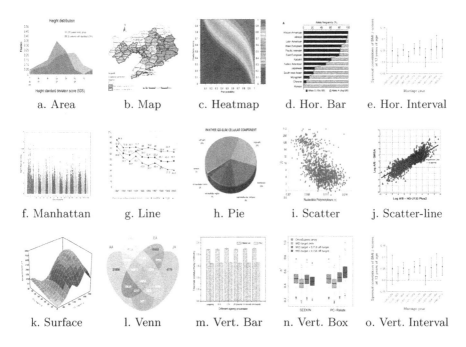

Fig. 1. Sample of chart images used in this study from UB-PMC [13] dataset

5 Experimental Protocol

5.1 Dataset

We use the ICPR2022 CHARTINFO UB PMC [13] dataset to perform our comparative study of deep learning models. The dataset is divided into training and testing sets. The number of chart images in the training and test set is 22,923 and 11,388, respectively. The ground truth values are annotated in JSON and XML formats. We further divide the provided training set into training and validation sets with an 80/20 ratio. The dataset contains charts of 15 categories: area, map, heatmap, horizontal bar, Manhattan, horizontal interval, line, pie, scatter, scatter-line, surface, Venn, vertical bar, vertical box, and vertical interval. Samples of each chart type present in the dataset are shown in Fig. 1.

5.2   Training and Testing Setup

We choose ResNet152, DenseNet121, Xception, and ConvNeXt CNN-based models and DeIT and Swin Transformers-based models for chart image classification. The CNN-based models were selected based on their performance in the existing literature on the ImageNet image classification task. The transformer-based models are chosen because they beat the CNN-based models. We use the pre-trained ImageNet weights of these models and fine-tune them for our chart classification task. The models are trained on a computer with an RTX 3090 video card with 24 GB memory. Pytorch [27] was used as the engine for our experiments. We use a batch size of 64 for CNN-based models and a batch size of 16 for transformer-based models. A learning rate of 10^{-4} is used to train each model for 100 epochs. Label Smoothing Cross Entropy Loss is used as a loss function. The evaluation measures the average over all classes and reports precision, recall, and F1-score.

Table 5. Comparative Performance of all the 13 Pre-trained Swin Transformer Models on ICPR2022 CHARTINFO UB PMC datase

Model	Precision	Recall	F1-measure
SwinT	0.929	0.924	0.922
SwinT_s3	0.931	0.923	0.922
SwinS	0.931	0.926	0.925
SwinS_s3	0.928	0.922	0.919
SwinB_224	0.933	0.926	0.925
SwinB_384	0.936	0.932	0.931
SwinB_224_in22k_ft1k	0.934	0.930	0.929
SwinB_384_in22k_ft1k	0.933	0.929	0.927
SwinB_s3	0.927	0.923	0.921
SwinL_224	**0.937**	**0.933**	**0.932**
SwinL_384	0.937	0.931	0.929
SwinL_224_in22k_ft1k	0.937	0.933	0.932
SwinL_384_in22k_ft1k	0.934	0.930	0.929

5.3   Comparative Results

The models were trained following the steps mentioned in the previous section and were tested on the UB-PMC test data set. We calculate all deep learning models' average precision, recall, and F1 score. Among CNN-based models, ResNet-152 and ConvNeXt provide the best results across all evaluation metrics. The ResNet-152 result is consistent with the results in [13] for CNN-based models. For Swin transformer we perform experiments on 13 models consisting Swin Tiny(SwinT), Swin Small(SwinS), Swin Base(SwinB) and Swin Larger(SwinL)

Table 6. Comparative performances of the CNN-based and Transformer-based models on ICPR2022 CHARTINFO UB PMC dataset

Model	Precision	Recall	F1-score
Resnet-152 [15]	0.905	0.899	0.897
Xception [8]	0.882	0.870	0.866
DenseNet-121 [17]	0.887	0.879	0.875
ConvNeXt [25]	0.906	0.898	0.896
DeIT [36]	0.888	0.879	0.874
Swin-Chart	**0.937**	**0.933**	**0.932**

Table 7. Comparison of Swin-Chart from Table 6 with models stated in [13] on the ICPR2022 CHARTINFO UB PMC dataset

Team	Precision	Recall	F1-score
Our (Swin-Chart)	**0.937**	**0.933**	**0.932**
IIIT_CVIT	0.926	0.901	0.910
UB-ChartAnalysis	0.900	0.881	0.886
six seven four	0.865	0.808	0.827
CLST-IITG	0.704	0.657	0.654

and their varients. SwinL with input image dimension 224 performs best with an F1-score of 0.932. So, **SwinL** model is further referred as **Swin-Chart**. The scores of all the Swin Transformer models are summarized in Table 5. The best performing CNN based models fail to compete with Swin-Chart for the chart classification task as it outperforms the other five models with an average F1-score of 0.932. The scores for the deep learning models are summarized in Table 6.

Furthermore, we compare our best-performing model(Swin-Chart) with the models reported in [13]. This comparison is summarized in Table 7. We note that Swin-Chart surpasses the winner of the ICPR 2022 CHART-Infographics competition with an average F1-score of 0.931.

6   Future Directions

Although there has been a significant increase in published articles on chart classification, several problems still need to be addressed.

6.1   Lack of Standard Benchmark Data Sets

The chart image classification problem has been extensively addressed in previous work. Efforts have been made to increase the size of chart image datasets that also cover a wide variety of charts [10,35]. With the growing literature in various domains, authors are finding creative ways to use different charts. This adds to the variety of chart types. Integrating such diverse chart types while creating chart datasets remains an open challenge. In addition, the popularity of charts such as bar, line, and scatter over others such as Venn, surface, and area adds to the problem of disparity between the number of samples in particular chart types.

6.2 Lack of Robust Models

Recent work makes some problematic assumptions in addressing this problem [11]. A lack of a diverse benchmark dataset adds to this problem, as there needs to be more consistency in model performance across publicly available datasets. The inherent intra-class dissimilarity and inter-class similarity of several chart types affect the model's performance.

6.3 Inclusion of Noise

Most of the work in the existing literature ignores the effect of noise. Different types of noise, such as background grids, low image quality, composite charts, and multiple components along with figures, lead to poor performance for models that perform exceptionally well on noiseless data [34]. In addition to the noiseless chart image dataset, if a small set of chart images could be provided that incorporates the noisy images, it would help fine-tune the models to work through the inclusion of noise and be invariant to the same.

7   Conclusion

We have provided a brief survey of existing chart classification techniques and datasets. We used a Transformer model to obtain state-of-the-art results. Although there has been a significant development both in terms of variety in models and in the size of datasets, we observe that the chart classification problem still needs to be solved, especially for noisy and low-quality charts. Our comparative study showed that Swin-Chart outperforms the other vision transformer and CNN-based models on the latest UB-PMC dataset. In the future, we plan to generalize the results of the Swin-Chart over other publicly available datasets and try to bridge the gap to a robust deep-learning model for chart image classification.

References

1. Amara, J., et al.: Convolutional neural network based chart image classification. In: International Conference in Central Europe on Computer Graphics, Visualization, and Computer Vision (2017)
2. Araújo, T., et al.: A real-world approach on the problem of chart recognition using classification, detection, and perspective correction. Sensors **20**(16), 4370 (2020)

3. Bajić, F., et al.: Data visualization classification using simple convolutional neural network model. Int. J. Electr. Comput. Eng. Syst. (IJECES) **11**(1), 43–51 (2020)
4. Bajić, F., Job, J.: Chart classification using siamese CNN. J. Imaging. **7**, 220 (2021)
5. Balaji, A., et al.: Chart-text: a fully automated chart image descriptor. ArXiv (2018)
6. Chagas, P., et al.: Evaluation of convolutional neural network architectures for chart image classification. In: International Joint Conference on Neural Networks (IJCNN), pp. 1–8 (2018)
7. Cheng, B., et al.: Graphical chart classification using data fusion for integrating text and image features. In: Proceedings of the International Conference on Document Analysis and Recognition, ICDAR (2013)
8. Chollet, F.: Xception: deep learning with depthwise separable convolutions. In: Proceedings of the IEEE Conference on Computer Vision and Pattern Recognition (2017)
9. Dai, W., et al.: Chart decoder: generating textual and numeric information from chart images automatically. J. Vis. Lang. Comput. **48**, 101–109 (2018)
10. Davila, K., et al.: ICDAR competition on harvesting raw tables from infographics (CHART-infographics). In: International Conference on Document Analysis and Recognition (ICDAR), pp. 1594–1599. IEEE, Sydney (2019)
11. Davila, K., et al.: Chart mining: a survey of methods for automated chart analysis. IEEE Trans. Pattern Anal. Mach. Intell. **43**(11), 3799–3819 (2021)
12. Davila, K., et al.: ICPR 2020 - competition on harvesting raw tables from infographics. In: Pattern Recognition. ICPR International Workshops and Challenges: Virtual Event, pp. 361–380 (2021)
13. Davila, K., et al.: ICPR: challenge on harvesting raw tables from infographics (CHART-infographics). In: 26th International Conference on Pattern Recognition (ICPR), pp. 4995–5001 (2022)
14. Gao, J., et al.: View: visual information extraction widget for improving chart images accessibility. In: 19th IEEE International Conference on Image Processing, pp. 2865–2868 (2012)
15. He, K., et al.: Deep residual learning for image recognition. In: Proceedings of the IEEE Conference on Computer Vision and Pattern Recognition (2016)
16. Howard, A.G., et al.: MobileNets: efficient convolutional neural networks for mobile vision applications (2017). http://arxiv.org/abs/1704.04861
17. Huang, G., et al.: Densely connected convolutional networks. In: Proceedings of the IEEE Conference on Computer Vision and Pattern Recognition (2017)
18. Jung, D., et al.: ChartSense: interactive data extraction from chart images. In: Proceedings of the 2017 CHI Conference on Human Factors in Computing Systems (2017)
19. Karthikeyani, V., Nagarajan, S.: Machine learning classification algorithms to recognize chart types in portable document format (PDF) files. IJCA **39**(2), 1–5 (2012)
20. Krizhevsky, A., et al.: ImageNet classification with deep convolutional neural networks. In: Advances in Neural Information Processing Systems (2012)
21. kv, J., et al.: DocFigure: a dataset for scientific document figure classification. In: International Conference on Document Analysis and Recognition Workshops (ICDARW) (2019)
22. Liu, X., et al.: Chart classification by combining deep convolutional networks and deep belief networks. In: 13th International Conference on Document Analysis and Recognition (ICDAR), pp. 801–805 (2015)

23. Liu, X., et al.: Data extraction from charts via single deep neural network. arXiv preprint arXiv:1906.11906 (2019)

24. Liu, Z., et al.: Swin transformer: hierarchical vision transformer using shifted windows. In: Proceedings of the IEEE/CVF International Conference on Computer Vision (2021)

25. Liu, Z., et al.: A ConvNet for the 2020s. In: Proceedings of the IEEE/CVF Conference on Computer Vision and Pattern Recognition (2022)

26. Luo, J., et al.: ChartOCR: data extraction from charts images via a deep hybrid framework. In: IEEE Winter Conference on Applications of Computer Vision (WACV), pp. 1916–1924. IEEE, Waikoloa (2021)

27. Paszke, A., et al.: PyTorch: an imperative style, high-performance deep learning library. In: Proceedings of the 33rd International Conference on Neural Information Processing Systems, pp. 8026–8037. Curran Associates Inc., Red Hook (2019)

28. Russakovsky, O., et al.: ImageNet large scale visual recognition challenge. Int. J. Comput. Vis. **115**, 211–252 (2015)

29. Savva, M., et al.: ReVision: automated classification, analysis and redesign of chart images. In: Proceedings of the 24th annual ACM symposium on User interface software and technology, pp. 393–402. Association for Computing Machinery, New York (2011)

30. Siegel, N., Horvitz, Z., Levin, R., Divvala, S., Farhadi, A.: FigureSeer: parsing result-figures in research papers. In: Leibe, B., Matas, J., Sebe, N., Welling, M. (eds.) ECCV 2016. LNCS, vol. 9911, pp. 664–680. Springer, Cham (2016). https://doi.org/10.1007/978-3-319-46478-7_41

31. Simonyan, K., Zisserman, A.: Very deep convolutional networks for large-scale image recognition (2015). http://arxiv.org/abs/1409.1556

32. Szegedy, C., et al.: Going deeper with convolutions. In: Proceedings of the IEEE Conference on Computer Vision and Pattern Recognition (2015)

33. Szegedy, C., et al.: Rethinking the inception architecture for computer vision. In: Proceedings of the IEEE Conference on Computer Vision and Pattern Recognition (2016)

34. Thiyam, J., et al.: Challenges in chart image classification: a comparative study of different deep learning methods. In: Proceedings of the 21st ACM Symposium on Document Engineering, pp. 1–4. Association for Computing Machinery, New York (2021)

35. Thiyam, J., et al.: Chart classification: an empirical comparative study of different learning models. Presented at the December 19 (2021)

36. Touvron, H., et al.: Training data-efficient image transformers & distillation through attention. In: Proceedings of the 38th International Conference on Machine Learning, pp. 10347–10357. PMLR (2021)

Segmentation-Free Alignment of Arbitrary Symbol Transcripts to Images

Pau Torras[1,2(✉)] , Mohamed Ali Souibgui[1] , Jialuo Chen[1] ,
Sanket Biswas[1,2] , and Alicia Fornés[1,2]

[1] Computer Vision Center (CVC), 08193 Bellaterra, Cerdanyola del Vallès, Spain
{ptorras,msouibgui,jchen,sbiswas,afornes}@cvc.uab.cat
[2] Department of Computer Science, Universitat Autònoma de Barcelona,
08193 Bellaterra, Cerdanyola del Vallès, Spain

Abstract. Developing arbitrary symbol recognition systems is a challenging endeavour. Even using content-agnostic architectures such as few-shot models, performance can be substantially improved by providing a number of well-annotated examples into training. In some contexts, transcripts of the symbols are available without any position information associated to them, which enables using line-level recognition architectures. A way of providing this position information to detection-based architectures is finding systems that can align the input symbols with the transcription. In this paper we discuss some symbol alignment techniques that are suitable for low-data scenarios and provide an insight on their perceived strengths and weaknesses. In particular, we study the usage of Connectionist Temporal Classification models, Attention-Based Sequence to Sequence models and we compare them with the results obtained on a few-shot recognition system.

Keywords: Historical Manuscripts · Symbol Alignment

1 Introduction

Historical ciphered manuscripts are an interesting source of insight into key events in human history [14]. Many historical ciphers employ arbitrary symbols to hide plain-text characters, making difficult their automated processing using regular text recognition systems. In this low-data scenario, either tailor-made recognition systems or generic few-shot models must be developed. In any case, for these to be successfully trained, a sufficient amount of annotated data is required.

In some cases, a limited subset of ciphered documents is manually transcribed by specialists. However, these transcriptions rarely contain symbol-level bounding boxes because computational decryption techniques only need the ciphertext;

Supported by the DECRYPT Project (grant 2018-0607).

M. Coustaty and A. Fornés (Eds.): ICDAR 2023 Workshops, LNCS 14193, pp. 83–93, 2023.
https://doi.org/10.1007/978-3-031-41498-5_6

any per-symbol annotation effort is usually redundant. Therefore, the scope of transcription techniques that can be employed are restricted to end-to-end segmentation-free recognition methods. Thus, it is necessary to devote efforts in symbol alignment to enable the full range of recognition methods.

In line with the state-of-the-art on aligning ancient manuscripts containing graphic symbols or characters, the research has evolved in mainly two different directions: image-to-image and text-to-image alignment. Classical approaches relied mostly on Dynamic Time Warping [11,15] or a collection of simple visual features to characterise and align text and image – usually Hidden Markov Models (HMMs). Yalniz, Feng and Mahnmatha [5,20] used the latter to align book scans with their corresponding transcriptions in order to perform automatic evaluation of Optical Character Recognition (OCR) systems. Other uses include the efforts of Indermühle *et al.* [8], which works on the IAM dataset and Toselli *et al.* [19]

More modern efforts for alignment devoted precisely to historical manuscripts are the recent works of de Gregorio *et al.* [3,4], which offer insight on a variety of methods ranging from the aforementioned HMMs and provide some additional ways of segmenting lines (a task that we abstract within this work).

The above described methods have been applied to textual documents and in many cases benefit of advances in OCR techniques. Thus, in this work we focus on the development, comparison and discussion of different alignment techniques for the particularities of cipher symbols. Although the work is focused on ciphers, the discussion is useful for any other low-resource scenario.

The rest of the paper is organized as follows. Section 2 describes the methods used for the comparisons. Section 3 describes the datasets used. Section 4 is devoted to present the experiments and analyze the advantages and disadvantages of each method. Finally, Sect. 5 concludes the paper and discusses future work.

2   Methods

In this section we describe the three methods that have been developed for symbol alignment.

2.1   Connectionist Temporal Classification Architectures

Connectionist Temporal Classification (CTC) [6] techniques span models trained using said loss function. A model trained using CTC has an image $I \in \mathbb{R}^{W \times H \times 3}$ as input, where W and H are its width and height. A neural network is used to convert this input into an output feature map $F \in \mathbb{R}^{W' \times C}$ where $W' <= W$ and C is the number of possible output classes. The idea of CTC is to generate a prediction for every vector in the output W' dimension and then, through the CTC decoding algorithm, produce the desired output sequence S.

The CTC decoding algorithm rules are as follows. Any sequence of repeated characters in the output maps into a single instance of the character. In order to distinguish different instances of the same character, a special 0 token must be interleaved between each subsequence of repeated symbols. Thus, the CTC loss tries to maximise the log likelihood of the target sequence by computing the joint probability of all possible decodings that map to it, assuming that each element along W' is an independent prediction. This is done efficiently through a dynamic programming algorithm [6].

In our use case we have the output sequences available at inference time; what we are interested in is the mapping of elements in W' to characters in the output sequence S. The alignment procedure we propose in this work is to train a CTC model for a regular transcription task, and then find the output decoding that maximises the probability of S during inference. Finding the optimal decoding is a computation of complexity $O(3^n)$. However, in practical settings, using a beam decoding algorithm [21] with a beam width of 30 seemed to yield robust results, partly due to the properties of the CTC loss [22]. Alternatives could be using a Monte-Carlo approach.

We employed two different architectures trained on the CTC loss; a Fully Convolutional (FC) model built on a standard Resnet 18 [7] with an upsampling step at the end to increase the size of W' and a simple Convolutional Recurrent Neural Network (CRNN) model with four convolutional layers and four LSTM layers inspired by [2].

2.2 Sequence to Sequence Architectures

Sequence to Sequence (Seq2Seq) architectures [17] are a family of models that map an arbitrary-length input sequence I into an arbitrary-length output sequence S. By interpreting an image as a sequence of pixel columns (or rather, a sequence of feature vectors after being processed by a neural network), they can also be used in transcription scenarios.

Canonical seq2Seq models are implemented as encoder-decoders; a stack of RNNs (the encoder) processes the input sequence into an intermediate representation, which is then used by a second stack of RNNs (the decoder) to produce as many output elements as required. We use a specific type of Seq2Seq models that use an Attention Mechanism to generate the input to the decoder at each inference step. The Attention mechanism computes a vector $\mathbf{a} \in \mathbb{R}^{W'} : \sum_{i=1}^{W'} a_i = 1$ that, when multiplied with the intermediate representation, generates a single vector that incorporates all of the necessary information for an inference step. This is useful for alignment because the model implicitly learns the correspondence of visual input features and output characters and models this mapping explicitly through $\mathbf{a}$. A representation of these $\mathbf{a}$ vectors can be found in Fig. 1.

After the recognition process finishes, a correction system is applied in order to ensure that all predictions generate the known output. In this case, the final output is corrected using the Levenshtein algorithm [12] to find the smallest edit path from the ground truth to the prediction. The character coordinates are then modified accordingly while keeping the alignment with the predicted tokens: that is, bounding boxes corresponding to characters that are false positives are

Fig. 1. First 5 log-attention masks on a Copiale line. Brighter outputs mean a higher activation.

removed, whereas blank space is inserted in those cases where there should be a character in the transcript. The blank space between characters is then divided equally among as many characters that need to be inserted. This comes from a rather strong assumption that the model will do well enough in order not to forget many characters in a row.

2.3   Few-Shot Recognition Architectures

The used few-shot text recognition method was proposed in [16]. This approach is adapting the few-shot object detection, which can be defined as finding one or several instance(s) of an object by providing a cropped example image, called a *support*, in an image that contains the object, called *query* image. For test recognition, the supports are the cropped alphabet set images and the query is the handwritten line image that is being transcribed. The model is used in two stages, symbol matching and similarity matrix decoding.

In the symbol matching stage, all the characters within the alphabet are passed with the handwritten line to the model to output the bounding boxes and similarity score. These are used to construct a similarity matrix between the alphabet and the line image. For the matrix decoding stage, the goal is to generate the sequence of characters as a final output. The method traverses the columns of the similarity matrix from left to right, and decides for each pixel column the final transcribed symbol class among the candidate symbols that construct the matrix columns (the alphabet). The choice of the symbols is based on the similarity score, i.e., choosing the symbol that is having the maximum similarity score.

For text alignment, the same idea as Seq2Seq is used; predict through a regular recognition task and then perform due corrections through edit distances.

3   Datasets

The datasets we tested the alignment systems on are two historical ciphered documents: the Borg cipher [1] and the Copiale cipher [10] (Fig. 2). The Copiale cipher is a manuscript that contains symbols from both the latin and greek alphabets, as well as some ideograms that represent special entities. Pages are scanned and binarised, while characters are noticeably distinct, making this dataset the easier of the two. The Borg cipher uses arbitrary slanted and touching symbols We also noticed the existence of pictures that include more than one handwritten line due to the shape of the symbols and the limited use of line spacing. This creates a further difficulty in the form of uneven crops and additional noise.

Since in both cases the amount of data is extremely limited (see Table 1), we devised a trick exploiting the fact that we do have a finely annotated subset of samples. We cropped all possible subwords of length higher than 16 characters and we shuffled the dataset. With this, we force the model to learn to recognise the same symbols in different positions within the input line and we also provide more distinct samples for training. We call this the Extended version of the datasets. We produced one version for all partitions in all datasets, but logged results employ the original versions of the datasets.

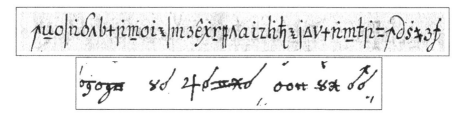

Fig. 2. Examples of ciphers: Copiale cipher (top) [10], Borg cipher (bottom) [1]. As it can be seen, some symbols are borrowed from widespread writing systems (mostly the greek and latin alphabets) but a sizeable subset are either invented or borrowed from other unrelated sources.

Table 1. Datasets description. From left to right, the columns are: the name of the cipher, whether the dataset is the subword exhaustive version, the number of images of the dataset in the training, validation and test partitions and the number of unique symbols in the vocabulary.

Cipher	Extended?	# Train	# Valid	# Test	Classes
Copiale		648	126	139	126
Copiale	✓	253,747	44,487	50,333	126
Borg		178	36	49	53
Borg	✓	2868	511	646	53

4   Experiments

In order to test our approaches, we have trained the aforementioned models using diverse setups. In particular, we have trained the three sequential models and used the publicly available weights for the Few Shot model. For training we have used the extended versions of the datasets and we validated the results epoch-wise using the normal partition.

Main results are logged in Average Intersection over Union (AIoU), which computes the degree of matching between the bounding boxes for the ground truth and the prediction. Considering one-dimensional coordinates, it is computed as $IoU = \frac{\min(x_2,x_2')-\max(x_1,x_1')}{\min(x_1,x_1')-\max(x_2,x_2')}$. The number of boxes with an IoU over a threshold are counted, with 25, 50 and 75 being the elected boundaries. The full results are shown in Table 2.

Table 2. Alignment results in Average Intersection Over Union and the number of characters with IoU over 25%, 50% and 75% in the Copiale and Borg datasets. As can be seen, the best performant is the CTC model trained on an RNN.

	Seq2Seq	CRNN+CTC	FC+CTC	Few Shot
AIoU (Copiale)	58.10%	70.53%	51.87%	0.981%
Hits@25	92.77%	96.97%	–	1.717%
Hits@50	64.57%	90.39%	–	0.572%
Hits@75	24.04%	44.99%	–	0.143%
AIoU (Borg)	20.56%	61.91%	–	48.68%
Hits@25	35.13%	82.17%	–	63.48%
Hits@50	15.82%	70.91%	–	49.58%
Hits@75	4.246%	46.07%	–	37.37%

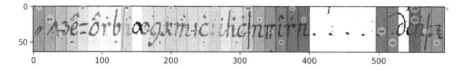

Fig. 3. A successful example on Copiale using Seq2Seq where the model manages to keep the alignment even with the presence of a rare symbol, something that used to be a problem. For this, training with much more real data and ensuring the batch size was small was the key to overcoming these such problems.

The best hyperparameters for the CTC models were found by running 30 instances of models with randomised hyperparameters and keeping those that obtained best Symbol Error Rate in validation. For the Fully Convolutional CTC models we initially only explored with Copiale since we quickly realised some major weaknesses; in particular, having an upsampling step causes the model to produce the same output for all feature weights related to the same upsampled region. Even if numeric alignment results were still quite promising after forcing a specific output sequence, we quickly realised that there was no practical advantage on these models over the CRNNs. In particular, in Fig. 4 an alignment output is presented that shows some of the issues hereby presented.

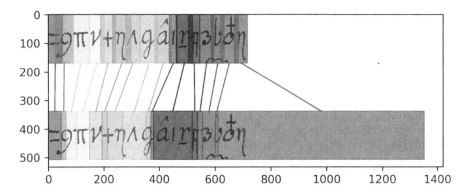

Fig. 4. Above, the sequence with the ground truth bounding boxes. Below, the resulting predicted bounding boxes using Convolutional CTC models. A recurrent issue of models with upsampling is that they tend to cause successive time steps to be tightly related, even if a specific output sequence is forced. The effect is that the character at the middle of such a sequence dominates over the others and causes their widths to be misrepresented. The trailing whitespace is caused by the exhaustive nature of CTC models.

In order to sidestep said weaknesses, we explored the idea of not reducing the width of the input images in order to keep as many output columns as possible while trying to keep them mostly independent. This originated the idea of using a simple five-layer CNN with a single max pooling on the horizontal dimension and as many vertical max pooling operations as required to reduce the height of the input to one.

The main conclusion that can be drawn from CRNN-based CTC models is that they seem to be the most promising by a significant margin. Overall, CTC-based models are better performing and substantially more lightweight than any of the other competitors. The CRNN in use here is around two million parameters big, which makes it relatively lightweight and easy to train. Finally, the CTC loss works well under low-data scenarios; during initial testing we found convergence using only the default datasets. Even with low-performance during recognition, acceptable results can be salvaged after forcing a specific decoding (although in this scenario there are specific failure cases that appear often – see Fig. 5).

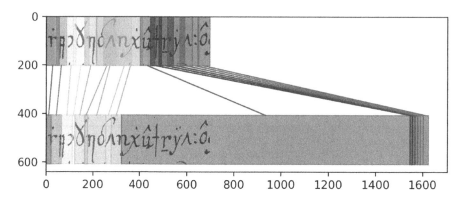

Fig. 5. An example of a real sequence with the ground truth bounding boxes. Below, the model's prediction. Forcing a specific decoding on a CTC model that is overconfident on its output while being wrong causes rogue predictions at the end of the image. This is due to the fact that a specific character or subsequence of characters have a massively low confidence at their real position, causing the model to place them in regions without characters (as is the case in this example). This is alleviated by forcing the decoding to be performed on a cropped image.

Seq2Seq models are quite decent at generating plausible alignments as long as they are trained to be significantly performant in recognition – our best results were obtained with a model with a 6% of symbol error rate. Moreover, in order to refine the outputs in those cases where there is no width correlation between the characters and the attention weights, a fairly significant amount of work is required, as seen in [18]. This is further proved here when applying the idea on the Borg cipher.

Another point worth considering is the difficulty of training a Seq2Seq model. Unlike CTC models, attention-based architectures seem sensitive to the output line length. Character co-occurrence influences the final output considerably more than in CTC models, which seem to quite aptly recognise characters even if the training and test partitions are different in terms of average length. For the case of Seq2Seq, a respectable amount of iteration was required in order to find a set of hyperparameters that suited the requirements of the task. Moreover, unless very small batch sizes were used, initial convergence without severe overfitting was very difficult. When these factors are combined, results improve from previous work [18] as can be seen in Fig. 3.

We also tried the newly released segment-anything architecture [9] without immediate success. Segmenting without prior information results in entire blocks of connected components being extracted from the input instead of individual symbols. Manually setting dots on symbols does help segmentation qualitatively. We speculate that providing the input dots from any of the models hereby presented could potentially improve results very substantially.

Finally, a comment on the results on the few-shot model using Copiale. This architecture seems very sensitive to the input aspect ratio of the images, which for this dataset has been tricky to find. This also makes it a little bit less appealing for this task in particular, considering the inference computational cost is quite high.

5   Conclusions and Future Work

In this work we have compared several alignment methods. Upon comparison, we have concluded that the most direct and better performing approaches for alignment of ciphered manuscripts seem to be the CTC loss-based approaches. These models provide an average IoU of 70.53% on the Copiale cipher and 61.91% in Borg. Moreover, these models are substantially easier to train and postprocess than the others presented while also being quite lightweight, making them an overall more appealing choice.

A relevant aspect of the models presented in this work is the fact that they are currently $ad - hoc$ solutions for specific ciphers. Nevertheless, a future line of research involves developing a generic alphabet with which to train a single model capable of aligning any cipher. The idea is to cluster a fixed set of symbols from examples of all available vocabularies; since the mapping from the actual cipher symbols and the generic alphabet is known, the only additional step needed to perform the post-processing and alignment strategies is converting the known transcription to the generic alphabet. The fact that transcriptions are known also allows the mapping between real and generic symbols to be non-injective.

We also have devised improvements on some of the models that we deem interesting to discuss. The first relevant improvement on CTC models is ensuring a certain degree of uncertainty on each bin in the output prediction. This is helpful when forcing a specific output, since CTC models tend to overfit or produce peaky output distributions [22], cases in which trying to produce a different decoding is, at best, harmless to the performance. To do so, the application of ideas such as label smoothing or regularising the output probabilities when computing the loss [13] seem worthwhile candidates.

Acknowledgements. This work has been partially supported by the Swedish Research Council (grant 2018-06074, DECRYPT), the Spanish projects PID2021-126808OB-I00 (GRAIL) and CNS2022-135947 (DOLORES), as well as the AGAUR Joan Oró FI grant 2023 FI-1-00324. The authors acknowledge the support of the Generalitat de Catalunya CERCA Program to CVC's general activities.

References

1. Aldarrab, N., Knight, K., Megyesi, B.: The Borg.lat.898 Cipher (2018). https://cl.lingfil.uu.se/bea/~borg/
2. Baró, A., Chen, J., Fornés, A., Megyesi, B.: Towards a generic unsupervised method for transcription of encoded manuscripts. In: Proceedings of the 3rd International Conference on Digital Access to Textual Cultural Heritage, DATeCH2019, pp. 73–78. Association for Computing Machinery, New York (2019). https://doi.org/10.1145/3322905.3322920

3. De Gregorio, G., Capriolo, G., Marcelli, A.: End-to-end transcript alignment of 17th century manuscripts: the case of Moccia code. J. Imaging **9**(1), 17 (2023). https://doi.org/10.3390/jimaging9010017, https://www.mdpi.com/2313-433X/9/1/17

4. De Gregorio, G., Citro, I., Marcelli, A.: Transcript alignment for historical handwritten documents: the MiM algorithm. In: Carmona-Duarte, C., Diaz, M., Ferrer, M.A., Morales, A. (eds.) IGS 2022. LNCS, vol. 13424, pp. 45–60. Springer, Cham (2022). https://doi.org/10.1007/978-3-031-19745-1_4

5. Feng, S., Manmatha, R.: A hierarchical, HMM-based automatic evaluation of OCR accuracy for a digital library of books. In: Proceedings of the 6th ACM/IEEE-CS Joint Conference on Digital Libraries, JCDL 2006, pp. 109–118. Association for Computing Machinery, New York (2006). https://doi.org/10.1145/1141753.1141776

6. Graves, A., Fernández, S., Gomez, F., Schmidhuber, J.: Connectionist temporal classification: labelling unsegmented sequence data with recurrent neural networks. In: Proceedings of the 23rd international conference on Machine learning, ICML 2006, pp. 369–376. Association for Computing Machinery, New York (2006). https://doi.org/10.1145/1143844.1143891

7. He, K., Zhang, X., Ren, S., Sun, J.: Deep residual learning for image recognition. In: Proceedings of the IEEE Conference on Computer Vision and Pattern Recognition (CVPR) (2016)

8. Indermühle, E., Liwicki, M., Bunke, H.: Combining alignment results for historical handwritten document analysis. In: 2009 10th International Conference on Document Analysis and Recognition, pp. 1186–1190 (2009). https://doi.org/10.1109/ICDAR.2009.19. iSSN 2379-2140

9. Kirillov, A., et al.: Segment anything (2023). https://doi.org/10.48550/arXiv.2304.02643, arXiv:2304.02643 [cs]

10. Knight, K., Megyesi, B., Schaefer, C.: The Copiale cipher. In: Proceedings of the 4th Workshop on Building and Using Comparable Corpora: Comparable Corpora and the Web, pp. 2–9. Association for Computational Linguistics, Portland (2011). https://aclanthology.org/W11-1202

11. Kornfield, E., Manmatha, R., Allan, J.: Text alignment with handwritten documents. In: 2004 Proceedings of First International Workshop on Document Image Analysis for Libraries, pp. 195–209 (2004). https://doi.org/10.1109/DIAL.2004.1263249

12. Levenshtein, V.I., et al.: Binary codes capable of correcting deletions, insertions, and reversals. In: Soviet Physics Doklady, vol. 10, pp. 707–710. Soviet Union (1966)

13. Liu, H., Jin, S., Zhang, C.: Connectionist temporal classification with maximum entropy regularization. In: Advances in Neural Information Processing Systems, vol. 31. Curran Associates, Inc. (2018). https://proceedings.neurips.cc/paper/2018/hash/e44fea3bec53bcea3b7513ccef5857ac-Abstract.html

14. Megyesi, B., et al.: Decryption of historical manuscripts: the decrypt project. Cryptologia **44**(6), 545–559 (2020)

15. Müller, M.: Dynamic time warping. In: Müller, M. (ed.) Information Retrieval for Music and Motion, pp. 69–84. Springer, Heidelberg (2007). https://doi.org/10.1007/978-3-540-74048-3_4

16. Souibgui, M.A., Fornés, A., Kessentini, Y., Megyesi, B.: Few shots are all you need: a progressive learning approach for low resource handwritten text recognition. Pattern Recognit. Lett. **160**, 43–49 (2022). https://doi.org/10.1016/j.patrec.2022.06.003, https://www.sciencedirect.com/science/article/pii/S016786552200191X

17. Sutskever, I., Vinyals, O., Le, Q.V.: Sequence to sequence learning with neural networks. In: Advances in Neural Information Processing Systems, vol. 27. Curran Associates, Inc. (2014). https://proceedings.neurips.cc/paper/2014/hash/a14ac55a4f27472c5d894ec1c3c743d2-Abstract.html

18. Torras, P., Souibgui, M.A., Chen, J., Fornés, A.: A transcription is all you need: learning to align through attention. In: Barney Smith, E.H., Pal, U. (eds.) ICDAR 2021. LNCS, vol. 12916, pp. 141–146. Springer, Cham (2021). https://doi.org/10.1007/978-3-030-86198-8_11

19. Toselli, A.H., Romero, V., Vidal, E.: Viterbi based alignment between text images and their transcripts. In: Proceedings of the Workshop on Language Technology for Cultural Heritage Data (LaTeCH 2007), pp. 9–16. Association for Computational Linguistics, Prague (2007). https://aclanthology.org/W07-0902

20. Yalniz, I.Z., Manmatha, R.: A fast alignment scheme for automatic OCR evaluation of books. In: 2011 International Conference on Document Analysis and Recognition, pp. 754–758 (2011). https://doi.org/10.1109/ICDAR.2011.157. iSSN 2379-2140

21. Zenkel, T., et al.: Comparison of decoding strategies for CTC acoustic models (2017). https://doi.org/10.48550/arXiv.1708.04469, arXiv:1708.04469 [cs]

22. Zeyer, A., Schlüter, R., Ney, H.: Why does CTC result in peaky behavior? (2021). https://doi.org/10.48550/arXiv.2105.14849, arXiv:2105.14849 [cs, eess, math, stat]

Optical Music Recognition: Recent Advances, Current Challenges, and Future Directions

Jorge Calvo-Zaragoza$^{(\boxtimes)}$, Juan C. Martinez-Sevilla, Carlos Penarrubia, and Antonio Rios-Vila

University Institute for Computing Research, University of Alicante, Alicante, Spain
jcalvo@dlsi.ua.es

Abstract. Optical Music Recognition (OMR) is an interdisciplinary field that aims to automate the process of transcribing sheet music into a digital format. Over the past few years, significant progress has been made in developing OMR systems that can recognize musical symbols with high accuracy. However, completing the pipeline of OMR remains a challenging endeavor due to the complexity and variability of music notation, and there are several open challenges that need to be addressed. In this position paper, we provide an overview of the current state-of-the-art in OMR through the two main lines of research. We include the problems that have been recently addressed and the techniques that have been considered. We then identify the key challenges that remain, such as learning to reconstruct the music notation, recognizing multiple voices, or dealing with artifacts such as lyrics. Finally, we suggest some possible directions for future research. We argue that addressing these challenges is crucial to making OMR a more practical and useful tool for musicians, scholars, and librarians alike.

Keyword: Optical Music Recognition

1 Introduction

Music is an important part of our cultural heritage. The digital humanities have played a crucial role in preserving and making music accessible to a wider audience. One area of research that has emerged in this context is Optical Music Recognition (OMR), which seeks to automate the process of transcribing written music sources into a digital format [5]. OMR represents an interdisciplinary field that draws on document image analysis, computer vision, and music theory to develop effective algorithms. The progress of this technology multiplies the

Work produced with the support of a 2021 Leonardo Grant for Researchers and Cultural Creators, BBVA Foundation. The Foundation takes no responsibility for the opinions, statements and contents of this project, which are entirely the responsibility of its authors.

options on which digital humanities can operate and, therein, lies its nowadays importance.

Recent advances in deep learning have led to significant improvements in OMR, and several state-of-the-art systems now rely on neural networks. In particular, deep learning has proven effective in addressing some of the challenges that have traditionally plagued OMR, such as staff-line removal [18] or music-object classification [23].

Despite these advances, OMR remains an open problem. In this paper, we provide a position statement on the state of the art in OMR and discuss the current use of deep learning techniques. We then identify the open challenges that need to be addressed and suggest some possible directions for future research. We argue that solving these challenges is crucial to advancing the field of OMR and making it a more practical tool for musicians, scholars, and enthusiasts alike.

For the sake of clarification, let us note that in the rest of the paper we will split the intended discussion into the two great paradigms that currently dominate OMR: the one based on a multi-step pipeline and the one focused on end-to-end formulations.

2 Background

Before elaborating on the aspects related to the state of the art in OMR, we introduce in this section the necessary background to understand the rest of the sections as regards how the task is approached from the two aforementioned paradigms.

2.1 Pipeline-Based Optical Music Recognition

The traditional approach to OMR involves a multi-stage workflow [12]. It consists of image preprocessing, music symbol identification, notation assembly, and encoding.

In the preprocessing step, the music score image is prepared for further analysis. This may include operations such as skew correction, binarization, and staff line removal. In the music symbol identification stage, individual symbols such as notes, rests, and accidentals are detected and classified. Then, notation assembly is performed, where the identified symbols are combined into larger structures such as compound symbols, measures, staves, and systems. Finally, in the encoding stage, the recognized notation is translated into a machine-readable format such as MusicXML or MIDI.

2.2 Holistic Optical Music Recognition

Holistic methods for OMR have been proposed that aim to transcribe entire sections of music notation at once, without explicitly identifying individual symbols. These methods typically involve a staff extraction step, where staves are identified within the whole page, followed by a staff-level end-to-end transcription, where an entire staff is transcribed as a single sequence of symbols.

This approach has the advantage of being more robust to variations in notation and layout, and can be applied to both printed and handwritten music. However, it also poses several challenges, such as the need to deal with overlapping staves and the difficulty of handling polyphonic music with multiple voices. Despite these challenges, holistic methods represent an active area of research in the field of OMR.

3   State of the Art

In this section, we outline the advances that have been taking over the publications in the OMR field for the last years.

3.1   Pipeline-Based Optical Music Recognition

Concerning the first stages of the pipeline, semantic segmentation has emerged as a promising method for OMR. This involves labeling each pixel of the image based on its layout category, such as staff, notes, rests, or lyrics. Recent works have shown success in this endeavor by considering deep learning models [9,33].

Direct music symbol identification, treated as an object detection task, has also received significant attention in recent years [16,21,22]. The idea is to detect and classify individual symbols directly from the music score image. Many researchers have proposed deep learning-based methods for this task. Furthermore, several datasets have been created to facilitate training and benchmarking of these methods [15,28,29].

Notation assembly, the process of combining identified symbols into larger structures such as compound symbols, measures, staves, and systems, has also seen a few learning-based approaches [20]. In these methods, the symbols are first identified, and then their relationships are modeled as a graph. Within this context, Graph Neural Networks (GNN) have also been used to learn the structure of the graph and assemble the symbols into the desired larger structures [3].

3.2   Holistic Optical Music Recognition

Stave detection, as a necessary preprocessing step for staff-line level recognition, has received much attention. This process involves identifying the location of staves as a specific region of the music score image. Stave detection is essential to achieving accurate results in subsequent stages, and many methods have been proposed to address this challenge [10].

Staff-level end-to-end recognition has been widely studied and achieved impressive results [31]. This approach involves recognizing the entire staff-level image and directly outputting the corresponding notation. This approach is particularly useful for old music, where monodic staves are common and music notation can be expressed as a plain sequence. Many publications have explored this topic, which can be categorized into two areas: Image-to-Sequence and CTC-based approaches. The former one takes the path of the sequence-to-sequence mechanisms proposed in the Machine Translation field, with works showing their effectiveness for OMR [4,24,30]; the latter resembles the Handwritten Text Recognition field by considering the CTC loss function [14], which has been demonstrated to perform accurately for OMR as well [6].

Recently, a step forward has been taken to extend the end-to-end paradigm to more complex music layouts. Several efforts have been proposed to transcribe music scores at a full-page level, from systems that combine layout analysis and staff-level recognition processes [8] to networks that learn to perform transcription in a single step [26]. A representation on how currently state-of-the-art methods address full page transcription can be found in Fig. 1 Although these advances are promising, they only cover currently monophonic music scores.

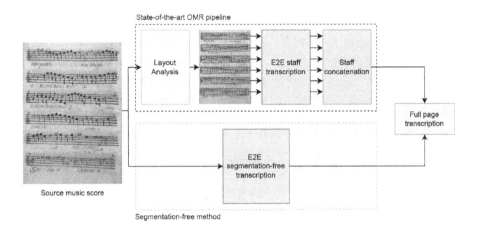

Fig. 1. General overview of the current OMR pipeline (top) in contrast to the holistic full-page approach (bottom), where a previous layout analysis is not needed to transcribe the music symbols in the score.

4 Current Challenges

As the field of OMR continues to evolve, researchers must face several new challenges. In this context, the open challenges differ greatly depending on the paradigm.

4.1 Pipeline-Based Optical Music Recognition

One of the current challenges in pipeline-based OMR is to improve the performance of music-object detection. Despite the advances mentioned above, it is still risky to assume that the algorithms work in any context, especially considering the less common musical symbols or the great variability in size between them as shown in Fig. 2. In many cases, the common object-detection metrics (e.g., mean Average Precision) do not necessarily reflect the goodness of the model for OMR.

Fig. 2. Variability in score formats and notational systems.

While the music-symbol detection stage can benefit from other advances in computer vision, both the notation assembly and the encoding stage are yet to be further developed because they are particular to OMR. For instance, it is not clear how the various music symbols should be generally related to one another, and how this information can be applied to generate a meaningful encoding. Additionally, the encoding stage itself has barely received attention from the literature, finding few works that address it [25]. One possible avenue is to keep on modeling the music notation as a graph and then use GNN to generate an encoding, but this has not been explored.

Furthermore, there is a lack of proper datasets that include the ground truth information required for both the notation assembly and encoding stages, which makes it difficult to train and evaluate full pipeline-based OMR systems.

4.2   Holistic Optical Music Recognition

The main challenge for holistic OMR is to move towards more complex music systems—such as quartets, orchestral scores or simultaneous lyric-accompanied music—as it is currently limited to single-staff and monophonic full-page recognition. Specifically, we refer to this challenge as OMR has to face scores where multiple melodic lines develop at the same time.

In this area, researchers have barely explored the use of language models to improve results [7,27]. These models can help predict the next note based on the context of the previous symbols, and thereby enhance the accuracy of the output. However, the improvement brought by language models is still limited, and more work is needed to explore their full potential in OMR, especially given the inherent properties of music as a language.

While current methods have shown promise in recognizing single staves, extending these methods to handle multiple staves and complex musical structures is still an open research problem. This requires developing algorithms that can separate and recognize multiple voices, handle overlapping and intersecting staves, and recognize complex musical symbols such as dynamic markings and articulations.

Another pressing challenge is the recognition and alignment of music notation and lyrics, which is essential for cultural heritage, where vocal music is prevalent. This challenge is particularly interesting because it requires the intersection of graphical recognition of written elements and underlying language processing to relate them appropriately. Just a few approaches have been considered [32], but the challenge is quite open for further research.

Additionally, there is a need to move towards actual holistic transcription, skipping the staff detection step and recognizing the music notation directly from the image. This has already begun in text transcription, where end-to-end methods have shown promising results.

Some efforts have been recently proposed in OMR. Specifically, to pianoform scores and aligned music and lyrics transcription. The most recent approach treats these scores like full-page handwritten text recognition paragraphs—thanks to the properties of well-known music encodings—and uses advanced neural network architectures to achieve impressive results. An example is shown in Fig. 3. This represents a significant advancement in OMR research and promises to enhance the capability of automated music transcription.

Finally, there are certain limitations on the music documents that OMR can effectively handle. For instance, there are only a limited number of techniques available to deal with handwritten modern music notation, which is mainly due to the scarcity of datasets for conducting experiments. As a result, OMR is currently biased towards being a tool for historic manuscripts, where although the number of available datasets is still limited, more data can be obtained.

This creates a more extensive challenge, which is the lack of adequate datasets that incorporate the ground truth information necessary to evaluate OMR systems, including different types of musical scores. Creating such datasets is a crucial step in advancing the field.

5 Future Directions

There are several exciting directions for future research in OMR, that can help address the current challenges and push the field forward.

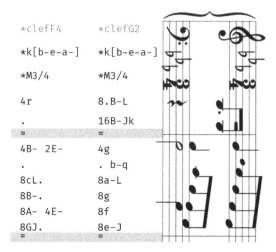

Fig. 3. Example on how a pianoform music excerpt can be aligned with its current digital music notation representation, in such a way that it could be read like a text paragraph.

5.1 End-to-End Music Notation Graph Retrieval

One promising direction is to merge the two branches of research in OMR, namely the pipeline-based and holistic approaches, and move towards end-to-end music-notation graph retrieval. This involves developing algorithms that can simultaneously perform staff extraction, music symbol identification, notation assembly, and encoding, using deep learning techniques to model the relationships between different music primitives in a single step. End-to-end approaches have shown promise in other computer vision tasks, and applying similar techniques to OMR can potentially improve recognition accuracy and reduce error propagation. This has only been considered on a very preliminary setting of compound symbols [13], as illustrated in Fig. 4.

5.2 Synthetic Data Generation and Data Augmentation

Another avenue for addressing the limitations of OMR is to generate synthetic data to augment existing datasets [2]. This can help overcome the shortage of appropriate data for various stages of the OMR process. Additionally, data augmentation can enhance the model generalizability and resilience. This trend is evident in other research fields, where deep learning solutions are trained on synthetic databases to develop a basic understanding of the task and then fine-tuned on specific datasets, resulting in impressive outcomes [11,17,19].

Generating realistic synthetic data for music notation is a challenging task, and it requires the development of techniques that can capture the complexity and diversity of various music systems.

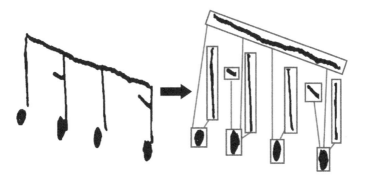

Fig. 4. Image to graph formulation for compound symbols.

5.3 Domain Adaptation Techniques

Domain adaptation techniques can complement the use of synthetic data by allowing models to generalize better to real-world data. Domain adaptation involves adapting models trained on synthetic or other sources of data to perform well on real-world data. This can be particularly useful in OMR, where the variations in notation style, font types, and scanning quality can significantly affect recognition accuracy.

5.4 Self-Supervised Learning

Self-Supervised Learning (SSL) is a machine learning technique that has gained attention in recent years due to its ability to learn useful representations from unlabeled data. Unlike supervised learning, where models are trained on labeled data, SSL involves training models on tasks that do not require explicit supervision. By leveraging large amounts of unlabeled data, SSL can help overcome the limitations of labeled datasets and improve model generalization and robustness.

In OMR, SSL has the potential to be a powerful technique for building more robust and generalizable models. By utilizing the vast amounts of unlabeled music data available on the internet, SSL can potentially help overcome the lack of labeled data for certain music systems or notation styles [1]. SSL also presents an opportunity to create general-purpose music transcription models that can transcribe music across various genres and styles, which can help capture the complexity and diversity of different music systems.

5.5 Foundational Models for OMR

As a crucial long-term objective, the creation of foundational models for OMR can establish a shared framework for the field, making it easier to compare different approaches. Foundational models would offer a cohesive representation of music notation that can be applied throughout various stages of the OMR pipeline and across diverse music systems or, indeed, produce general-purpose holistic solutions to perform music scores transcription. This can enhance recognition accuracy, diminish the spread of errors, and propel the field toward more universal and resilient OMR systems.

References

1. Bardes, A., Ponce, J., LeCun, Y.: Vicreg: variance-invariance-covariance regularization for self-supervised learning. In: The Tenth International Conference on Learning Representations, ICLR 2022, Virtual Event, 25–29 April 2022 (2022)
2. Baró, A., Riba, P., Calvo-Zaragoza, J., Fornés, A.: From optical music recognition to handwritten music recognition: a baseline. Pattern Recognit. Lett. **123**, 1–8 (2019)
3. Baró, A., Riba, P., Fornés, A.: Musigraph: optical music recognition through object detection and graph neural network. In: Porwal, U., Fornés, A., Shafait, F. (eds.) Frontiers in Handwriting Recognition - 18th International Conference, ICFHR 2022, Proceedings. Lecture Notes in Computer Science, Hyderabad, India, 4–7 December 2022, vol. 13639, pp. 171–184. Springer, Heidelberg (2022). https://doi.org/10.1007/978-3-031-21648-0_12
4. Baró, A., Badal, C., Fornés, A.: Handwritten historical music recognition by sequence-to-sequence with attention mechanism. In: 2020 17th International Conference on Frontiers in Handwriting Recognition (ICFHR), pp. 205–210 (2020)
5. Calvo-Zaragoza, J., Jr, J.H., Pacha, A.: Understanding optical music recognition. ACM Comput. Surv. (CSUR) **53**(4), 1–35 (2020)
6. Calvo-Zaragoza, J., Toselli, A.H., Vidal, E.: Handwritten music recognition for mensural notation with convolutional recurrent neural networks. Pattern Recogn. Lett. **128**, 115–121 (2019)
7. Calvo-Zaragoza, J., Toselli, A.H., Vidal, E.: Hybrid hidden Markov models and artificial neural networks for handwritten music recognition in mensural notation. Pattern Anal. Appl. **22**(4), 1573–1584 (2019)
8. Castellanos, F.J., Calvo-Zaragoza, J., Inesta, J.M.: A neural approach for full-page optical music recognition of mensural documents. In: Proceedings of the 21st International Society for Music Information Retrieval Conference, pp. 558–565. ISMIR, Montreal (2020)
9. Castellanos, F.J., Calvo-Zaragoza, J., Vigliensoni, G., Fujinaga, I.: Document analysis of music score images with selectional auto-encoders. In: Proceedings of the 19th International Society for Music Information Retrieval Conference, pp. 256–263 (2018)
10. Castellanos, F.J., Garrido-Munoz, C., Ríos-Vila, A., Calvo-Zaragoza, J.: Region-based layout analysis of music score images. Expert Syst. Appl. **209**, 118211 (2022)
11. Coquenet, D., Chatelain, C., Paquet, T.: Dan: a segmentation-free document attention network for handwritten document recognition. IEEE Trans. Pattern Anal. Mach. Intell. **45**, 8227–8243 (2023)
12. Fujinaga, I., Vigliensoni, G.: The art of teaching computers: the SIMSSA optical music recognition workflow system. In: 27th European Signal Processing Conference, EUSIPCO 2019, A Coruña, Spain, 2–6 September 2019, pp. 1–5. IEEE (2019)
13. Garrido-Munoz, C., Ríos-Vila, A., Calvo-Zaragoza, J.: A holistic approach for image-to-graph: application to optical music recognition. Int. J. Doc. Anal. Recognit. **25**(4), 293–303 (2022)

14. Graves, A., Fernández, S., Gomez, F.J., Schmidhuber, J.: Connectionist temporal classification: labelling unsegmented sequence data with recurrent neural networks. In: Proceedings of the Twenty-Third International Conference on Machine Learning, (ICML 2006), Pittsburgh, Pennsylvania, USA, 25–29 June 2006, pp. 369–376 (2006)

15. Hajic, J., Pecina, P.: The MUSCIMA++ dataset for handwritten optical music recognition. In: 14th IAPR International Conference on Document Analysis and Recognition, ICDAR 2017, Kyoto, Japan, 9–15 November 2017, pp. 39–46. IEEE (2017)

16. Huang, Z., Jia, X., Guo, Y.: State-of-the-art model for music object recognition with deep learning. Appl. Sci. **9**(13), 2645 (2019)

17. Kang, L., Riba, P., Rusiñol, M., Fornés, A., Villegas, M.: Pay attention to what you read: non-recurrent handwritten text-line recognition. Pattern Recogn. **129**, 108766 (2022)

18. Konwer, A., et al.: Staff line removal using generative adversarial networks. In: 2018 24th International Conference on Pattern Recognition (ICPR), pp. 1103–1108. IEEE (2018)

19. Li, M., et al.: Trocr: transformer-based optical character recognition with pre-trained models (2021). arXiv preprint arXiv:2109.10282

20. Pacha, A., Calvo-Zaragoza, J., Hajic Jr., J.: Learning notation graph construction for full-pipeline optical music recognition. In: Proceedings of the 20th International Society for Music Information Retrieval Conference, ISMIR 2019, Delft, The Netherlands, 4–8 November 2019, pp. 75–82 (2019)

21. Pacha, A., Choi, K.Y., Coüasnon, B., Ricquebourg, Y., Zanibbi, R., Eidenberger, H.: Handwritten music object detection: open issues and baseline results. In: 2018 13th IAPR International Workshop on Document Analysis Systems (DAS), pp. 163–168. IEEE (2018)

22. Pacha, A., Hajič, J., Jr., Calvo-Zaragoza, J.: A baseline for general music object detection with deep learning. Appl. Sci. **8**(9), 1488 (2018)

23. Paul, A., Pramanik, R., Malakar, S., Sarkar, R.: An ensemble of deep transfer learning models for handwritten music symbol recognition. Neural Comput. Appl. **34**(13), 10409–10427 (2022)

24. Ríos-Vila, A., Iñesta, J.M., Calvo-Zaragoza, J.: On the use of transformers for end-to-end optical music recognition. In: Pattern Recognition and Image Analysis: 10th Iberian Conference, IbPRIA 2022, Aveiro, Portugal, 4–6 May 2022, Proceedings, pp. 470–481. Springer, Heidelberg (2022). https://doi.org/10.1007/978-3-031-04881-4_37

25. Ríos-Vila, A., Esplà-Gomis, M., Rizo, D., Ponce de León, P.J., Iñesta, J.M.: Applying automatic translation for optical music recognition's encoding step. Appl. Sci. **11**(9), 3890 (2021)

26. Ríos-Vila, A., Inesta, J.M., Calvo-Zaragoza, J.: End-to-end full-page optical music recognition for mensural notation. In: Proceedings of the 23rd International Society for Music Information Retrieval Conference, pp. 226–232. ISMIR, Bengaluru (2022)

27. Torras, P., Baró, A., Kang, L., Fornés, A.: On the integration of language models into sequence to sequence architectures for handwritten music recognition. In: Proceedings of the 22nd International Society for Music Information Retrieval Conference, pp. 690–696 (2021)

28. Tuggener, L., Elezi, I., Schmidhuber, J., Pelillo, M., Stadelmann, T.: Deepscores-a dataset for segmentation, detection and classification of tiny objects. In: 24th International Conference on Pattern Recognition, ICPR 2018, Beijing, China, 20–24 August 2018, pp. 3704–3709. IEEE Computer Society (2018)

29. Tuggener, L., Satyawan, Y.P., Pacha, A., Schmidhuber, J., Stadelmann, T.: The deepscoresv2 dataset and benchmark for music object detection. In: 25th International Conference on Pattern Recognition, ICPR 2020, Virtual Event/Milan, Italy, 10–15 January 2021, pp. 9188–9195. IEEE (2020)
30. van der Wel, E., Ullrich, K.: Optical music recognition with convolutional sequence-to-sequence models. In: Cunningham, S.J., Duan, Z., Hu, X., Turnbull, D. (eds.) Proceedings of the 18th International Society for Music Information Retrieval Conference, pp. 731–737 (2017)
31. Wen, C., Zhu, L.: A sequence-to-sequence framework based on transformer with masked language model for optical music recognition. IEEE Access 10, 118243–118252 (2022)
32. Wick, C., Puppe, F.: Experiments and detailed error-analysis of automatic square notation transcription of medieval music manuscripts using cnn/lstm-networks and a neume dictionary. J. New Music Res. 50(1), 18–36 (2021)
33. Wick, C., Hartelt, A., Puppe, F.: Staff, symbol and melody detection of medieval manuscripts written in square notation using deep fully convolutional networks. Appl. Sci. 9(13), 2646 (2019)

Reconstruction of Power Lines
from Point Clouds

Alexander Gribov$^{(\boxtimes)}$ and Khalid Duri

Esri, 380 New York Street, Redlands, CA 92373-8100, USA
{agribov,kduri}@esri.com

Abstract. This paper proposes a solution for constructing line features modeling each catenary curve present within a series of points representing multiple catenary curves. The solution can be applied to extract power lines from lidar point clouds, which can then be used in downstream applications like creating digital twin geospatial models and evaluating the encroachment of vegetation. This paper offers an example of how the results obtained by the proposed solution could be used to assess vegetation growth near transmission power lines based on freely available lidar data for the City of Utrecht, Netherlands [1].

Keywords: Lidar point cloud · Power lines · Digital twin · Vegetation encroachment · Catenary curve fitting · Finding closest point

1 Introduction

This paper proposes a solution of constructing line features modeling each catenary curve that is present within a series of points representing multiple catenary curves. Utility companies can leverage this solution to extract power lines from lidar surveys and identify specific attributes. The extracted power lines can be used to assess the locational accuracy of the power lines and determine whether the amount of sag (see calculation of sag in [2]) indicates any need for maintenance of the support structure. Additionally, the extracted power lines can be incorporated into downstream operations like mitigating fire risk from vegetation encroachment. The power lines can help define the vegetation clearance zone surrounding them, as demonstrated later in this paper. The importance of maintaining power lines and their surrounding vegetation was underscored by the spate of fires in California caused by power lines (see Table 1). Of the fires related to power lines in 2020, vegetation contact accounted for nearly 93 percent of the area damaged [3].

Lidar provides a cost-effective solution for surveying the location of power lines and other objects of interest with great measures of accuracy. The points obtained from a lidar survey can be classified using a variety of techniques—from deterministic, rule-based operations and manual editing to deep learning. Once the points from power lines are classified, they can be processed with the proposed solution to produce a polyline representation for each power line.

© The Author(s), under exclusive license to Springer Nature Switzerland AG 2023
M. Coustaty and A. Fornés (Eds.): ICDAR 2023 Workshops, LNCS 14193, pp. 105–119, 2023.
https://doi.org/10.1007/978-3-031-41498-5_8

Table 1. History of California wildfires caused by power lines. [3–7]

Year	Incident count	Incident count of acres burned ≥ 10	Acres burned	Dollar damage	% of total burned area
2021	284	16	$207,983$	$\$2,777,873$	74%
2020	335	31	$59,334$	$\$52,001,282^1$	4%
2019	304	37	$83,729$	$\$388,843,293$	65%
2018	297	26	$246,873$	$\$2,001,803,168$	23%
2017	408	65	$249,501$	$\$12,046,544,839$	53%

[1] Fires from undetermined causes resulted in $\$2,121,798,631$ in damages.

The resultant polylines can overcome gaps in the lidar survey and ensure the connectivity of each power line span.

The approach can be summarized as follows:

1. **Clusterization Phase:** Use the minimum spanning tree clustering algorithm to connect points and cut them into sequences of *combined* points (Sect. 2).
2. **Segmentation Phase:** Apply the dynamic programming approach to ensure the separation of each catenary curve (Sect. 3). The dynamic programming approach uses a penalty function based on fitting the catenary curve to points (Sect. 5).
3. **Refinement and Curve Fitting Phase:** Use the k-mean clustering algorithm to improve partitioning. In k-mean iterative steps, the update step refits catenary curves to the newly found clusters (Sect. 5) and the assignment step uses the algorithm to find the closest points along the catenary curves (Sect. 4).

An example of the solution is shown in Sect. 6.

2   Clustering Catenary Curves

This section describes the process of clustering catenary curves. The process involves defining a "combined point" as a nonempty set of points and a "combined polyline" as a list of combined points. A combined polyline is considered small if it contains no more than a predetermined minimum number of combined points and not small if it has more combined points than a predetermined maximum number of points. The prolongated form, or extended shape, is discovered for all combined polylines with a point count within a predetermined range. This form is found by evaluating the ratio of the median axis to the largest axis of an ellipse constructed from the covariance matrix of the coordinates for each point in the combined polyline. This measure is used to determine if the combined polyline has exceeded a predetermined threshold value.

The steps of the algorithm to cluster the catenary curves are as follows:

1. For the source points, apply the minimum spanning tree (MST) clustering algorithm using edges whose length does not exceed the user-specified threshold for the maximum allowed gap. See algorithms for the minimum spanning tree in [8–11]. Note the following: *"Out of many available algorithms to solve MST, the Borůvka's algorithm is the basis of the fastest known algorithms"* [12].
2. Assign each node from the network, constructed in the previous step using the MST algorithm, to a unique set of combined polylines. Initially, each of these sets is empty, meaning it does not contain any combined polylines.
3. Save all non-small combined polylines for each end node with more than one non-small combined polyline, and discard all other combined polylines. Add the empty combined polyline to the neighboring node. To prevent merging combined polylines through this node, mark that empty combined polyline as not small. Perform the following for all other end nodes:
   a. Find the longest combined polylines by the number of combined points, giving preference to non-small combined polylines.
   b. If there is only one longest combined polyline, take all points from other combined polylines, add these points and the node point as a combined point to the longest combined polyline, and add this modified combined polyline to the neighboring node.
   c. When two are the longest, construct a new combined polyline with a combined point containing all points from all combined polylines in the node and the node point, and add this combined polyline to the neighboring node.
4. Delete all processed end nodes and repeat step 3 until no end nodes remain.
5. At this stage, all nodes are isolated. Save all non-small combined polylines for each node with more than two non-small combined polylines. Otherwise, merge non-small combined polylines through the combined point with all points from other combined polylines and the node point.
6. For all saved combined polylines, apply the division algorithm described in the next section and use the resultant partitioning as a starting solution for the k-mean clustering algorithm [13]. The k-mean clustering algorithm is then applied by alternating steps for fitting catenary curves to points and reallocating points to the closest catenary curves. An essential intermediate step is merging similar catenary curves.

The results identify all points that belong to the same catenary curve. Consideration must be taken for the density of the points and expected length of the nominal catenary curve when determining appropriate values for the minimum and maximum number of points.

3 Dividing a Combined Polyline

Dividing combined polylines involves breaking down a large combined polyline into smaller segments, with each segment representing a unique catenary curve.

The optimal division aims to minimize the number of partitions and the associated penalty within each partition. Each partition should fit the underlying data, and new partitions should be created when they significantly improve the overall fit.

We found the following empirical penalties for each partition: $-\dfrac{\sum \epsilon_i^2}{2 \cdot n \cdot T^2}$, where ϵ_k is the deviation from the fitted catenary curve, n is the number of points in the partition, and T is the maximum deviation from the catenary curve specified by the user. Fitting is performed without removing any points even if they violate the maximum deviation. Also, all points in a combined point cannot be divided.

And for the partitions not larger than the predetermined number of points, the penalty is $\log (1/2)$, plus the total number of partitions times $\log (1/2)$.

From a probabilistic point of view, this penalty formulation can be treated in the following way:

- Each partition has a probability $f_{exp}\left(\dfrac{\sum \epsilon_i^2}{2 \cdot n \cdot T^2} \right)$, where f_{exp} is the probability density function of the exponential distribution with an intensity equal to 1,
$$f_{exp}(x) = \begin{cases} e^{-x}, & x \geq 0; \\ 0, & \text{otherwise.} \end{cases}$$
- And for the partitions not larger than the user-specified number of points, the probability is $1/2$. The probability for each partition is $1/2$, or in other words, the probability to have extra partition is $1/2$ and the probability not to have any more partitions is $1/2$. Therefore, the number of partitions follows the geometric distribution with probability equaling $1/2$.

The total probability is equal to the product of probabilities.

To find the optimal division, we use the dynamic programming approach described in [14].

4   Finding the Closest Point on the Catenary Curve

The catenary curve equation is

$$y = c + a \cdot \cosh \left(\frac{x - m}{a} \right). \tag{1}$$

Finding the point C on the catenary curve closest to some point A in three-dimensional space is the same as finding the point C on the catenary curve closest to the point B, where B is a projection of point A onto the catenary curve's plane—see Fig. 1. Because point B is a projection of point A onto the catenary curve's plane and point C is the closest point to point B, it follows that point A is located on the tangent plane of the catenary curve constructed at point C. Therefore, finding the closest point in three-dimensional space can be done by projecting the point onto the catenary curve's plane, then determining the closest point on the catenary curve for the projected point.

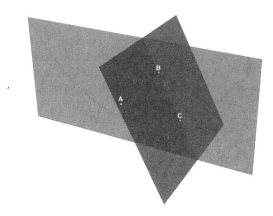

Fig. 1. The catenary curve (red line) is defined in a plane (green). Point B is a projection of point A into the catenary curve's plane. Point C is the point on the catenary curve closest to point B. The tangent plane (blue) for the catenary curve is constructed for point C. (Color figure online)

To find the point closest to point $\mathbf{p} = [x_p, y_p]^\top$ on the catenary curve defined by (1), we first apply the coordinate system's transformation: $x' = \dfrac{x - m}{a}$, $y' = \dfrac{y - c}{a}$.

After this transformation, the catenary curve (1) becomes

$$y' = \cosh(x'), \tag{2}$$

and point $\mathbf{p}$ transforms to $\mathbf{p}' = [x_p', y_p']^\top$, where $x_p' = \dfrac{x_p - m}{a}$, $y_p' = \dfrac{y_p - c}{a}$.

Because this transformation is translation and dilation, it follows that if the point $\mathbf{p}_c' = [x_c', y_c']^\top$ on the catenary curve (2) closest to the point is $\mathbf{p}'$, then $\mathbf{p}_c = [x_c, y_c]^\top$, where $x_c = a \cdot x_c' + m$, $y_c = a \cdot y_c' + c$ is the location on the catenary curve (1) closest to the point $\mathbf{p}$.

Therefore, without loss of generality, it is sufficient to solve the task of finding the closest point on the catenary curve (1) by considering only the case when $a = 1$, $m = 0$, and $c = 0$; therefore, we will refer to the catenary curve as a function:

$$y = \cosh(x).$$

The derivative at point x is $\sinh(x)$, and it is strictly an increasing function. The normal to the catenary curve is $[\tanh(x), -\operatorname{sech}(x)]^\top$ (see Fig. 2), and it will intersect the y-axis at

$$y = \cosh(x) + \frac{x}{\sinh(x)}. \tag{3}$$

Equation (3) is an even function, strictly increasing for positive x, and strictly decreasing for negative x; see proof in Appendix. From these properties, it follows that the closest point is unique for any location except the line from point $[0, 2]^\top$

to $[0, +\inf]^{\top}$, shown in blue. While the point $[0, 2]^{\top}$ has a unique closest point $[0, 1]^{\top}$, all other points on the blue line have two equally distant closest points. There is an area of numerical instability for finding the closest point for points close to the point $[0, 2]^{\top}$. In this paper, for all points on the line from point $[0, 2]^{\top}$ to $[0, +\inf]^{\top}$ and around point $[0, 2]^{\top}$, point $[0, 1]^{\top}$ will be assigned as the closest.

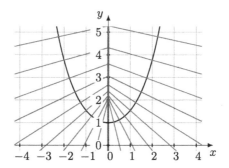

Fig. 2. The relationship between the points and the closest points on the catenary curve. The black line is a catenary curve; the red lines are normal lines to the catenary curve; the blue line is where two points on the catenary curve have the same distance; and the blue dot is an area of instability for finding the closest point on the catenary curve. (Color figure online)

Let's start by partitioning the area by normal lines from the catenary curve. Without loss of generality, it is sufficient to consider only half of the plane $x > 0$. Choosing $x_i = k \cdot i$, $i \in \mathbb{N}_0$, construct lines passing through points $[x_i, \cosh(x_i)]^{\top}$ with directions $[\tanh(x_i), -\operatorname{sech}(x_i)]^{\top}$. The example of partitioning the area for the case when $k = 1/4$ is shown in Fig. 2.

Finding the partition is performed by a binary search. The bounds can be narrowed by finding the intersection of horizontal and vertical lines drawn from the point $\mathbf{p}$ with the catenary curve.

While this algorithm can be used for finding the catenary curve's closest point by subdividing on each iteration, many steps are needed to converge. However, if performance is not a concern, it can be the chosen algorithm for this task.

In the next step, using the found partition, the approximate location for the closest point is found by maintaining the same proportion of distances between the points and the region bounds. See Fig. 3, for the point $\mathbf{p}$, the distances are measured as the shortest distances to the region's bounds. For the point on the catenary curve $\mathbf{p}_c$, the distances are measured along the catenary curve to the region's bounds. Therefore, $\dfrac{\ell(\mathbf{p}'_a, \mathbf{p}')}{\ell(\mathbf{p}'_a, \mathbf{p}'_b)} = \dfrac{d(\mathbf{p}_a, \mathbf{p})}{d(\mathbf{p}_a, \mathbf{p}) + d(\mathbf{p}, \mathbf{p}_b)}$, where $\ell(\mathbf{p}_0, \mathbf{p}_1) = |\sinh(x_1) - \sinh(x_0)|$ is the distance along the catenary curve between points

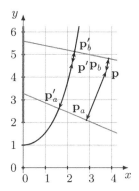

Fig. 3. Example of partitioning by normal lines to the catenary curves (red lines) at points $\mathbf{p}'_a$ and $\mathbf{p}'_b$. $\mathbf{p}_a$ and $\mathbf{p}_b$ are the points on these normal lines closest to the point $\mathbf{p}$. $\mathbf{p}'$ approximates the point closest to the point $\mathbf{p}$ on the catenary curve. See the text for details. (Color figure online)

$\mathbf{p}_0 = [x_0, \cosh(x_0)]^\top$ and $\mathbf{p}_1 = [x_1, \cosh(x_1)]^\top$; and $\mathrm{d}(\mathbf{p}_0, \mathbf{p}_1)$ is the Euclidean distance between points $\mathbf{p}_0$ and $\mathbf{p}_1$.[1]

In the final step of finding the closest point, we will use an iterative algorithm. We consider two approximations: one by the osculating circle, and another by the osculating parabola.

In point $\mathbf{p}_c = [x_c, y_c]^\top$, where $y_c = \cosh(x_c)$, we will approximate the catenary curve as an osculating circle; see Fig. 4. The osculating circle will have the radius $\cosh^2(x_c)$ and center at $[x_c - \sinh(x_c) \cdot \cosh(x_c), y_c + \cosh(x_c)]^\top$.[2] Point $\mathbf{p}'$ is the projection of point $\mathbf{p}$ onto the osculating circle. Measuring the distance on the circle from point $\mathbf{p}_c$ to $\mathbf{p}'$ and finding point $\mathbf{p}_1$ at the same distance from point $\mathbf{p}_c$ measured on the catenary curve will produce the next approximation of the closest point on the catenary curve.

Let $\mathbf{p}_c = [x_c, \cosh(x_c)]^\top$ be the current approximation of the point closest to the point $\mathbf{p}$. The normal vector to the catenary curve at the point $\mathbf{p}_c$ will be $v_n = (x_n, y_n) = (\tanh(x_c), -\operatorname{sech}(x_c))$. Rotate vector $\mathbf{p} - \mathbf{p}_c$ in the opposite direction of vector v_n. This should rotate vector $\mathbf{v}_n$ to $[1, 0]^\top$.

$$\mathbf{v}_r = \begin{bmatrix} x_r \\ y_r \end{bmatrix} = \begin{bmatrix} x_n & y_n \\ -y_n & x_n \end{bmatrix} \cdot (p - p_c).$$

The radius of the osculating circle is $r = \cosh^2(x_c)$.

---

[1] Because partitioning lines change directions, they are not parallel and, therefore, will intersect. Another way to find an approximate location of the closest point is by evaluating proportion from formed angles. This approach is less robust for larger values of x.

[2] Notice that the x-coordinate of the osculating circle center is of the opposite sign of x_c unless x_c is equal to 0; and the y-coordinate is always twice the y_c.

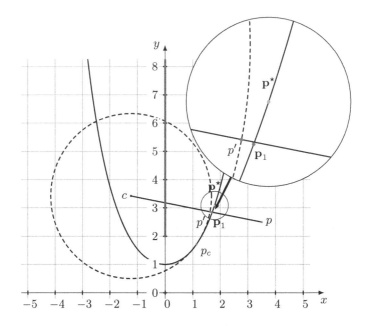

Fig. 4. Approximation of a catenary curve (black line) by an osculating circle (dashed line) at the point $\mathbf{p}_c$. This represents one step of the algorithm for finding the closest point on the catenary curve for the point $\mathbf{p}$ as point $\mathbf{p}_1$. The point $\mathbf{p}'$ is the projection of the point $\mathbf{p}$ onto the osculating circle. The point $\mathbf{p}_1$ is found to have the same distance along the catenary curve as the point $\mathbf{p}'$ along the circle from the point $\mathbf{p}_c$. The point $\mathbf{p}^\star$ is the location on the catenary curve closest to the point $\mathbf{p}$.

The chordal distance between $\mathbf{p}_c$ and $\mathbf{p}'$ is

$$l_c = \sqrt{\left(r \cdot \left(1\Big/\sqrt{\left(1+\left(\frac{y_r}{r+x_r}\right)^2\right)}-1\right)\right)^2 + \left(y_r\Big/\sqrt{\left(1+\frac{x_r}{r}\right)^2+\left(\frac{y_r}{r}\right)^2}\right)^2}.$$

In this equation, dividing first by $r + x_r$ and r improves the stability of the numeric evaluation.

$$l = \begin{cases} 2 \cdot r \cdot \arcsin\left(\dfrac{l_c}{2 \cdot r}\right), & \text{if } 2.149\,119\,332\,890\,820\,95 \times 10^{-8} < \dfrac{l_c}{2 \cdot r}; \\ l_c, & \text{otherwise.} \end{cases}$$

In this equation, to prevent underflow, the constant $2.149\,119\,332\,890\,820\,95 \times 10^{-8}$ was used to skip evaluation of arcsin for small values. This constant is the largest number for which arcsin does not modify the number; however, it depends on the implementation of the arcsin function. In the case where r overflows, $l = |y_r|$.

Taking the distance l in the direction of y_r from point $\mathbf{p}_c$ along the catenary curve will yield point $\mathbf{p}_1$, which is the next approximation for the point on the catenary curve closest to point $\mathbf{p}$.

For a more in-depth exploration of the approximation by osculating parabola, please refer to the preprint [15], which covers this topic in detail. Both approximation algorithms work well and can be used to find the closest point on the catenary curve. While approximating by the osculating parabola produces a more precise solution than using the osculating circle, the latter produces a faster solution. About three iterations are sufficient to reach a relative precision of 10^{-6}.

5   Fitting a Catenary Curve to Points

In the previous section, the task of finding the closest point on the catenary curve was solved. This section will describe the algorithm for fitting a catenary curve to a set of points using a nonlinear least squares algorithm where the parameters of the curve are optimized to minimize the sum of minimum distances between the source points and the curve.

Let's define a signed distance from a point $\mathbf{p} = [x_p, y_p]^{\top}$ to the catenary curve as a scalar product of the normal to the catenary curve at the point $\mathbf{p}_c$ and $\overrightarrow{\mathbf{p}_c, \mathbf{p}}$,

$$
F(c, a, m | \mathbf{p}, \mathbf{p}_c) = \left(\tanh\left(\frac{x_c - m}{a}\right), - \operatorname{sech}\left(\frac{x_c - m}{a}\right) \right) \cdot
$$
$$
\cdot \left(x_p - x_c, y_p - (c + a \cdot \cosh\left(\frac{x_c - m}{a}\right) \right),
$$

where $\mathbf{p}_c = [x_c, y_c]^{\top}$, $y_c = c + a \cdot \cosh\left(\frac{x_c - m}{a}\right)$, is the point on the catenary curve closest to $\mathbf{p}$. Point $\mathbf{p}_c$ is found using the algorithm described in Sect. 4.

This equation will be used in the trust region algorithm [16–20] to find optimal parameters c, a, and m by minimizing the sum of $\left\| F(c, a, m | \mathbf{p}_i, \mathbf{p}_{c,i}) \right\|^2$ over a set of points $\mathbf{p}_i$, where $\mathbf{p}_{c,i}$ is the point on the catenary curve closest to the point $\mathbf{p}_i$. The algorithm requires an evaluation of the function and its first derivatives,

$$
F(c, a, m | p, p_c) = a + \operatorname{sech}\left(\frac{x_c - m}{a}\right)\left((x_p - x_c) \cdot \sinh\left(\frac{x_c - m}{a}\right) - (y_p - c)\right),
$$
$$
F_c'(c, a, m | p, p_c) = \operatorname{sech}\left(\frac{x_c - m}{a}\right), F_a'(c, a, m | p, p_c) = \frac{x_c - m}{a} \cdot F_m'(c, a, m | p, p_c)
$$
$$
+ 1, F_m'(c, a, m | p, p_c) = -\frac{\operatorname{sech}^2\left(\frac{x_c - m}{a}\right) \cdot \left((y_p - c) \cdot \sinh\left(\frac{x_c - m}{a}\right) + (x_p - x_c)\right)}{a}.
$$

While the x-coordinate of the closest point location is not moving when catenary curve parameters change, the scalar product to the normal keeps only the orthogonal distance to the linear approximation at location $\mathbf{p}_c$; see examples in Fig. 5. This approach is essentially a linear approximation compared to the distance calculated using the fixed point on the catenary curve.

Fig. 5. Example of approximation of the shortest distance from the point to the catenary curve when the catenary curve is moving. The distance from the point $\mathbf{p}$ to the catenary curve (thin black line) is approximated as the shortest distance $\overrightarrow{\mathbf{p},\mathbf{q}}$ (dotted line) to the tangent line (dashed line) constructed at the point on the catenary curve $\mathbf{p}'_c$. That distance can be found as the absolute value of the scalar product of the normal to the catenary curve at point $\mathbf{p}'_c$ and $\overrightarrow{\mathbf{p},\mathbf{p}'_c}$. The point $\mathbf{p}'_c$ on the catenary curve is found to have the same x-coordinate as the point $\mathbf{p}_c$ on the source catenary curve (thick black line) closest to the point $\mathbf{p}$.

The domain of parameter a is $(0; +\inf)$. To apply the trust region algorithm without constraint, we will make the following substitution: $a = e^{a_e}$. The first derivative for a_e is $F'_{a_e}(c, e^{a_e}, m|\mathbf{p}, \mathbf{p}_c) = F'_a(c, e^{a_e}, m|\mathbf{p}, \mathbf{p}_c) \cdot e^{a_e}$.

The steps of the algorithm are the following:

1. Find the vertical plane that best fits the original points in terms of the least squares distance from the plane.
2. Define a coordinate system on the fitted plane with the x-axis being horizontal and the y-axis being vertical. The x-axis can be found by projecting the normal vector to the horizontal plane, normalizing it, and rotating clockwise by 90 degrees. The y-axis can be found by the vector product of the normal and the found x-axis direction.
3. Project source points onto the plane. Let projected points be $\mathbf{p}_i = [x_i, y_i]^\top$, where $i = \overline{1..n}$, and n is the number of points.
4. Fit the parabola to the points.
5. Check if the parabola is close to the straight line or concave. If it is, the catenary curve does not represent the points.
6. Find the center of mass for the projected points.
7. Find the closest point on the parabola for the found center. It is done through solving a cubic equation [21].
8. Find catenary curve parameters a_0 and m_0 by matching the first and second derivatives of the catenary curve at the closest point on the parabola. The parameter c_0 is found to minimize the sum of squared deviations between the catenary curve and the data (average difference between the points and the catenary curve). This will produce initial estimates for catenary curve parameters c_0, a_0, and m_0.

9. Apply the trust region algorithm.[3]

We have noticed that some power lines captured in lidar surveys are slightly rotated by the wind. In such cases, in step 1, the optimal plane's orientation can be found to account for the influence of the wind in lieu of fitting to a vertical plane. We assume that the wind blows in one direction with consistent strength.[4]

This algorithm is applied to real data that often contains outliers introduced by the following:

- Sensor noise producing improper points during the lidar survey
- Misclassified points that do not belong to any catenary curve
- Points from a different catenary curve during the clustering process

Consequently, a user-specified maximum deviation threshold from the catenary curve can be used to tailor the algorithm for the data being processed. To make this algorithm robust to outliers, points exceeding this threshold need to be removed, and the algorithm needs to be reapplied to the remaining points.

Note that fitting a plane in step 1 does not find the optimal plane for the catenary curve but rather a close approximation. Finding an optimal plane requires simultaneous optimization of parameters for the catenary curve and the plane. The impact of windage yaw was eliminated by fixing the endpoints of the fitted catenary curve and determining a new catenary curve with the same length, which hangs freely.

6   Example

This solution can be demonstrated using a lidar survey of approximately 57 million points collected over Utrecht, the Netherlands [1]. The points representing power lines were labeled using a point cloud classification model (PointCNN), resulted in about 100 thousand classified points; see Fig. 6a. Line features modeling each power line's catenary curve were obtained by using the Extract Power Lines From Point Cloud tool in ArcGIS Pro [22], which implemented the solution discussed in this paper. The processing took 26 seconds on an Intel Xeon W-2265 CPU @ 3.50 GHz and led to the reconstruction of 163 distinct power lines.

1. The Extract Power Lines From Point Cloud tool was executed with the following parameters: power line class code 14, point tolerance 80 cm, wire separation distance 1 m, maximum wire sampling gap 15 m, output line

---

[3] Note that the result of this algorithm does not depend on the distance of the points to the fitted catenary curve plane (projection plane) because the squared distance to any point on the projection plane is the sum of the squared distance to the projection plane and the squared distance from the projected point to that point. However, squared distances from the points to the projection plane are fixed.

[4] Strong or inconsistent winds acting on a long power line can produce irregular distortions of its shape that create noncatenary curves. No examples of this were discovered when surveying available data for power lines and therefore were not considered in this paper.

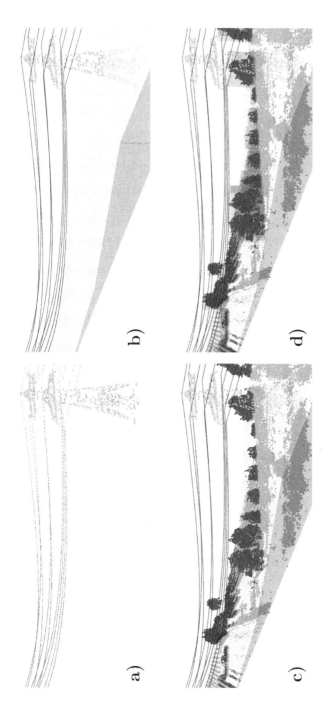

Fig. 6. Screenshots of results obtained in ArcGIS Pro when applying the catenary curve modeling solution to a workflow for managing vegetation growth around power lines. a) Lidar points classified as power lines are drawn in purple, and transmission towers are drawn in gray. b) This screenshot shows line features constructed from the power line classified points and the clearance surface constructed from the line features. c) Vegetation points are displayed against a semitransparent clearance surface. Vegetation points within the clearance surface are tinted gray. d) Vegetation points exceeding the clearance surface are enclosed within a red bounding box. (Color figure online)

tolerance 1 cm, adjust for wind distortion *enabled*, eliminate wind *enabled*, minimum span for wind correction 60 m, maximum deviation angle 10°, end point search radius 10 m, and minimum wire length 5 m.

2. The tool generated a unique catenary curve for each power line; see Fig. 6b.
3. The resultant power line was used to create a surface modeling a 15-m horizontal clearance from the outermost power lines and a 9-m vertical clearance from the lowest power lines; see Fig. 6c.
4. This clearance surface was then used alongside a surface model of vegetation cover to evaluate the volume and area of the trees that exceed the clearance surface; see Fig. 6d. These results provide a basis for determining the quantity of vegetation that has exceeded the clearance zone.

The aforementioned example demonstrates how the solution presented in this paper can be used to extract power lines as part of a downstream application for evaluating vegetation encroachment.

7 Conclusion

This paper describes a solution for accurately extracting unique catenary curves from an array of points representing multiple catenary curves. The solution can be used to extract power lines from lidar surveys to support downstream applications. An example illustrating the potential use of this solution was provided in the examination of vegetation encroachment around transmission power lines.

The clearance zone refers to the area surrounding the power lines that must be kept clear of any objects or structures to prevent contact with the power line and reduce the risk of electrical hazards. Knowing the position of power lines without wind is crucial in determining the appropriate clearance zone for safe operation. Without knowledge of the power line's position without wind, it can be difficult to accurately determine the appropriate clearance zone, as the line may sway or shift during periods of high wind. This can lead to clearance zones that are over-estimated in one direction and under-estimated in another, which can potentially create safety hazards or limit the use of the surrounding land. Therefore, it's important to have accurate information on the position of power lines without wind to ensure that appropriate clearance zones are established and maintained for safe and reliable operation.

Also in this paper, a new, efficient algorithm to find the closest point on the catenary curve is introduced, along with a new algorithm for fitting a catenary curve to a set of points based on the shortest distance. This algorithm fits a catenary curve to points by approximating the shortest distance as a scalar product of a normal and the vector; see Sect. 5. Such an approach provides a general framework that can be applied to fitting other curves. The clustering algorithm is based on a minimum spanning tree, a dynamic programming approach, and the k-mean clustering technique.

Acknowledgments. The authors would like to thank Lois Stuart and Linda Thomas for proofreading this paper; and Manoj Lnu, senior software engineer, 3D analyst at

Esri, for helpful discussions about the initial fitting of a catenary curve to a set of points.

Appendix   Equation (3)

$$y' = \sinh(x) + \frac{\sinh(x) - x\cosh(x)}{\sinh^2(x)} = \frac{\sinh^3(x) + \sinh(x) - x\cosh(x)}{\sinh^2(x)} =$$

$$= \frac{\sinh(x)\left(\sinh^2(x) + 1\right) - x\cosh(x)}{\sinh^2(x)} = \frac{\sinh(x)\cosh^2(x) - x\cosh(x)}{\sinh^2(x)} =$$

$$= \frac{\cosh(x)}{\sinh^2(x)}\left(\sinh(x)\cosh(x) - x\right) = \frac{\cosh(x)}{\sinh^2(x)}\left(\frac{\sinh(2x)}{2} - x\right). \quad (4)$$

Because $\sinh(x) - x$ is a strictly increasing odd function, it follows that (4) is positive for all positive values and negative for all negative values; therefore, (3) is strictly increasing for all positive values and strictly decreasing for all negative values.

The minimum point of the intersection of the y-axis is always above 2, because

$$\lim_{x \to 0}\left(\cosh(x) + \frac{x}{\sinh(x)}\right) = 2.$$

References

1. AHN3 downloads. Actueel Hoogtebestand Nederland (current height file, Netherlands). https://app.pdok.nl/ahn3-downloadpage/
2. Hatibovic, A.: Derivation of equations for conductor and sag curves of an overhead line based on a given catenary constant. Periodica Polytechnica Electr. Eng. Comput. Sci. **58**(1), 23–27 (2014). https://doi.org/10.3311/PPee.6993
3. 2020 Wildfire Activity Statistics. California Department of Forestry and Fire Protection. https://www.fire.ca.gov/our-impact/statistics
4. 2021 Wildfire Activity Statistics. California Department of Forestry and Fire Protection. https://www.fire.ca.gov/our-impact/statistics
5. 2019 Wildfire Activity Statistics. California Department of Forestry and Fire Protection. https://www.fire.ca.gov/our-impact/statistics
6. 2018 Wildfire Activity Statistics. California Department of Forestry and Fire Protection. https://www.fire.ca.gov/our-impact/statistics
7. 2017 Wildfire Activity Statistics. California Department of Forestry and Fire Protection. https://www.fire.ca.gov/our-impact/statistics
8. Borůvka, O.: O jistém problému minimálním. In: Práce Moravské přírodovědecké společnosti, Czech, German, vol. III, no. 3, pp. 37–58 (1926). http://hdl.handle.net/10338.dmlcz/500114
9. —. Příspěvek k otázce ekonomické stavby elektrovodných sítí. In: Elektrotechnický obzor, Czech, German, vol. 15, pp. 153–154 (1926). http://hdl.handle.net/10338.dmlcz/500188

10. Kruskal, J.B.: On the shortest spanning subtree of a graph and the traveling sales-man problem. Proc. Am. Math. Soc. **7**(1), 48–50 (1956). https://doi.org/10.1090/S0002-9939-1956-0078686-7

11. Prim, R.C.: Shortest connection networks and some generalizations. Bell Syst. Tech. J. **36**(6), 1389–1401 (1957). https://doi.org/10.1002/j.1538-7305.1957.tb01515.x

12. Nešetřil, J., Milková, E., Nešetřilová, H.: Otakar Borůvka on minimum spanning tree problem: translation of both the 1926 papers, comments, history. Disc. Math. **233**(1–3), 3–36 (2001). https://doi.org/10.1016/S0012-365X(00)00224-7

13. Bock, H.-H.: Origins and extensions of the k-means algorithm in cluster analysis. J. Electronique d'Histoire des Probabilités et de la Statistique/Electron. J. Hist. Probabil. Stat. **4**(2), 1–18 (2008). http://www.jehps.net/Decembre2008/Bock.pdf

14. Gribov, A., Bodansky, E.: A new method of polyline approximation. In: Fred, A., Caelli, T.M., Duin, R.P.W., Campilho, A.C., de Ridder, D. (eds.) SSPR /SPR 2004. LNCS, vol. 3138, pp. 504–511. Springer, Heidelberg (2004). https://doi.org/10.1007/978-3-540-27868-9_54

15. Gribov, A., Duri, K.: Reconstruction of power lines from point clouds, pp. 1–15. ArXiv e-prints (2022). https://arxiv.org/abs/2201.12499

16. Conn, A.R., Gould, N.I.M., Toint, P.L.: Trust Region Methods. Society for Indus-trial and Applied Mathematics (2000). https://doi.org/10.1137/1.9780898719857

17. Levenberg, K.: A method for the solution of certain non-linear problems in least squares. Q. Appl. Math. **2**, 164–168 (1944). https://doi.org/10.1090/qam/10666

18. Marquardt, D.W.: An algorithm for least-squares estimation of nonlinear param-eters. J. Soc. Ind. Appl. Math. **11**(2), 431–441 (1963). https://doi.org/10.1137/0111030

19. Ranganathan, A.: The Levenberg-Marquardt algorithm (2004). http://www.ananth.in/Notes_files/lmtut.pdf

20. Gavin, H.P.: The Levenberg-Marquardt method for nonlinear least squares curve-fitting problems (2020). https://people.duke.edu/~hpgavin/ce281/lm.pdf

21. Press, W.H., Teukolsky, S.A., Vetterling, W.T., Flannery, B.P.: Numerical Recipes, Third Edition: The Art of Scientific Computing. Cambridge University Press, Cam-bridge (2007). www.cambridge.org/9780521880688

22. Extract power lines from point cloud (3D Analyst). Retrieved Octo-ber 2021. https://pro.arcgis.com/en/pro-app/latest/tool-reference/3d-analyst/extract-power-lines-from-point-cloud.htm

KangaiSet: A Dataset for Visual Emotion Recognition on Manga

Ruddy Théodose$^{(\boxtimes)}$ (ID) and Jean-Christophe Burie (ID)

L3i Laboratory, SAIL joint Laboratory, La Rochelle Université,
17042 La Rochelle CEDEX 1, France
{ruddy.theodose,jean-christophe.burie}@univ-lr.fr

Abstract. This paper presents KangaiSet, a dataset for facial emotion recognition on Manga. This dataset is based on the japanese manga books available in the Manga109 dataset and focuses on visual aspects. While faces are considered as the main emotional information carrier, artists employ various graphic cues through the panels to highlight different aspects of a character's emotional states. Consequently, for each face image, pictures corresponding to the character's body and the whole panel are also included so they can be studied simultaneously.

Keywords: Emotion Recognition · Manga · Comics Analysis · Document Analysis

1 Introduction

Affective computing is an active area of research with application fields including human-robot interaction and customer/tester satisfaction evaluation. To the best of our knowledge, the field has been mainly explored on real data captured in different forms (photos, videos, sound, text, physiological signals), through either unimodal or multimodal approaches. Concerning visual data, two main paradigms are studied : Facial Expression Recognition (FER) that focuses on face attributes and Gesture Recognition (GR) that focuses on the body movements or gestures.

While they are considered more expressive and intimate medias, the visual arts haven't drawn much attention to this task. The found works revolve essentially around paintings [13,27] and include modalities outside to the images themselves such as textual descriptions or titles. Among the artistic mediums, comics, defined as "Juxtaposed pictorial and other images in deliberate sequence, intended to convey information and/or to produce an aesthetic response in the viewer." [17], have become an important part of popular cultures around the world.

Research into automated comics book processing began in the last decade. Various topics were addressed, from the detection of main elements in comic pages such as panels [9,14,22], speech bubbles [6,10,15], characters [3,19,24] to

M. Coustaty and A. Fornés (Eds.): ICDAR 2023 Workshops, LNCS 14193, pp. 120–134, 2023.
https://doi.org/10.1007/978-3-031-41498-5_9

layout analysis, for example, to automatically detect the reading order in atypical panel arrangements [2,4,5]. However, while these methods mostly focus on the form, to our knowledge, no study have dived into the automated understanding of the narrated content. In this sense, character analysis can become a major topic in order to represent the stakes of each panel.

The study of the drawings and graphic elements specific to comics has almost never been explored for the automated recognition of emotions. Faces tend to be the privileged vector of information about emotions. On the facial aspect, the drawn human and anthropomorphic faces tend to rely on mechanisms that move real faces, so that the reader can grasp points of reference between reality and the drawings. Artists employ multiple graphic cues that can amplify or even contradict the emotion apparent on the characters' face. Symbols, background effects, onomatopoeia, speech bubbles and so on are often used to illustrate or highlight concepts that are not supposed to be visible in real life.

In this paper, we present KangaiSet, a new dataset for emotion recognition in Manga. At this stage, the dataset relies exclusively on data of the Manga109 dataset that includes pictures from manga of various genres and eras. The goal of this dataset is to gather for each emotion, the different forms of expression and tools used by artists through their artistic styles. KangaiSet can be seen as an extension of Manga109, so its use requires the original Manga109 dataset and must respect the rules defined on the website[1]. We defined the metrics and establish baselines on this dataset at different scales of study.

The proposed definition of a comic does not imply the need to include textual information, although most of the comic books uses text mainly for visualizing what is supposed to be heard by the reader. Even if the understanding of expressions depends on both the image and the text, we explore in this paper exclusively the graphical aspect. If omitting a modality induces a consequent loss of information, it allows to focus on the different tools used by artists to convey the emotions, regardless the difference of artistic styles.

Our contributions are :

- a dataset for emotion recognition based on Manga109 dataset, manually annotated and focused on the graphical cues. Conscious that emotional information may not be located only on faces, we linked the thumbnail image of each face to the panel image it comes from, and the sub-image including the body of the character ;
- a set of baselines on the three scales (face, body, panel) and an analysis of how these baselines react to the other scales.

2   Emotion Modeling

The definition of an emotion is often debated through the psychology community. The American Psychology Association defines an emotion as "a complex reaction

---
[1] http://www.manga109.org/en/.

| Anger | Disgust | Fear | Joy | Neutral | Sadness | Surprise |

Fig. 1. Samples from KangaiSet dataset. Each face was assigned a single label. Anger : ©Ono Yasuyuki, ©Saijo Shinji ; Disgust : ©Ishioka Shoei, ©Nakanuki Eri ; Fear : ©Hukuyama Kei, ©Aida Mayumi ; Joy : ©Sato Harumi, ©Deguchi Ryusei ; Neutral : ©Miyauchi Saya, ©Aida Mayumi ; Sadness : ©Minamisawa Hishika, ©Satonaka Machiko ; Surprise : ©Miyauchi Saya, ©Kuriki Shoko

pattern, involving experiential, behavioral, and physiological elements, by which an individual attempts to deal with a personally significant matter or event". The problem of emotion recognition have been approached through various models which can be grouped into two main paradigms : (1) Categorical or discrete models (Table 1) aim at representing emotions as a finite and discrete set. They allow a simpler processing, transforming the emotion recognition problem into a classification problem. However, these models limit the representation spectrum as they do not allow the expression of more complex states or emotions defined outside their scope, except through combinations. (2) Dimensional/Continuous models (Fig. 2) illustrate the emotional spectrum as a multidimensional continuous space, where each axis is assigned to one specific aspect of an emotion. A character's emotional state is then represented as a single point in this space. Such models can also allow the definition of distances to better assess similarities between signals. However, using a continuous space is much more complex as it requires to assign to an emotion some sort of quantitative value (Fig. 1).

In this study we decided to choose the categorical model in order to allow the participation of non-expert annotators. The Ekman model [7] is one of the most used and documented. It divides the emotional spectrum into 6 basic emotions : Anger, Joy, Disgust, Fear, Sadness and Surprise. We added the neutral state as a seventh emotion even though it can be debated as the absence of the six other ones.

Table 1. Discrete emotion models

Main Author	Basic emotions
Ekman [7]	Anger, Disgust, Fear, Joy, Sadness, Surprise
Plutchik [23]	Anger, Anticipation, Disgust, Fear, Joy, Sadness, Surprise, Trust
Shaver [26]	Anger, Fear, Joy, Love, Sadness, Surprise
Tomkins [28]	Anger, Contempt, Disgust, Distress, Fear, Interest, Joy, Shame, Surprise

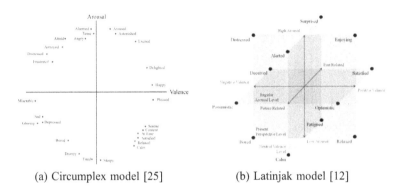

(a) Circumplex model [25]   (b) Latinjak model [12]

Fig. 2. Dimensional emotion models

3   Comic and Art Related Datasets

3.1   Element Detection

eBDTheque. eBDtheque [8] is one of the first published dataset about element detection in comic pages. The dataset contains 100 comic pages from America, Europe and Japan. Annotations include characters, panels and text lines as bounding boxes, speech bubbles represented as closed polygons and links between speech bubbles and characters. On the last version of the dataset, there is 1620 bounding boxes assigned to characters.

DCM77. The Digital Comic Museum is an online library that has collected public domain books from the golden age of comic books (1938–1956). The built dataset [20] contains 772 pages from 27 comic books, each book being issued by a different editor in order to foster a variety of artistic choices. All annotated entities (panels, characters' body and characters' face) are represented as bounding boxes. Unlike eBDTheque dataset that includes any speaking characters in the annotations whatever its form, the built dataset's labels only concern human characters. At the time, the dataset contains 3740 panels, 4638 faces and 7049 bodies, some faces and bodies corresponding to the same characters.

Manga109. Manga109 [1, 16] is currently the biggest dataset on comic elements detection. The latest version of the dataset is composed of 10602 pages from 109 japanese mangas written by 94 authors, and contains 118593 faces and 157234 bodies (Fig. 3).

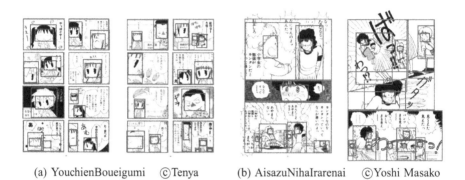

(a) YouchienBoueigumi   ©Tenya   (b) AisazuNihaIrarenai   ©Yoshi Masako

Fig. 3. Samples from Manga109, panels in red, bodies in green, faces in blue. (Color figure online)

3.2   Emotion Recognition

EmoRecCom. The EmoRecCom dataset [21] has been built upon the COMIC dataset [11] that contains books from the Digital Comic Museum. This dataset was used for a Multimodal Emotion Recognition challenge for ICDAR 2021 and has two specificities. First, the images are proposed at panel scale, i.e. for one image, a whole scene is represented, possibly with multiple characters. The second point is that the challenge encourages the use of both modalities, image and text line transcriptions.

Wikiart Emotion. The Wikiart Emotion dataset [18] does not include comic-related images but is composed of 4000 images from the WikiArt.org collection containing paintings from various eras and western artistic movements. Some of the pictures were abstract works and do not represent human figures. Annotators were asked for each work, taking into account the image and its title, which emotion its was inspiring them and how they would rate the images. Consequently, the dataset is more oriented towards the viewer's interpretation rather than on the piece of art itself and the graphical tricks of the authors that are causing the viewer's reaction.

4   KangaiSet

In this section, we introduce our dataset, KangaiSet. We first describe how the data was processed. Then we provide details about the statistics about the annotated images.

4.1   Data Construction

As aforementioned, our dataset rests exclusively on the Manga109 dataset. That means that the dataset is focused on japanese mangas and information produced may not be valid for other forms of comics from different origins and cultures.

Our dataset was, at the beginning, focused on faces. In fact, the most informative area is the face, through the facial expressions. Drawn faces are often represented according to mechanisms based on real faces. Consequently, the expressions on drawn and real faces tend to show similarities, such as mouth muscles activation for smiling when a character is happy. On the same aspect, body language and gestures can provide supplementary information on the character's state. Furthermore, other graphical tricks that are more specific to the medium are often used by the artists such as symbols that are graphical icons associated to ideas and are mostly drawn on or around the characters. Similarly, background effects can be applied on the whole panel to illustrate or amplify a mood that can be related to multiple characters in the drawn scene. Consequently, conscious that the whole drawn scene can provide information on the emotional state of the studied character (for example the reason that provoked a reaction), we wanted to include contextual information for each face and then link them to larger regions of interest.

The Manga109 dataset already provides bounding boxes for panels, body, faces, speech balloons. Each box has a unique identifier. Hence, we didn't need to draw bounding boxes. However, no explicit relationship between the categories is defined in the original dataset, for example in which panel is located a given face. To establish those relationships, we computed the Intersection over Union (IoU) between the face boxes and the panel/body boxes and assigned to each face box the panel/body box with the highest IoU. Not all bodies were annotated, so we limited the study to the face boxes with both body and panel boxes.

The original Manga109 dataset also includes an ID for the characters drawn in each box. We decided to keep this information because it can be used to understand how one character's face can vary through the different drawings.

A single face can express complex emotions that may not be covered by the spectrum of Ekman model's. In this dataset, we selected only faces that we considered conveying single emotions defined by the Ekman model. Figures considered too complex or ambiguous enough to cause divergent opinions were excluded and the annotation was done by assigning a single emotion to each image. However, due to the subjectivity of the topic, complex expressions may have been added under our supervision. Consequently, we do not expect a strict one-label classification for the dataset.

To summarize, each sample is described with the following attributes:

- Face Box ID: the bounding box for the focused face;
- Body Box ID: the bounding box of the body related to the focused face;
- Panel Box ID: the bounding box of the panel containing the focused face;
- Character ID: unique identifier linked to the identity of the depicted character;

– Source Page: relative path of the comic page, contains the book title and the page number;
– Emotion (as a single label): Values belong to the set {Anger, Disgust, Fear, Joy, Neutral, Sadness and Surprise}.

Figure 4 gives an overview of the images extracted with the three types of box: face, body and panel. In the second column, the body and panel pictures are almost identical. In fact, like movie directors, authors have to decide which type of shot they want, which information they want to show, conditioning what is drawn inside the panel. In that case, this is a close shot focusing on the character reading a newspaper. The chosen type of shot also determines which part of the body is visible, meaning that gestures or body poses are not always available.

Fig. 4. Illustration of the different scales of study, rows: face, body and panel. Columns in order: Anger - Appare Kappore ©Kanno Hiroshi, Neutral - Hanzai Kousyounin Minegishi Eitarou ©Ki Takashi, Anger - Jiji Baba Fight ©Nishikawa Shinji, Joy - Karappo Highschool ©Takaguchi Satosumi

4.2   Data Statistics

The whole dataset is composed of 9387 annotated samples extracted from 98 series of mangas. Table 5 shows how the emotion labels are scattered across the dataset.

The first observation is that the classes are highly imbalanced. In fact, images come from pages of published books. The characters are integrated inside a story or a specific context, which means that every emotion has to be justified in some way. Characters are not asked to realise expression but react to events that happen to them. Consequently, some emotions like disgust are barely visible because the situations causing them are much rarer, while neutral and joyful expressions are more common reactions, even in real life.

	Anger	Disgust	Fear	Joy	Neutral	Sadness	Surprise	Total
#	1348	44	372	2428	3288	794	1113	9387
%	14.36	0.47	3.96	25.865	35.02	0.84	11.86	

Fig. 5. Distribution of the different emotions in the training and testing sets

5 Baselines

5.1 Experiment Settings

Data Management. Our dataset is split into training, validation and testing subsets. Both training and validation subsets were built around the same 92 series with a ratio 80:20 while the samples of the testing subset come from the 6 remaining series of album. In fact, each authors has its own method to design a face. Some of them tend to get closer from real faces by drawing realistic proportions while others can choose more simplified constructions or more expressive figures. The objective of the proposed testing set is to evaluate the performance of a classifier that has been trained on different art styles and determine its ability of our system to find similarities in images that depict faces with drastically different designs. Table 2 describe how the labels are scattered on the different subsets.

Table 2. Emotion distribution in subsets

Subset	# Series	Anger	Disgust	Fear	Joy	Neutral	Sadness	Surprise	Total
Train	92	921	30	282	1723	2348	608	766	6678
Val	92	231	7	71	431	587	152	191	1670
Test	6	196	7	19	274	353	34	156	1039

Training Setup. In the following paragraphs, the term "size" refers to the number of pixels in the input image while the term "scale" will refer to the content of the image (face, body or scale). The used network is a Resnet50 that is trained with Adam optimizer with a fixed learning of 1×10^{-4} for 100 epochs with a batch size of 24. All images are automatically resized to 128×128 pixels. In order to evaluate the importance of the observed content for the recognition of emotion, we selected four data setups that depends on the training scale: face, body, panel and face+body (concatenation). Moreover, as aforementioned, our classes are heavily imbalanced. Thus, for each data setup, we conducted three experiments:

– the first run is a basic training and will be the reference;
– for the second training, a weighted resampling with a class weight defined as $w_{cls} = \frac{1}{\#_{cls}}$ was applied so rare labels appear more frequently during the training;

– for the third training, an additional linear output was added to the network to generate identity related features. The computed loss is $L = L_{emo} + 0.1L_{id}$ where L_{emo} is a focal loss for emotion classification and L_{id} is a triplet loss applied to the additional output. We selected a identity feature size of 256.

To sum up, we conducted 12 experiments that differ from the scale used for the training and the way the network is trained.

Metrics. In the following sections, we chose to display three metrics to take into account the class imbalance in the dataset. Results are presented with the form "micro/macro/weighted" where:

– micro: F1 score is computed once for all classes by summing all the true positives, false positives and false negatives;
– macro: F1 scores are computed for each class and averaged without considering class support;
– weighted: F1 scores are computed for each class and weighted considering class support

5.2   Results Analysis

Base Experiment. Table 3 displays F1-scores on the different scales while Figs. 7 and 8 illustrate some predictions on the test set from the model trained on the "face" scale.

Table 3. Base network. Each value represents, in order, the micro F1, the macro F1 and the Weighted F1 scores. Bold cells : best for each test scale

Train	Test		
	Face	Body	Panel
Face	70.93/40.95/69.71	30.80/17.48/30.65	23.87/13.71/24.21
Body	**71.41/46.28/70.85**	54.19/33.32/54.42	36.77/22.04/37.18
Panel	56.30/37.93/59.04	50.72/31.95/51.29	**41.19/25.20/41.17**
Face+Body	69.78/45.64/70.30	**56.30/36.05/56.62**	39.27/25.29/40.19

First, scales of study either on training or testing phases does not provide equivalent results. Evaluations on panels scales lead to the worst results, whatever the training scale, whereas face scale tend to produce the best results. In fact, some scenes drawn in panels may include multiple characters, causing the confusion of any classification system. A more appropriate task would be more to recognize a general mood in the drawn scene than trying to estimate the state of one of the characters without indication on which one. Moreover, depending on the shot, characters may cover small regions of a panel, this phenomenon being amplified by the resizing of the image for the network. On the other hand, faces are often considered as one of the most critical medium when it comes to emotional communication. To that extent, the "face" scale is focused on the minimal and necessary information. This also justifies the fact that face-to-face results provide better results than both body-to-body and panel-to-panel combinations.

However, for the testing set, the network trained on "body" scale provides equivalent results on the "face" testing set as the "face" trained network. Unlike body whose regions of interest may greatly vary in the samples depending on the type of shots used by the artists, from full body representation to very close shots, the quantity of information tends to remain fixed for the face scale. That explains why networks trained on face scale have their performances dropping faster than the other networks on unknown scales because the additional information was never processed by the network and then considered as noise.

While the micro F1-score provides interesting results, the macro F1-score that averages F1 scores from all classes drops significantly, meaning that some of the classes have much lower accuracy. Table 4 and Fig. 6 respectively illustrate the F1-scores for each class and the confusion matrix. "Anger", "Joy" and "Neutral" classes which are the largest classes have high values while none of the pictures labeled as "Disgust" or "Fear" were accurately predicted. The scores are also linked to the small size of the corresponding categories, which means that each error has more impact on the F1-score produced.

Table 4. F1 scores per class for the "face" base network on "face" testing data.

	Training #	Testing #	F1 score
Anger	921	196	66.53
Disgust	30	7	0.0
Fear	282	19	0.0
Joy	1723	274	**84.37**
Neutral	2348	353	79.55
Sadness	608	34	4.76
Surprise	766	156	51.47

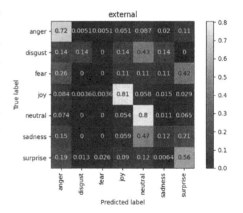

Fig. 6. Confusion Matrix for "face" network on "face" testing set. All values were normalized across the ground truths.

Weight Resampling. F1-scores from models trained with weight resampling are presented in Table 5. Although there are slight improvements on macro F1 scores (+2.14 on "face-to-face", +0.26 on "body-to-body"), the macro and

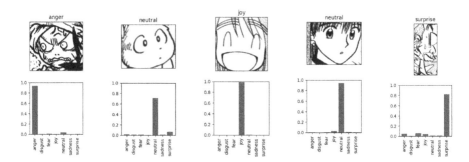

Fig. 7. Good predictions from the "face" testing subset, base network trained on the "face" scale.

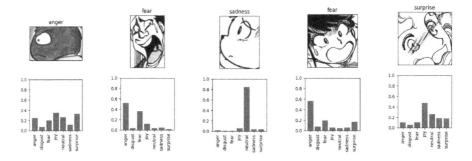

Fig. 8. Error cases from the "face" base network on the "face" testing subset

weighted scores tend to globally drop, which means that the gap between large and small classes has reduced but the large classes see their accuracy drop. The few samples in the smaller classes such as "disgust" were observed more often than some "joy" or "neutral" images. However, the few repeated samples may lead to overfitting due to the lack of diversity for these classes. To conclude, the weighted resampling have negative effect on the results when applied to this dataset.

Identity and Triplet Loss. The goal of this experiment is to evaluate if providing identity related cues can improve the performance of the network on emotion recognition by expliciting the features specific to the graphic representation of the characters. Table 6 presents the results obtained by the third experiment. The scores for studies on the same scale have slightly decreased due to the added complexity during the training. An interesting point is that the "body" network have become more accurate on the "panel" scale. We suppose that the network became better at separating characters on wide shots through the identity features, instead of averaging the whole image.

Table 5. Results from the Weight resampling training. In each cell, the second row is the difference with the "base" training on the same scale setup, green color means improvement

Training	Testing		
	Face	Body	Panel
Face	65.06/43.09/67.30	19.83/15.04/22.26	9.34/8.25/11.52
	−5.87/ /−2.41	−10.97/−2.44/−8.39	−14.53/−5.46/−12.69
Body	63.72/41.98/65.75	51.20/33.58/53.35	33.69/22.92/36.76
	−7.69/−4.3/−5.1	−2.99/ /−1.07	−3.08/ /−0.42
Panel	47.26/33.13/51.48	41.29/29.57/44.46	34.46/22.58/35.62
	−9.04/−4.8/−7.56	−9.43/−2.38/−6.83	−6.73/−2.62/−5.55
Face+Body	65.83/43.09/68.97	17.23/13.47/19.85	13.47/10.05/14.55
	−3.95/−2.55/−1.33	−39.07/−22.58/−36.77	−25.8/−15.24/−25.64

Table 6. Results from the Triplet loss training. In each cell, the second row is the difference with the "base" training on the same scale setup, green color means improvement.

Training	Testing		
	Face	Body	Panel
Face	69.20/40.07/67.29	29.74/18.41/30.37	21.75/12.68/21.92
	−1.73/−0.88/−2.42	−1.06/ /-0.28	−2.12/−1.03/−2.29
Body	65.74/36.50/63.13	56.40/31.17/54.14	44.66/23.35/41.42
	−5.67/−9.78/−7.72	/−2.15/−0.28	
Panel	46.01/21.43/38.75	49.18/26.16/45.97	42.64/21.59/39.01
	−10.29/−16.5/−20.29	−1.54/−5.79/−5.32	/−3.61/−2.16
Face+Body	71.22/40.42/69.32	50.91/28.24/48.91	39.65/19.83/35.55
	/−5.22/-0.98	−5.39/−7.81/−7.71	/−5.46/−4.64

As we employed a triplet loss, so without any explicit label, we want to check if the generated features have been correctly built. In Fig. 9, we illustrated some of the sample distributions through T-SNE visualization algorithm applied on the features used by the triplet loss. The samples are grouped through visual attributes such as large eyes (on the left) or dark hair ratio on image (on the bottom). However, different characters were often grouped together only because they looked similar in certain specific pictures. To conclude, while our approach generate additional information on characters' appearance, it did not improve the accuracy, so the problem is closer to the data distribution rather than to how networks are trained.

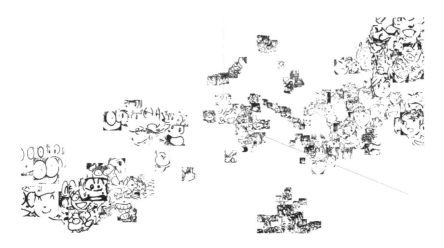

Fig. 9. T-SNE on the features used for the triplet loss

6   Conclusion

We introduced KangaiSet, a new dataset for emotion recognition in Manga. Whereas textual information can provide supplementary cues, our dataset focuses on purely visual standpoint so translations of transcriptions are not included. The main targets are faces, however various indications can be drawn around each character so corresponding body and panel images are also included. So users will be able to exploit the level of detail they want. We then defined and analyzed some baselines to determine the impact of the specificities of our created dataset. Our classes are heavily imbalanced as it samples the distribution of basic emotions reproduced in comics albums. This disparities between classes tend to affect the performances on the testing set, and despite our strategies to mitigate class imbalance, some classes remain badly evaluated, suggesting that the smaller classes are not diverse enough for the training of machine learning algorithms. Multiple evolutions of the dataset are studied. First, images were annotated with single labels. However, a face expression can suggest multiple feelings simultaneously. Consequently. We plan to extend the dataset to multi-label classification. Moreover, the current version of the dataset was created by only one person. As the process was highly subjective, we wish to organize, for future versions of the dataset, an annotation campaign at larger scale to gather as much interpretations as possible. Furthermore, all the images currently in the dataset were only extracted from mostly black and white Japanese comics. Future work may include comic books from other sources, especially from western countries as each culture tend to develop its own tools and language to represent the same concepts. Finally, some emotions remain under-represented in comics albums, so image synthesis for comics might become a viable research topic.

Acknowledgements. This work is supported by the Research National Agency (ANR) in the framework of the 2017 LabCom program (ANR 17-LCV2-0006-01) and the Region Nouvelle Aquitaine in the framework of the EmoRecCom project.

References

1. Aizawa, K., et al.: Building a manga dataset "manga109" with annotations for multimedia applications. IEEE Multimedia **27**(2), 8–18 (2020)
2. Bach, B., Wang, Z., Farinella, M., Murray-Rust, D., Riche, N.H.: Design patterns for data comics. In: Proceedings of the 2018 Chi Conference on Human Factors in Computing Systems, pp. 1–12 (2018)
3. Chu, W.T., Li, W.W.: Manga FaceNet: face detection in manga based on deep neural network. In: Proceedings of the 2017 ACM on International Conference on Multimedia Retrieval, Bucharest Romania, pp. 412–415. ACM (2017)
4. Cohn, N.: Navigating comics: an empirical and theoretical approach to strategies of reading comic page layouts. Front. Psychol. **4**, 186 (2013)
5. Cohn, N.: The architecture of visual narrative comprehension: the interaction of narrative structure and page layout in understanding comics. Front. Psychol. **5**, 680 (2014)
6. Dubray, D., Laubrock, J.: Deep cnn-based speech balloon detection and segmentation for comic books. In: 2019 International Conference on Document Analysis and Recognition (ICDAR), pp. 1237–1243. IEEE (2019)
7. Ekman, P., Friesen, W.V.: Constants across cultures in the face and emotion. J. Pers. Social Psychol. **17**(2), 124–129 (1971)
8. Guérin, C., et al.: ebdtheque: a representative database of comics. In: 2013 12th International Conference on Document Analysis and Recognition, pp. 1145–1149. IEEE (2013)
9. He, Z., et al.: An end-to-end quadrilateral regression network for comic panel extraction. In: Proceedings of the 26th ACM International Conference on Multimedia, MM 2018, New York, NY, USA, pp. 887–895. Association for Computing Machinery (2018)
10. Ho, A.K.N., Burie, J.C., Ogier, J.M.: Panel and speech balloon extraction from comic books. In: 2012 10th IAPR International Workshop on Document Analysis Systems, Gold Coast, Queenslands, TBD, Australia, pp. 424–428. IEEE (2012)
11. Iyyer, M., et al.: The amazing mysteries of the gutter: drawing inferences between panels in comic book narratives. In: Proceedings of the IEEE Conference on Computer Vision and Pattern recognition, pp. 7186–7195 (2017)
12. Latinjak, A.T.: The underlying structure of emotions: a tri-dimensional model of core affect and emotion concepts for sports. Revista Iberoamericana de Psicología del Ejercicio y el Deporte **7**(1), 71–88 (2012)
13. Li, J., Chen, D., Ning, Yu., Zhao, Z., Lv, Z.: Emotion recognition of Chinese paintings at the thirteenth national exhibition of fines arts in china based on advanced affective computing. Front. Psychol. **12**, 741665 (2021)
14. Li, L., Wang, Y., Tang, Z., Gao, L.: Automatic comic page segmentation based on polygon detection. Multimedia Tools Appl. **69**(1), 171–197 (2014)
15. Liu, X., Li, C., Zhu, H., Wong, T.-T., Xuemiao, X.: Text-aware balloon extraction from manga. Visual Comput. Int. J. Comput. Graph. **32**(4), 501–511 (2016)
16. Matsui, Y., et al.: Sketch-based manga retrieval using manga109 dataset. Multimedia Tools Appl. **76**(20), 21811–21838 (2017)

17. McCloud, S.: Understanding Comics: The Invisible Art, Northampton, vol. 7, p. 4 (1993)
18. Mohammad, S., Kiritchenko, S.: Wikiart emotions: an annotated dataset of emotions evoked by art. In: Proceedings of the Eleventh International Conference on Language Resources and Evaluation (LREC 2018) (2018)
19. Nguyen, N.H., Rigaud, C., Burie, J.C.: Comic characters detection using deep learning. In: 2017 14th IAPR International Conference on Document Analysis and Recognition (ICDAR), Kyoto, November, pp. 41–46. IEEE (2017)
20. Nguyen, N.-V., Rigaud, C., Burie, J.-C.: Digital comics image indexing based on deep learning. J. Imaging 4(7), 89 (2018)
21. Nguyen, N.-V., Vu, X.-S., Rigaud, C., Jiang, L., Burie, J.-C.: ICDAR 2021 competition on multimodal emotion recognition on comics scenes. In: Lladós, J., Lopresti, D., Uchida, S. (eds.) ICDAR 2021. LNCS, vol. 12824, pp. 767–782. Springer, Cham (2021). https://doi.org/10.1007/978-3-030-86337-1_51
22. Pang, X., Cao, Y., Lau, R.W.H., Chan, A.B.: A robust panel extraction method for manga. In: Proceedings of the 22nd ACM International Conference on Multimedia, Orlando Florida USA, pp. 1125–1128. ACM (2014)
23. Plutchik, R.: The nature of emotions: human emotions have deep evolutionary roots, a fact that may explain their complexity and provide tools for clinical practice. Am. Sci. 89(4), 344–350 (2001)
24. Qin, X., Zhou, Y., He, Z., Wang, Y., Tang, Z.: A faster R-CNN based method for comic characters face detection. In: 2017 14th IAPR International Conference on Document Analysis and Recognition (ICDAR), vol. 01, pp. 1074–1080 (2017)
25. Russell, J.: A circumplex model of affect. J. Pers. Social Psychol. 39, 1161–1178 (1980)
26. Shaver, P., Schwartz, J., Kirson, D., O'connor, C.: Emotion knowledge: further exploration of a prototype approach. J. Pers. Social Psychol. 52(6), 1061 (1987)
27. Tashu, T.M., Hajiyeva, S., Horvath, T.: Multimodal emotion recognition from art using sequential co-attention. J. Imaging 7(8), 157 (2021)
28. Tomkins, S.S.: Affect theory. Appr. Emot. 163(163–195), 31–65 (1984)

MuraNet: Multi-task Floor Plan Recognition with Relation Attention

Lingxiao Huang[(✉)], Jung-Hsuan Wu, Chiching Wei, and Wilson Li

Foxit Software, Fremont, CA 94538, USA
{lingxiao_huang,matt_wu,jeremy_wei,wensheng_li}@foxitsoftware.com

Abstract. The recognition of information in floor plan data requires the use of detection and segmentation models. However, relying on several single-task models can result in ineffective utilization of relevant information when there are multiple tasks present simultaneously. To address this challenge, we introduce MuraNet, an attention-based multi-task model for segmentation and detection tasks in floor plan data. In MuraNet, we adopt a unified encoder called MURA as the backbone with two separated branches: an enhanced segmentation decoder branch and a decoupled detection head branch based on YOLOX, for segmentation and detection tasks respectively. The architecture of MuraNet is designed to leverage the fact that walls, doors, and windows usually constitute the primary structure of a floor plan's architecture. By jointly training the model on both detection and segmentation tasks, we believe MuraNet can effectively extract and utilize relevant features for both tasks. Our experiments on the CubiCasa5k public dataset show that MuraNet improves convergence speed during training compared to single-task models like U-Net and YOLOv3. Moreover, we observe improvements in the average AP and IoU in detection and segmentation tasks, respectively. Our ablation experiments demonstrate that the attention-based unified backbone of MuraNet achieves better feature extraction in floor plan recognition tasks, and the use of decoupled multi-head branches for different tasks further improves model performance. We believe that our proposed MuraNet model can address the disadvantages of single-task models and improve the accuracy and efficiency of floor plan data recognition.

Keywords: Floor plan · Unified backbone · Attention mechanism · Multi-head branches · Multi-task recognition

1 Introduction

Architectural floor plan data are standardized data that are used to make the design, construction, and other related work convenient. There are stringent requirements for the recognition accuracy of objects with different design components when automatically identifying design drawing data, to reduce errors between different works around the same floor plan data. There are several factors that makes the floor plan recognition difficult such as: (1) Special data

© The Author(s), under exclusive license to Springer Nature Switzerland AG 2023
M. Coustaty and A. Fornés (Eds.): ICDAR 2023 Workshops, LNCS 14193, pp. 135–150, 2023.
https://doi.org/10.1007/978-3-031-41498-5_10

characteristics: These data are composed of simple lines; thus, few features exist in aspects such as texture and color, and most features focus on the shape, structure, and relationships among different parts. (2) Complex semantic relationships exist between different objects. (3) Approaches such as detection and segmentation tasks usually work best for subsets of objects differently.

By looking into the literatures, classical convolutional neural network (CNN) models have primarily been used for different visual tasks on floor plan datasets. For example, the fully convolutional network (FCN), U-Net [11], HRNet [26], DeepLab [27], and Mask R-CNN [28] have been applied for mainframe and layout parsing, whereas YOLO and Faster R-CNN [29] have been used for component detection. However, these models do not consider the characteristics of floor plan data, including few low-level features and strong correlations among different objects at a high level. In recent years, attention mechanisms have been applied to vision tasks. Several state-of-the-art models have achieved success on many well-known public vision datasets, such as DETR [24], ViT [8], and the Swin Transformer [25]. Attention mechanisms can focus on the relationships among the data and improve model accuracy, which is beneficial for the complex relationships between different object types and recognition tasks in floor plan data. Because of the high-level features of floor plan data, a strong correlation exists between the different components, which makes it possible to use attention mechanisms in recognition tasks on floor plan data.

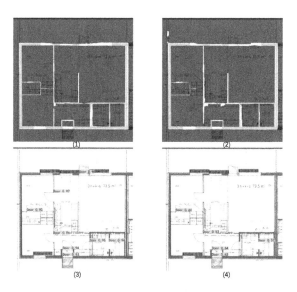

Fig. 1. Visualization results of a. MuraNet and b. U-Net + YOLOv3. (1) visualizes the MuraNet segmentation results, (2) visualizes the segmentation results with U-Net(D5), (3) visualizes the MuraNet detection results, (4) visualizes the detection results with YOLOv3. We can see that the segmentation results of MuraNet have fewer misidentifications and are more accurate than those of U-Net(D5). MuraNet is able to accurately and completely detect each door and window, but YOLOv3 misses two doors.

In this study, we propose MuraNet, which is an attention-based multi-task model for segmentation and detection tasks in floor plan analysis. We adopt a unified encoder with the MURA module as the backbone. The improved segmentation decoder branch and decoupled detection head branch from YOLOX [14] are used for the segmentation and detection tasks, respectively.

Our contributions are twofold. First, our proposed MuraNet jointly considers the wall pixel-level segmentation and doors and windows vector-level detection at the same time. Second, we add attention mechanism for the model to leverage the correlations between walls, doors and windows to enhance the accuracy. Because walls, doors, and windows usually jointly constitute the main frame of a floor plan's architecture, we believe the joint training will provide advantages. The performance of our model is excellent in terms of experimental values and intuitive visualization, as illustrated in Fig. 1.

This article is structured as follows: First, we introduce classical models for the recognition task and their applications to floor plans in the related work section. We also analyze the feasibility of applying the attention mechanism and explain its function and principle in this context. Next, in the methods section, we provide a detailed description of the MuraNet model architecture, including the attention module, the unified backbone, and multi-head branches. Then, in the experimental section, we present comparative experiments with classical models and conduct ablation experiments for the attention module and decoupled branch. Finally, we conclude by summarizing the contributions of this work and proposing future research directions while also acknowledging the current limitations.

2   Related Works

2.1   Recognition Tasks

In recognition tasks, traditional CNN models are widely used and stable. Different models are often applied to different recognition tasks in floor plans: scene segmentation models are applied to mainframe and layout parsing, object detection models are applied to component detection and recognition, and other models are used for specific tasks. For example, the graph convolutional network model [23] has been used for the analysis of vector data such as CAD and PDF, whereas generative adversarial network models have been used to analyze the main structure of drawings in a data-generated manner.

Detection Models. Many graphic symbols, such as doors, sliding doors, kitchen stoves, bathtubs, sinks, and toilets, need to be recognized in floor plans. Faster R-CNN includes an end-to-end model, and region proposal-based CNN architectures for object detection are often used to identify the parts and components in floor plans [1,3]. The YOLO model, which is a fast one-stage convolutional method that automatically detects structural components from scanned CAD drawings, was proposed in [21]. Several improved versions of the YOLO model, such as YOLOv2, have also been applied extensively [22].

Segmentation Models. FCNs have exhibited effective performance in segmentation tasks in the early stages [1]. U-Net [11] and the variants of U-Net have been designed based on this concept and have been proven to be outstanding in image segmentation tasks. Mask R-CNN is an instance segmentation model that is used to detect building and spatial elements [4].

In some works, they use separate detection and segmentation models to identify different objects. In some multi-task works [5], they use same backbone such as VGG-16 for feature extraction, a decoder such as U-Net [11] for segmentation, and a detector such as SSD for detection to achieve detection and segmentation tasks simultaneously. But recognition tasks are not independent of one another, and a relationship exists among recognition targets because they constitute a complete floor plan. In some multi-task works [7,18], the outputs of these models mentioned are pixel-level segmentation maps, but the outputs of our model are pixel-level segmentation maps and vector-level detection coordinates. None of these studies considered the interrelationships among different recognition targets.

2.2 Attention Networks

Attention mechanisms enable a network to focus on important parts adaptively. Recent models with attention mechanisms have exhibited high performance in image classification and dense prediction tasks such as object detection and semantic segmentation—for example, DETR [24], ViT [8], and the Swin Transformer [25] based on self-attention, as well as HRNet+OCR, VAN [9], and SegNeXt [10] based on convolution attention. Several state-of-the-art models have used a similar approach on many well-known public computer vision datasets.

Self-Attention. Self-attention mechanisms are primarily concerned with the overall relationships of all data. The self-attention mechanism first appeared in the transformer model in the natural language processing (NLP) field and is currently an indispensable component of NLP models. Several self-attention-based methods perform similarly to or even better than CNN-based methods. These models break down the boundaries between the computer vision and NLP fields, thereby enabling the development of unified theoretical models.

Convolution Attention. Convolution attention mechanisms are primarily concerned with the relationships among data that are processed by different convolution kernels. Convolution attention includes two categories: spatial attention and channel attention. Spatial attention focuses on the mutual relationships based on spatial information, whereas channel attention aims to select important objects for the network adaptively. In CNN models, a large kernel convolution layer is used to build both spatial and channel attention, which is known as a large kernel attention (LKA) mechanism. Both VAN [9] and SegNeXt [10] use the LKA mechanism, with SegNeXt being the most relevant reference for our model.

3 Our Method

3.1 Model Architecture

In this section, we describe the architecture of the proposed MuraNet in detail. Our model consists of a backbone with attention module, segmentation decoder and decoupled detection head. The backbone consists of 4 down-stages, and each down-stage consists of a set of downsampling convolutions and MURA in attention module. The segmentation decoder consists of 4 upsample-convolutional layers, and the features of last three stages will be kept and processed by the topmost convolutional layer after being aggregated. The decoupled detection head divides branches into classification and regression, and the regression branch divides branches into coordinate and IoU. We adopt a unified architecture, as illustrated in Fig. 2, which incorporates an attention-based relation attention module and uses multi-head branches for multiple tasks.

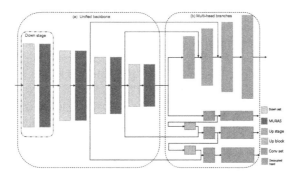

Fig. 2. MuraNet architecture uses (a) a unified backbone with the attention module as the encoder and (b) multi-head branches for multiple tasks.

Relation Attention Module. We adopt a similar structure to that of ViT [8] and SegNeXt [10] to build our unified backbone as the encoder, but instead of multi-branch depth-wise strip convolutions, a unified multi-scale attention module is designed to focus on the relationships among multi-scale features. This module is known as the multi-scale relation attention (MURA) module. As depicted in Fig. 3, the MURA module uses a series of 3 × 3 convolution kernels instead of large-kernel depth-wise strip convolutions to avoid the sensitivity of specific-shaped convolution kernels to the detection target shape. Although most walls, doors, and windows are elongated in a floor plan, multi-angle directions exist, and most frames and components, such as house types, room shapes, and furniture, are square. The MURA module also uses skip add connections to aggregate multi-scale features [11]. Thus, the parameters can be shared and the associations among multi-scale features can be strengthened.

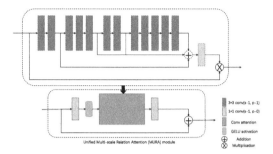

Fig. 3. MURA architecture uses (a) a series of 3 × 3 convolution kernels instead of large-kernel depth-wise strip convolutions and (b) unified skip add connections to aggregate multi-scale features.

Unified Backbone as the Encoder. Most components of floor plan data, such as walls, doors, and windows, are composed of simple lines, which means that few features exist in aspects such as texture and color. Furthermore, most features focus on the shape, structure, and relationships among different parts. As low-level feature information is insufficient, multi-level structured models and multi-scale feature aggregation are required to extract high-level features at different scales. We adopt the multi-level pyramid backbone structure for our encoder, following most previous studies [12,13]. Figure 4 depicts a stage of the encoder block with the MURA module. We use two residual connections in one stage of the encoder block to avoid the vanishing gradient problem as the model depth increases.

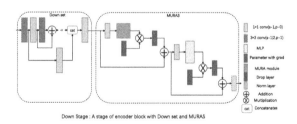

Fig. 4. Down Stage : A stage of encoder block with Down set and MURAS: two residual connections are used in one stage of the encoder block.

We adopt a robust four-level pyramid hierarchy that contains a downsampling block in each stage for the overall model architecture. The downsampling block uses a two-stride 3 × 3 kernel convolution to decrease the spatial resolution, which is followed by a batch normalization layer. The model architecture is illustrated in Fig. 5. The four resolutions are $\frac{H \times W}{4 \times 4}$, $\frac{H \times W}{8 \times 8}$, $\frac{H \times W}{16 \times 16}$, and $\frac{H \times W}{32 \times 32}$, where $H \times W$ is the shape of the input data.

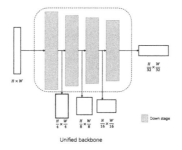

Fig. 5. Unified encoder architecture

Multi-Head Branches. The input data are processed by a unified multi-level pyramid encoder to generate features of four different scales. The features of these different stages must be processed by heads that correspond to different tasks to obtain required data. We believe that it is necessary to use different head branches to process the corresponding tasks immediately following feature extraction through the unified backbone, because of the significant differences in the segmentation and detection tasks. We adopt multi-head branches for different recognition tasks, a decoder for the segmentation task, and a detector for the detection task.

Segmentation Decoder. It is necessary for the decoder to aggregate and process multi-scale features in segmentation tasks. Thus, we adopted a powerful decoder based on SegNeXt [10] and U-Net [11]. SegNeXt uses a lightweight Hamburger, which has been proven as an effective decoder, that can process features aggregated from the last three stages. Only the features of the last three stages are aggregated because the features from stage 1 contain excessive low-level information, which degrades the performance. The characteristics of the Hamburger decoder are consistent with those of the floor plan data; that is, few low-level features exist. In U-Net, the encoder–decoder structure is a symmetrical U-shape, and the model structure and skip connections that can connect multi-scale features have been demonstrated to be effective and stable in segmentation tasks. As illustrated in Fig. 6, our decoder aggregates the features from the last three stages, upsamples symmetrically with the encoder, and forms skip connections with multi-scale features, thereby forming an encoder–decoder architecture similar to that of U-Net.

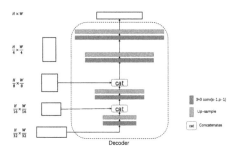

Fig. 6. Segmentation decoder

Detection Head. The conflict between the classification and regression tasks is a well-known problem in object detection [15,16]. In YOLOX [14], a lite decoupled detection head separates the classification and regression tasks for end-to-end training, which enables fast fitting for training and improves model performance. As illustrated in Fig. 7, for each feature stage, this head adopts a 1 × 1 conv layer to reduce the feature channels to 256 and then adds two parallel branches, each with two 3 × 3 conv layers, for the classification and regression tasks, respectively. An IoU branch is added to the regression branch. This decoupled detection head is critical in complex multi-task models.

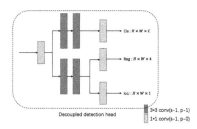

Fig. 7. Decoupled detection head

3.2   Model Training

Public Floor Plan Datasets. CVC-FP includes 122 high-resolution images in four different drawing styles, and the image resolution of the dataset ranges from 1098 × 905 to 7383 × 5671 pixels [2]. R2V [6] consists of 870 ground-truth floor plan images of urban residences that were collected from various regions in Japan. R3D [17] consists of 214 original images, in which most room shapes are irregular with a nonuniform wall thickness. CubiCasa5k [7] is a large-scale public dataset that consists of 5000 ground-truth Finnish floor plan images and is annotated using the SVG vector graphics format. In this study, we selected CubiCasa5k as the experimental dataset. We think the results of CubiCasa5k is sufficient because of its rich-in-type, high-quality, and close-to-real-world-data nature.

Multi-task Loss. Studies [18] have defined reasonable weights for the loss function from the perspective of the number of labels for segmentation tasks. These loss weights represent the distribution of different target categories in the overall dataset, which can stabilize the loss value during training and enable the model to perform normal iterative training.

Segmentation-Weighted Loss. We define the task-weighted loss in an entropy style as follows:

$$L_{segmentation} = \sum_{i=0}^{C} -w_i y_i log p_i \tag{1}$$

where y_i is the label of the i-th floor plan segmentation element in the floor plan, C is the number of floor plan segmentation elements in the task, when C is equal to 0, it represents the background category, in our experiments in this work, when C is equal to 1, it represents the wall category, and p_i is the prediction label of the pixels for the i-th element ($p_i \in [0, 1]$). Furthermore, w_i is defined as follows:

$$w_i = \frac{\hat{N} - \hat{N}_i}{\sum_{j=0}^{C}(\hat{N} - \hat{N}_j)} \tag{2}$$

where $\hat{N}_i$ is the total number of ground-truth pixels for the i-th floor plan segmentation element in the floor plan and $\hat{N} = \sum_{i=0}^{C} \hat{N}_i$ represents the total number of ground-truth pixels over all C floor plan segmentation elements. The W_i in the loss function is calculated based on the training-set. Multiple experiments will divide different datasets, the W_i will also be recalculated. The specific value will be different, and the effect of the experiments is stable and consistent. In our experiments in this work, W_0 is approximately 0.2, W_1 is approximately 0.8.

Detection Loss. A widely used loss function combination [14,19] ensures effective model training in detection tasks. In this study, we use the detection loss functions of YOLOv3 [20]. The overall detection loss consists of the bbox, class, and objectness losses. The segmentation and detection losses are combined as the overall loss during model training to strengthen the relevance of the segmentation and detection task targets.

$$L_{total} = L_{segmentation} + L_{detection} \tag{3}$$

4   Experiments

We evaluated our method on the public dataset CubiCasa5k. We divided the dataset into 4000 images for training, 500 images for validation, and 500 images for testing. The size of the image data ranged from 430 × 485 to 6316 × 14304 with few repeating dimensions, and the average size was 1399 × 1597. It was necessary for the data to be divisible by 32 to fit the data size change of the model architecture. Therefore, we directly resized the input images to 1536 × 1536, and the aspect ratio will change, but not change greatly. When the input images need to be down-sampled, we use the area interpolation algorithm, and when the input images need to be up-sampled, we use the cubic interpolation algorithm. The training parameters: the batch size is 10 by default to typical 10-GPU devices, the max lr is 0.01, the initial lr is 0.0001, the min lr is 0.000001, the weight decay is 0.0005 and the SGD momentum is 0.937, and the cosine lr schedule, total epoch is 1000, 50 epochs linear warm-up. After experiments, these parameters can achieve the best training effect in the experimental models.

CubiCasa5k contains over 80 categories, due to the nature that walls, doors and windows usually jointly constitute the main frame of the architecture of a floorplan, we believe the joint training will have advantages. In our experiments, the segmentation task was performed on the walls, and the detection task was performed on the doors and windows. All models, including the comparison models, were trained on a node with 10 RTX 3090 GPUs. Because of the particularity of the floor plan data, pretrained models were not used in any of our experiments. That is, all models were trained from scratch. We adopted the AP (@0.5 and @[.5:.95]) and IoU as evaluation metrics for detection and segmentation, respectively. Furthermore, we recorded the convergence speed of the model during the training process. Regarding the convergence epoch, we define that when the first accuracy of the training epoch reaches 99.9% of the final, and the subsequent accuracy fluctuations do not exceed 0.2% of the final, the training curve also indicate the convergence speed.

4.1 Comparison with Single-Task Models

U-Net has often been used in segmentation tasks because of its high performance. It is also used for the recognition task of the main architecture in floor plans. Models in the YOLO series are frequently used for fast component detection. We use U-Net and YOLOv3 as the comparative test models to show the performance of MuraNet.

Comparison with U-Net. We trained our MuraNet model and U-Net model with base(Fig. 8), 5, and 6 depth structures using the same training parameter configurations, and the IoU of the wall was considered as the primary evaluation metric. Since U-Net cannot fit the data well, the data of U-Net 5-stage(D5) which has 5 downsampling modules, one more than the standard U-Net is added in Fig. 8. As indicated in Table 1, the wall IoU of MuraNet was generally higher than that of U-Net, which means that MuraNet achieved higher accuracy and the attention-based MURA module can assist the backbone network in improving the segmentation feature extraction. The convergence speed of MuraNet training was faster than that of U-Net at similar performance.

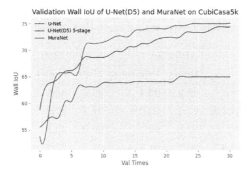

Fig. 8. Validation wall IoU of MuraNet and U-Net(D5)

Table 1. Comparison of MuraNet and U-Net counterparts in terms of IoU (%) on CubiCasa5k. All models were tested with a resolution of 1536 × 1536 .

Model	Wall IoU (%)	Convergence epochs
U-Net base	65.5	6
MuraNet base(+12.9)	78.4	8
U-Net 5-stage	74.4	10
MuraNet 5-stage(+1.3)	75.7	9
U-Net 6-stage	75.8	11
MuraNet 6-stage(+0.6)	76.4	11

Comparison with YOLOv3. We designed MuraNet base(Fig. 9) and MuraNet-spp, according to the structures of YOLOv3 and YOLOv3-spp, respectively. We trained these models using the same training parameter configuration and used the AP (@0.5 and @0.5:0.95) as the primary evaluation metrics. As indicated in Table 2, the convergence speed of MuraNet training was generally faster than that of YOLOv3, which means that the decoupled detection head caused the model to converge faster during training. According to the AP value, MuraNet could achieve the same accuracy as YOLOv3 with rapid convergence, and spp-layer has little effect on model performance.

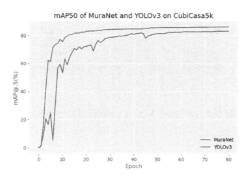

Fig. 9. Validation mAP50 of MuraNet and YOLOv3 on CubiCasa5k

Table 2. Comparison of MuraNet and YOLOv3 counterparts in terms of AP (%) on CubiCasa5k. All models were tested with a resolution of 1536×1536 .

Model	AP50(%)			AP@[.5:.95](%)		
	Doors	Windows	Mean	Doors	Windows	Mean
YOLOv3 base	89.2	90.1	89.6	43.6	55.4	49.5
MuraNet base(+4.3)	91.2	92.2	91.7	47.9	59.7	53.8
YOLOv3+spp	89.8	90.2	90.0	43.8	55.6	49.7
MuraNet+spp(+3.7)	90.9	91.5	91.2	47.7	59.1	53.4

4.2 Ablation Experiments

MURA. We designed three comparative experiments to verify the functions of our proposed attention-based MURA module in the backbone for feature extraction from floor plan data. In the first experiment, we compared MuraNet model with and without MURA modules. In the second experiment, we compared the model with U-Net and only added MURA modules to the backbone, without changing the U-Net structure. In the third experiment, we compared the model with SegNeXt and replaced the MSCA module, which uses depth-wise strip convolutions to compute the attention, with MURA modules that use a series of 3×3 convolution kernels. We also trained these models using the same training parameter configuration on CubiCasa5k. As shown in Fig. 10) and Table 3, MuraNet, U-Net and SegNeXt with the MURA modules achieved higher wall category accuracy on the CubiCasa5k dataset, and the convergence epochs of the training cycle were approximately the same. Thus, the attention module can extract feature information better, and a series of 3×3 convolution kernels is more suitable for floor plan data than depth-wise strip convolutions.

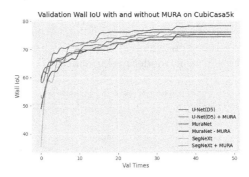

Fig. 10. Validation wall IoU of MuraNet, U-Net(D5) and SegNeXt with MURA and without MURA

Table 3. Comparison of MuraNet, U-Net and SegNeXt with and without MURA modules and counterparts in terms of IoU (%) on CubiCasa5k. All models were tested with a 1536 × 1536 resolution.

Model	Wall IoU (%)	Convergence epochs
MuraNet base	78.4	8
MuraNet-MURA base(-3.1)	75.3	7
MuraNet 5-stage	75.7	9
MuraNet-MURA 5-stage(-0.7)	75.0	10
MuraNet 6-stage	76.4	11
MuraNet-MURA 6-stage(-0.2)	76.2	11
U-Net base	65.5	6
U-Net+MURA base(+9.58)	75.1	8
U-Net 5-stage	74.4	10
U-Net+MURA 5-stage(+1.7)	76.1	9
U-Net 6-stage	75.8	11
U-Net+MURA 6-stage(+0.42)	76.2	10
SegNeXt base	75.8	12
SegNeXt+MURA base(+0.38)	76.2	12
SegNeXt small	75.3	11
SegNeXt+MURA small(+0.71)	76.0	10
SegNeXt large	76.3	10
SegNeXt+MURA large(-0.02)	76.3	10

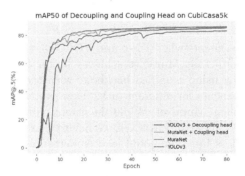

Fig. 11. Validation mAP50 of Decoupling and Coupling Head on CubiCasa5k

Decoupled Detection Head. We only used decoupled and coupled detection heads on MuraNet and YOLOv3 to verify the effect of the decoupled head in the detection task(Fig. 11). We also trained these models using the same parameter configuration on CubiCasa5k. Table 4 indicates that the decoupled detection head could accelerate the model fitting without compromising on the accuracy in the normal YOLO detection models.

Table 4. Comparison of MuraNet and YOLOv3 with and without decoupled detection head and counterparts in terms of AP (%) on CubiCasa5k. All models were tested with a resolution of 1536 × 1536.

Model	AP50 (%)			AP@[.5:.95] (%)		
	Doors	Windows	Mean	Doors	Windows	Mean
MuraNet base+coupled head	89.9	90.1	90.0	43.9	56.7	50.3
MuraNet base+decoupled head(+3.5)	91.2	92.2	91.7	47.9	59.7	53.8
MuraNet+spp+coupled head	89.7	90.1	89.9	44.0	56.2	50.1
MuraNet+spp+decoupled head(+3.3)	90.9	91.5	91.2	47.7	59.1	53.4
YOLOv3 base+coupled head	89.2	90.1	89.6	43.6	55.4	49.5
YOLOv3 base+decoupled head(+0.4)	89.8	90.1	90.0	43.9	55.9	49.9
YOLOv3+spp+coupled head	89.8	90.2	90.0	43.8	55.6	49.7
YOLOv3+spp+decoupled head(+0.9)	90.0	90.1	90.0	45.0	56.2	50.6

5   Conclusions

This paper proposes MuraNet, an attention-based multi-task model for segmentation and detection tasks in floor plan data, designed to jointly train the model on both detection and segmentation tasks. MuraMet adopt a unified encoder called MURA module as the backbone, an improved segmentation decoder branch for the segmentation task, and a YOLOX-based decoupled detection head branch for the detection task. The two key contributions of our work is (1) our proposed model integrates the pixel-level segmentation and vector-level detection at the same time. (2) we add attention mechanism for the model to leverage the correlations between walls, doors and windows to enhance the accuracy.

Comparative experiments with U-Net and YOLOv3 on the CubiCasa5k public dataset demonstrate that MuraNet achieves better feature extraction and higher performance by using a unified backbone with the attention mechanism and different head branches for different tasks. However, we believe that there is room for improvement in MuraNet. While the attention mechanism is currently only utilized in the backbone of MuraNet, we believe that its use in other parts, such as the head or loss function, could lead to better learning of relationships among different recognition targets, improving recognition tasks in floor plan data. This is a direction for future exploration.

References

1. Dodge, S., Xu, J., Stenger, B.: Parsing floor plan images. In: MVA, pp. 358–361 (2017). https://doi.org/10.23919/MVA.2017.7986875
2. de las Heras, L.P., Fernández, D., Valveny, E., Lladós, J., Sánchez, G.: Unsupervised wall detector in architectural floor plans. In: ICDAR, pp. 1245–1249 (2013). https://doi.org/10.1109/ICDAR.2013.252

3. Surikov, I.Y., Nakhatovich, M.A., Belyaev, S.Y., et al.: Floor plan recognition and vectorization using combination UNet, faster-RCNN, statistical component analysis and Ramer-Douglas-Peucker. In: COMS2, pp. 16–28 (2020)

4. Wu, Y., Shang, J., Chen, P., Zlantanova, S., Hu, X., Zhou, Z.: Indoor mapping and modeling by parsing floor plan images. Int. J. Geogr. Inf. Sci. **35**(6), 1205–1231 (2021)

5. Lu, Z., Wang, T., Guo, J., et al.: Data-driven floor plan understanding in rural residential buildings via deep recognition. Inf. Sci. **567**, 58–74 (2021)

6. Liu, C., Wu, J., Kohli, P., Furukawa, Y.: Raster-to-vector: revisiting floorplan transformation. In: ICCV, pp. 2195–2203 (2017)

7. Kalervo, A., Ylioinas, J., Häikiö, M., Karhu, A., Kannala, J.: CubiCasa5K: a dataset and an improved multi-task model for floorplan image analysis. In: Felsberg, M., Forssén, P.-E., Sintorn, I.-M., Unger, J. (eds.) SCIA 2019. LNCS, vol. 11482, pp. 28–40. Springer, Cham (2019). https://doi.org/10.1007/978-3-030-20205-7_3

8. Dosovitskiy, A., et al.: An image is worth 16×16 words: Transformers for image recognition at scale. In: International Conference on Learning Represent (2020)

9. Guo, M.H., Lu, C.Z., Liu, Z.N., Cheng, M.M., Hu, S.M.: Visual Attention Network. arXiv preprint arXiv:2202.09741 (2022)

10. Guo, M.H., et al.: SegNeXt: rethinking convolutional attention design for semantic segmentation. arXiv preprint arXiv:2209.08575 (2022)

11. Ronneberger, O., Fischer, P., Brox, T.: U-Net: convolutional networks for biomedical image segmentation. In: MICCAI (2015)

12. Xie, E., Wang, W., Yu, Z., Anandkumar, A., Alvarez, J.M., Luo, P.: Segformer: simple and efficient design for semantic segmentation with transformers. Adv. Neural Inf. Process. Syst. **34**, 12077–12090 (2021)

13. Chen, L.C., Papandreou, G., Kokkinos, I., Murphy, K., Yuille, A.L.: Deeplab: semantic image segmentation with deep convolutional nets, Atrous convolution, and fully connected CRFs. IEEE Trans. Pattern Anal. Mach. Intell. **40**(4), 834–848 (2018)

14. Ge, Z., Liu, S., Wang, F., Zeming, L., Jian, S.: YOLOX: exceeding YOLO series in 2021. arXiv preprint arXiv:2107.08430 (2021)

15. Song, G., Liu, Y., Wang, X.: Revisiting the sibling head in object detector. In: CVPR (2020)

16. Wu, Y, Chen, Y., Yuan, L. et al.: Rethinking classification and localization for object detection. In: CVPR (2020)

17. Liu, C., Schwing, A., Kundu, K., Urtasun, R., and Fidler, S.: Rent3D: floor-plan priors for monocular layout estimation. In: CVPR (2015)

18. Zeng, Z., Li, X., Yu, Y.K., Fu, C.W.: Deep floor plan recognition using a multi-task network with room-boundary-guided attention. In: ICCV, pp. 9095–9103 (2019)

19. Ge, Z., Liu, S., Li, Z., Yoshie, O., and Sun, J.: OTA: optimal transport assignment for object detection. In CVPR, pp. 303–312 (2021)

20. Redmon, J., Farhadi, A.: YOLOv3: an incremental improvement. arXiv preprint arXiv:1804.02767 (2018)

21. Zhao, Y., Xueyuan, D., Huahui, L.: A deep learning-based method to detect components from scanned structural drawings for reconstructing 3D models. Appl. Sci. **10**(6), 2066 (2020)

22. Rezvanifar, A., Cote, M., and Albu, A.B.: Symbol spotting on digital architectural floor plans using a deep learning-based framework. In: CVPRW (2020)

23. Fan, Z., Zhu, L., Li, H., et al.: FloorPlanCAD: a large-scale CAD drawing dataset for panoptic symbol spotting. In: ICCV (2021)

24. Nicolas, C., Francisco, M., Gabriel, S., Nicolas, U., Alexander, K., Sergey, Z.: End-to-end object detection with transformers. arXiv:2005.12872 (2020)
25. Ze, L., Yutong, L., Yue, C., et al.: End-to-end object detection with transformers. In: ICCV (2021)
26. Wang, J., Sun, K., Cheng, T., et al.: Deep high-resolution representation learning for visual recognition. IEEE Trans. Pattern Anal. Mach. Intell. **43**(10), 3349–3364 (2020)
27. Liang-Chieh, C., George, P., Iasonas, K., Kevin, M., Alan, L.Y.: Semantic image segmentation with deep convolutional nets and fully connected CRFs. arXiv preprint arXiv:1412.7062 (2014)
28. He, K., Gkioxari, G., Dollár, P., Girshick, R.: Mask R-CNN. In: ICCV(2017)
29. Ren, S., He, K., Girshick, R., Sun, J.: Faster R-CNN: towards real-time object detection with region proposal networks. IEEE Trans. Pattern Anal. Mach. Intell. **39**, 1137–1149 (2017)

Automatic Detection of Comic Characters: An Analysis of Model Robustness Across Domains

Javier Lucas, Antonio Javier Gallego$^{(\boxtimes)}$ ⓘ, Jorge Calvo-Zaragoza ⓘ, and Juan Carlos Martinez-Sevilla ⓘ

Department of Software and Computing Systems, University of Alicante, Alicante, Spain

jla45@alu.ua.es, {jgallego,jcalvo}@dlsi.ua.es, jcmartinez.sevilla@ua.es

Abstract. The popularity of comics has increased in the digital era, leading to the development of several applications and platforms. These advancements have opened up new opportunities for creating and distributing comics and experimenting with new forms of visual storytelling. One of the most promising research areas in this field is the use of deep learning techniques to process comic book images. However, one of the main challenges associated with the use of these models is adapting them to different domains because comics greatly vary in style, subject matter, and design. In this paper, we present a study on the problem of generalization across different domains for the automatic detection of characters in comics. We evaluate the performance of state-of-the-art models trained in different domains and analyze the difficulties and challenges associated with generalization. Our study provides insights into the development of more robust deep-learning models for processing comics' characters and improving their generalization to new domains.

Keywords: Comic Analysis · Character Detection · Domain Shift

1 Introduction

A *comic* is a means of artistic expression that combines text and images to tell a story. The history of comics goes back to the end of the 19th century, when the first publications of comic strips appeared in the press. Since then, comics have evolved and diversified, encompassing a wide variety of genres and styles, from science fiction and fantasy to horror and drama, among others.

In the digital era, comics have undergone a significant transformation owing to new technologies and the growing popularity of mobile phones and tablets. Currently, there are several applications and digital platforms that allow access to a wide variety of comics instantly. This has opened up new opportunities for reading, creating, and distributing comics, as well as experimenting with new forms of visual storytelling.

M. Coustaty and A. Fornés (Eds.): ICDAR 2023 Workshops, LNCS 14193, pp. 151–162, 2023.
https://doi.org/10.1007/978-3-031-41498-5_11

Within this context, the development of applications for comics has gained great importance in recent years, as for instance interactive reading and reading assistance for people with functional diversity [15]. Interactive reading allows the reader to explore the story in a more dynamic and immersive way, while assistive-reading applications help people with visual or learning disabilities enjoy the comic in a more accessible and personalized way.

The use of deep learning techniques to process comic book images is one of the most promising research areas nowadays to develop new applications. However, the associated challenges encompass several difficulties, such as the complexity of the illustrations, the variability in the drawing style, the composition of the strips, or the non-linear narrative [8]. To address these, several approaches based on computer vision have been proposed (cf. Sect. 2).

Furthermore, in addition to the challenges inherent in any task (as in comics here), one of the main drawbacks associated with the use of deep learning models is the difficulty of adapting the models trained for a particular comic to different domains. This is because comics can be highly variable in terms of style, subject matter, design, and other factors that influence the appearance and layout of the panels. This problem of learning neural networks that generalize across different domains is one of the main challenges faced by researchers in this area.

In this work, we present a study on the problem of generalization across different domains for the automatic detection of characters in comics. This is important because comics feature a wide range of artistic styles and character designs, making it challenging for a recognition system trained on one comic to perform well on another. Whether existing approaches can be well generalized to other domains in comics is a key piece for future strategies, not just to a task but general to the field of comic book processing itself.

In particular, we here evaluate the performance of state-of-the-art models trained in different domains of comics for character recognition, and we analyze the difficulties and challenges associated with adapting the models to other domains. Our work demonstrates improved cross-domain character detection through multi-dataset training.

2   Background

Compared to other graphics recognition fields, the automatic processing of comic sources is barely explored [11]. The new contributions usually focus on one specific task within the full spectrum of challenges.

In particular, Iyyer et al [8] explored whether computers can understand the implicit narrative conveyed by comic book panels, which often rely on readers to infer unseen actions. The authors built a dataset containing over 1.2 million panels with automatic textbox transcriptions. They asked deep models to predict narrative and character-centric aspects of a panel given context from preceding panels. In general, the automatic models underperformed human baselines, suggesting that comics present fundamental challenges for both vision and language understanding for current technology.

Nguyen et al. [13] described a method for indexing digital comic images. They considered deep learning to automatically split images into panels, and encode and index them through XML-based formats. The authors evaluated their method on a dataset consisting of online library content and proposed a new public dataset.

Concerning the automatic detection or recognition of characters—which is the scope of the present work—there also exists some previous work. Nguyen et al. [12] considered an object-detection model trained with a custom dataset called *Sequencity*, comprising a total of 612 pages. They tested the resulting models against different datasets, such as Sun60, Ho42, or Fahad18. Their method reported better results than those achieved until that date. Furthermore, Dutta and Biswas [2] not only trained a model to perform character recognition, but also panel detection. To accomplish these tasks, they applied Transfer Learning on a deep learning model using a new dataset they created: the Bengali Comic Book Image Dataset (BCBId). They tested the resulting model on different available datasets, such as eBDtheque, Manga109, or DCM, obtaining successful results.

Similarly to the approaches mentioned above, we also propose addressing character recognition in comics using deep learning. In our case, however, we particularly focus on studying the generalization capacity of these approaches to domains other than the one used for training.

3   Method

Object detection stands for the task of locating and delimiting within a bounding box the different elements present in an image for eventually classifying them [18]. Conceptually, these techniques model a continuous function that predicts the bounding boxes along with a discrete predictor that infers their associated labels.

A model that has stood out in object detection—which will be the architecture considered in this work—is the YOLO (You Only Look Once) algorithm. YOLO was proposed by Redmon et al. [16], and since then several versions have been developed that improve both the speed and accuracy of the method. The main idea behind YOLO is to look at the image only once (as its name indicates), as opposed to other proposals of the state of the art, such as DPM v5 [3], Fast-RCNN [5], Faster-RCNN [17], or Mask-RCNN [7], which perform this task in two steps: first, they detect the bounding boxes and then they perform a classification step. The reader is referred to the work by [19] that provides a comprehensive review of the existing formulations and architectures.

YOLO is composed of a backbone architecture that extracts image features and a neck that generates bounding boxes for objects and the associated class probabilities. It first divides the whole image into a $S \times S$ grid. Every cell in this grid will generate B bounding boxes, each represented by the spatial coordinates of the box, the confidence of containing an object, and the conditional probability of the object belonging to a particular class. The loss function used is a combination of these components: the sum of the mean squared error (MSE) between the predicted and true coordinates, the binary cross-entropy loss for the confidence score, and the cross-entropy loss for the class probabilities.

In our particular case, we will configure this architecture for the prediction of a single class: comic book characters. Since this type of data contains a wide variability of drawing styles, colors, types of characters, etc., and given the relatively limited amount of data available, we will resort to the use of transfer learning techniques for model initialization and data augmentation processes to improve the results and generalization capabilities of the models obtained (described in detail in Sect. 4.2).

Transfer learning consists of taking advantage of the knowledge learned for a domain with sufficient labeled data and applying it to another through a process of fine-tuning—i.e., using the weights of the trained model and using it as initialization for a new model. In our case, we will freeze the first part of the network (i.e. the backbone), so that these weights do not change during training. This solution helps to reduce the amount of labeled data needed for training, to make training converge faster, and to obtain models that generalize better.

In addition, since it is intended to analyze the performance of the model when training at the panel level as well as with the complete pages of the comics, it was necessary to perform a preprocessing of the datasets (see Fig. 1). First, the annotations available in the datasets were used to extract the panels. The characters were then associated with the panel in which they were contained, readjusting the coordinates of the panels in some cases. Finally, the character coordinates were recalculated to make them relative to the panel and to conform to YOLO's format (center x, center y, width and height, all rescaled in the range $[0, 1]$, so that the values are independent of the size of the image).

4   Experimental Setup

This section describes the configuration followed during the experimentation process, including the considered datasets, the details of the training process, and the evaluation metrics. All the experiments were performed using an Intel(R) Xeon(R) CPU @ 2.00 GHz with 12 GB RAM and an Nvidia Tesla T4 GPU with 12 GB of RAM.

4.1   Datasets

For the experimentation, we considered two datasets from the state of the art: eBDtheque [6] and Manga109 [4]. A description of them is provided below. Table 1 includes a summary of their characteristics and some random image examples can be consulted in Fig. 2, which shows the great variability of the characters and the appearance of the comics considered.

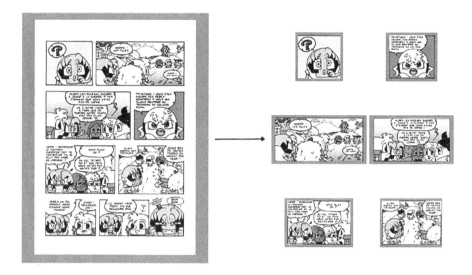

Fig. 1. Graphic representation of the panel extraction process from comic book pages.

eBDtheque. This dataset contains color images of hundreds of comics, most of them from French-Belgian authors. The ground truth includes labels for panels, text lines, and the main characters. We will use both the annotations for panels and characters.

Manga109. It contains 109 grayscale Japanese comics, with more than 21,100 pages in total. It includes labels for panels and characters, differentiating between the face and the whole body. In our experiments, we will use 25 % of all the data available in Manga109, both for pages and panels. We made this decision due to the resources available, as this helps us to reduce training time without notably reducing the results obtained.

Table 1. Summary of the characteristics of the datasets considered, including the number of samples and the minimum, maximum, and average resolution, both at the page level and at the panel level.

Dataset	Labeling level	# samples	Resolution (width×height px.)		
			Avg.	Min.	Max.
eBDtheque	Pages	96	$1,841 \times 2,230$	800×301	$5,320 \times 3,632$
	Panels	715	644×571	121×58	$4,959 \times 3,422$
Manga109	Pages	2,517	$1,666 \times 1,179$	$1,654 \times 1,170$	$2,960 \times 2,164$
	Panels	20,669	404×365	60×62	$2,224 \times 1,470$

Fig. 2. Random examples from the datasets considered for the experimentation.

For the experimentation, we used 80% of the available data to train the models, 10% for validation, and the remaining 10% for testing. Since there are many more samples from Manga109 than from eBDtheque, we applied *oversampling* to generate the models that combine both datasets during training. This technique consists of duplicating the images of the dataset with fewer samples until the number of samples is balanced, so that the model is trained with the same amount of images of each of them. This approach eliminates the possible bias towards a type of data and also ensures that the result obtained cannot be attributed to the use of more or less data for a specific dataset.

4.2 Network Configuration

In our experiments, we used the YOLO v5 version of the algorithm [9], which provides several network sizes (from small to large) with different numbers of parameters. We selected the medium size (with a total of 21.2 million parameters) since after several preliminary tests we determined that it presented a good balance between efficiency (in the resources available) and efficacy.

The original model was initialized using the pre-trained weights obtained for the Common Objects in COntext (COCO) corpus [10] for object detection. From this initialization, we carry out the fine-tuning process for the datasets considered in this work. In this process, the first part of the network (i.e. the backbone) was frozen and only the final layers were trained during 100 epochs, which, as will be seen, was sufficient for the convergence of the model due to the good initialization. The learning rate was fixed to 10^{-3}, with a weight decay regularization of 10^{-4}, and considering a warm-up process of 3 epochs with a learning rate of 10^{-1}. Stochastic Gradient Descent [1] was selected as the optimization function with a mini-batch size of 32 samples.

Regarding the data augmentation, we considered a collection of transformations that were randomly applied to each training image: horizontal flips (with a 50% of probability), mosaics, translations of the images by -10% to 10% on both axes, axis-independent scale changes within -50% and 50% of their original size, and color alterations in the Hue-Saturation-Value (HSV) color space by a fraction of 0.015, 0.7, and 0.4, respectively to each of these channels.

4.3   Metrics

In terms of evaluation, we considered different figures of merit commonly used in the assessment of object detection methods [14]: Average Precision (AP), F-measure (F_1), Precision (P), and Recall (R).

Considering a predicted character $\hat{m}$ and its associated m ground-truth annotation, we first calculate the Intersection over Union (IoU) as the ratio between their overlapping area and the total surface covered by their union. Based on this indicator, a character is considered correctly detected *iff* IoU $\geq \delta$, where $\delta \in [0,1]$ represents an evaluation threshold that relates to the severity of the assessment.

The figures of merit considered build upon this definition using the IoU value obtained for one or more thresholds. F_1, P, and R are defined as:

$$P = \frac{TP}{TP + FP} \tag{1}$$

$$R = \frac{TP}{TP + FN} \tag{2}$$

$$F_1 = \frac{2 \cdot P \cdot R}{P + R} = \frac{2 \cdot TP}{2 \cdot TP + FP + FN} \tag{3}$$

where TP, FP, and FN denote the True Positives (number of correctly detected characters), False Positives (number of incorrectly detected characters), and False Negatives (number of non-detected or missed characters) for a given δ threshold, for which we will use $\delta = 0.5$ since it is the value commonly used for this type of tasks.

Regarding the AP metric, it measures the area under the precision-recall curve, representing the average of the individual ratios between the number of correctly estimated characters and the number of elements to retrieve. Since this metric also depends on the δ threshold, we resort to the values usually reported in related works: $\delta = 0.5$ (denoted as $AP^{0.50}$) as well as the average of the AP scores (referred to as AP^m) when considering 11 equispaced threshold values in the range $\delta \in [0.5, 1]$.

5   Results

In this section, the proposed method is evaluated using the experimental setup previously described.

In the first place, we analyze the training process of the proposed architecture in order to assess the model's learning capacity on the training data both at the page and panel level, and the generalization made with the validation set. Figures 3a to 3d show the evolution of the loss values across the epochs for the training and validation sets of the two datasets considered. As can be seen, the models manage to converge in all cases and do not perform overfitting. In addition, learning at the panel level seems to be easier, since the error is reduced faster and a lower error rate is reached.

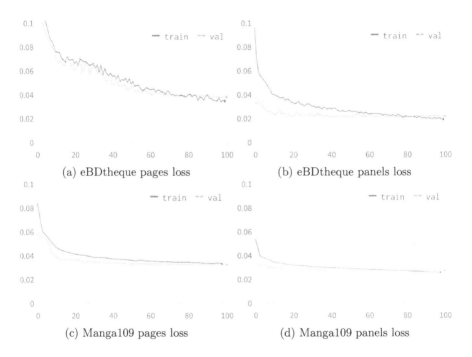

(a) eBDtheque pages loss

(b) eBDtheque panels loss

(c) Manga109 pages loss

(d) Manga109 panels loss

Fig. 3. Evolution of the bounding box loss (vertical axis) across the epochs (horizontal axis) for the training and validation sets of eBDtheque and Manga109.

We now analyze the results obtained with the models trained for the different scenarios contemplated: train at the page and panel level and carry out the evaluation both in the same data domain (i.e. intra-domain case) and in the other domain considered (i.e. inter-domain case). Besides, the case of combining both domains for training is also studied. The results of these experiments can be seen in Table 2, which details the dataset used for training (*source* column) and for the evaluation (*target* column), and whether training was conducted at the page or panel level.

Analyzing these results in general, it is observed that training at the page level and the panel level achieves a similar average performance for all the metrics, which slightly improves or worsens depending on the dataset used. Looking in detail at the case of training with a single domain and performing the intra-domain evaluation, eBDtheque obtains a better result at the page level and Manga109 at the panel level, although with fairly close figures for all metrics.

In the case of the inter-domain evaluation, it also depends, since when training with Manga109 and evaluating with eBDtheque (Manga109→eBDtheque) a better result is reported at the panel level, while for eBDtheque→Manga109 a certain improvement is obtained at the page level. In both cases, the results are quite good considering that the models were not trained with data from the same domain. It is important to remember that the styles are very different, one in color and the other in grayscale.

Table 2. Results of the experimentation carried out considering the different possibilities of domain changes for training (source column) and evaluation (target column), as well as the combination of data domains for training. The intra-domain evaluation case for each scenario is underlined. The best results for each metric, labeling level, and target test set are highlighted in bold.

Source	Target	AP	$AP^{0.5}$	P	R	F_1
Pages						
eBDtheque	eBDtheque	**50.3**	**85.3**	**88.8**	**78.9**	**83.6**
	Manga	38.0	70.2	76.4	63.7	69.5
Manga	eBDtheque	17.7	42.0	61.1	39.2	47.8
	Manga	**58.2**	**86.4**	86.3	**79.1**	**82.5**
eBDtheque + Manga	eBDtheque	48.7	81.9	80.8	76.0	78.3
	Manga	56.9	86.1	**86.6**	78.1	82.1
Panels						
eBDtheque	eBDtheque	41.6	77.4	**81.9**	**77.2**	**79.5**
	Manga	36.4	68.2	70.8	68.7	69.7
Manga	eBDtheque	26.3	61.5	63.7	59.3	61.4
	Manga	**61.2**	**86.2**	**83.0**	**84.7**	**83.8**
eBDtheque + Manga	eBDtheque	**42.4**	**79.2**	78.4	75.9	77.1
	Manga	61.0	**86.2**	82.7	84.4	83.5

Finally, if we analyze the case of combining domains for training (eBDtheque + Manga109), quite similar results are also obtained at the page and panel levels, being slightly better for eBDtheque at the page level and for Manga109 at the panel level. Remarkably, the result obtained equals or even improves in some cases the individual training (see P for Manga109 at the page level or $AP^{0.5}$ at the panel level, and, in the case of eBDtheque, the AP and $AP^{0.5}$ metrics at the panel level). If we compare these results with the case of the inter-domain evaluation when training with a single domain, a remarkable improvement is observed (an average improvement of 22.7% of AP and 18.2% of F_1). Therefore, the proposed strategy for the combination of domains is much more appropriate when it is intended to use a model in several domains.

5.1 Discussion

As seen in the experimentation carried out, the training at the page level and the panel level have reported quite similar results, not the case for the resources and time used for training. Specifically, training the model with eBDtheque + Manga109 using panels required 51 h and 23 min, whereas training at the page level took only 13 h and 18 min. Therefore, it is more convenient to train at the page level, since having more individual images does not bring any improvement. It seems that having a larger variability of characters within a single sample outweighs the benefits of having more samples but with fewer characters.

Another relevant finding is the competitive results obtained when combining domains for training. This approach allows obtaining models that are usable in multiple domains, increasing their generalizability and range of application, which is very convenient for their practical use.

The task of detecting characters in comics represents a great challenge due to the wide variability of elements that we can find, which is only limited by the imagination. This variability not only makes labeling and model building difficult, but can lead to other unexpected problems. For example, we have detected that in some comics there are characters that are parts of the body, like the one that can be seen in Fig. 4a, in which the characters are eyes. This has caused the model to learn to detect the eyes of some characters as characters (see Fig. 4b).

   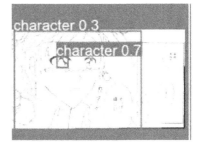

(a) Ground truth   (b) Prediction

Fig. 4. Examples of panels from eBDtheque with characters that are eyes. The model predicts the eye of a character as an actual character.

Another problem related to the aforementioned variability is that we can find many secondary characters, which are often not labeled in these datasets, either because they are not important, they do not participate in the main action, or simply because the labeling is not complete. However, the generated models are capable of detecting these missing characters (see Fig. 5). This can help improve the quality of available datasets, as well as make it easier to label new ones. Therefore, it would be of great help in the tedious and laborious task of labeling.

Fig. 5. Examples of incomplete labeling in the eBDtheque and Manga109 ground truth (first row) and how the generated model exhaustively predicts all characters even if they were not labeled (second row).

6   Conclusions

This work has presented a study on the problem of generalization in different domains for the automatic detection of characters in comics. The results of the experimentation show that when the model is trained with characters from different datasets, the model has a higher precision for the detection of characters in different domains. Moreover, training the model using panels instead of pages is not worth it at all, because the results obtained are similar, but the time it takes to train the model is significantly longer.

To keep on the development of automatic character recognition in comics, there are several ways to explore in future work. One is to try other architectures that take into account the context of the panels, for which the Transformers could be very effective. The datasets used for the training and evaluation of the models can also be expanded, in order to obtain more robust and representative results, as well as the combination of data coming from different domains, such as comics of different styles and genres, to further assess the generalization capacity of the trained models.

References

1. Bottou, L.: Large-scale machine learning with stochastic gradient descent. In: Lechevallier, Y., Saporta, G. (eds.) Proceedings of COMPSTAT'2010, pp. 177–

186. Springer-Verlag, Berlin, Heidelberg (2010). https://doi.org/10.1007/978-3-7908-2604-3_16

2. Dutta, A., Biswas, S.: CNN based extraction of panels/characters from Bengali comic book page images. In: 2019 International Conference on Document Analysis and Recognition Workshops (ICDARW), vol. 1, pp. 38–43. IEEE (2019)

3. Felzenszwalb, P.F., Girshick, R.B., McAllester, D., Ramanan, D.: Object detection with discriminatively trained part based models. IEEE Trans. Pattern Anal. Mach. Intell. **32**(9), 1627–1645 (2010)

4. Fujimoto, A., Ogawa, T., Yamamoto, K., Matsui, Y., Yamasaki, T., Aizawa, K.: Manga109 dataset and creation of metadata. In: Proceedings of the 1st International Workshop on Comics Analysis, Processing and Understanding, pp. 1–5 (2016)

5. Girshick, R.: Fast r-cnn. In: 2015 IEEE International Conference on Computer Vision (ICCV), pp. 1440–1448 (2015). https://doi.org/10.1109/ICCV.2015.169

6. Guérin, C., et al.: eBDtheque: a representative database of comics. In: 2013 12th International Conference on Document Analysis and Recognition, pp. 1145–1149. IEEE (2013)

7. He, K., Gkioxari, G., Dollár, P., Girshick, R.: Mask r-cnn. In: IEEE International Conference on Computer Vision, pp. 2980–2988 (2017)

8. Iyyer, M., et al.: The amazing mysteries of the gutter: drawing inferences between panels in comic book narratives. In: Proceedings of the IEEE Conference on Computer Vision and Pattern recognition, pp. 7186–7195 (2017)

9. Jocher, G.: YOLOv5 by Ultralytics, May 2020. https://doi.org/10.5281/zenodo.3908559, https://github.com/ultralytics/yolov5

10. Lin, T.-Y., et al.: Microsoft COCO: common objects in context. In: Fleet, D., Pajdla, T., Schiele, B., Tuytelaars, T. (eds.) ECCV 2014. LNCS, vol. 8693, pp. 740–755. Springer, Cham (2014). https://doi.org/10.1007/978-3-319-10602-1_48

11. Lladós, J.: Two decades of GREC workshop series. Conclusions of GREC2017. In: Fornés, A., Lamiroy, B. (eds.) GREC 2017. LNCS, vol. 11009, pp. 163–168. Springer, Cham (2018). https://doi.org/10.1007/978-3-030-02284-6_14

12. Nguyen, N.V., Rigaud, C., Burie, J.C.: Comic characters detection using deep learning. In: 2017 14th IAPR International Conference on Document Analysis and Recognition (ICDAR), vol. 3, pp. 41–46. IEEE (2017)

13. Nguyen, N.V., Rigaud, C., Burie, J.C.: Digital comics image indexing based on deep learning. J. Imaging **4**(7), 89 (2018)

14. Padilla, R., Passos, W.L., Dias, T.L., Netto, S.L., Da Silva, E.A.: A comparative analysis of object detection metrics with a companion open-source toolkit. Electronics **10**(3), 279 (2021)

15. Rayar, F.: Accessible comics for visually impaired people: challenges and opportunities. In: 14th IAPR International Conference on Document Analysis and Recognition (ICDAR), vol. 3, pp. 9–14. IEEE (2017)

16. Redmon, J., Divvala, S., Girshick, R., Farhadi, A.: You only look once: unified, real-time object detection. In: Proceedings of the IEEE Conference on Computer Vision and Pattern Recognition, pp. 779–788 (2016)

17. Ren, S., He, K., Girshick, R., Sun, J.: Faster R-CNN: towards real-time object detection with region proposal networks. IEEE Trans. Pattern Anal. Mach. Intell. **39**(6), 1137–1149 (2017)

18. Xiao, Y., et al.: A review of object detection based on deep learning. Multimed. Tools Appl. 23729–23791 (2020). https://doi.org/10.1007/s11042-020-08976-6

19. Zhao, Z.Q., Zheng, P., Xu, S.t., Wu, X.: Object detection with deep learning: a review. IEEE Trans. Neural Netw. Learn. Syst. **30**(11), 3212–3232 (2019)

FPNet: Deep Attention Network for Automated Floor Plan Analysis

Abhinav Upadhyay[⊠], Alpana Dubey, and Suma Mani Kuriakose

Accenture Labs, Bangalore, India
{k.a.abhinav,alpana.a.dubey,suma.mani.kuriakose}@accenture.com

Abstract. In this work, we propose a deep neural network, FPNet, for parsing and recognizing floor plan elements. We develop a multi-task deep attention network to recognize room boundaries and room types in CAD floor plans. We evaluate our network on multiple datasets. We perform quantitative analysis along three metrics - Overall accuracy, Mean accuracy, and Intersection over union (IoU) to evaluate the efficacy of our approach. We compare our approach with the existing baseline and significantly outperform on all these metrics.

Keywords: Floor plan · segmentation · CAD plan

1 Introduction

Floor plan image analysis or understanding has long been a research topic in automatic document analysis, a branch of computer vision. Floor plans are scalable drawings that show the structure of buildings or apartments as seen from a top view. Their purpose is to convey the structural information and related semantics to viewers. Key elements in a floor plan are rooms, walls, doors, windows, and fixed furniture. However, they can also provide technical information such as building materials, electrical wiring, and plumbing lines.

Floor plan images created by professional designers and architects are initially drawn using CAD softwares such as AutoCAD, Sketchup, and HomeStyler. The output from these CAD softwares is rendered into images in vector graphics format. These vector-graphic images of architectural floor plans are rasterized before printing or publication to digital media for marketing purposes. These rasterized images are utilized for assisting users to decorate rooms, design furniture layouts, and remodel indoor spaces. However, this rasterization process leads to a loss of geometric and semantic information, making it unfit for applications such as 3D real estate virtual tours and floor plan-based 3D model creation. Recovering the lost information from a rasterized floor plan image is not trivial.

Traditional approaches rely on processing low-level features and heuristics, which is error-prone. Traditional methods do not generalize well, are difficult to fully automate, and require a large amount of manual effort. The recognition of floor plans needs to consider various information such as room structure, type,

M. Coustaty and A. Fornés (Eds.): ICDAR 2023 Workshops, LNCS 14193, pp. 163–176, 2023.
https://doi.org/10.1007/978-3-031-41498-5_12

symbols, text, and scale. Due to the advancement in deep vision techniques, there is a shift from feature engineering to methods relying on learning from data [11,21].

We propose a deep neural network, FPNet, for parsing and recognizing floor plan elements. We develop a multi-task deep attention network to recognize room-boundary and room-type elements in CAD floor plans. We evaluate our network on multiple floor plan datasets. We perform quantitative analysis along three metrics - Overall accuracy, Mean accuracy, and Intersection over union (IoU), to evaluate the efficacy of our approach. We compare our approach with the existing baseline and significantly outperform on all these metrics.

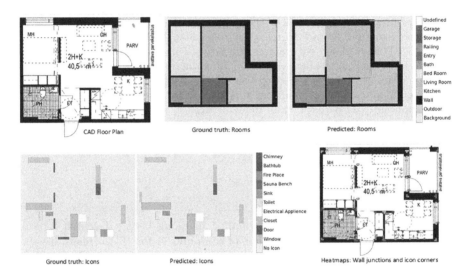

Fig. 1. Outputs predicted by our network for rooms and icons along with their ground truths.

2   Related Work

Floor plan analysis and recognition is an active area of research. Traditional approaches recognize elements in floor plans based on low-level image processing algorithms and heuristics. Ryall et al. [16] propose a semi-automatic method for room segmentation that makes use of a proximity metric for delineating partially or fully bounded regions of a scanned bitmap that depicts a building floor plan. Some approaches [1,5] locate rooms, walls, and doors by detecting graphical shapes in the layout, e.g., lines, arcs, and small loops. In recent years, there has been a shift towards applying deep learning for automatic floor plan parsing. Dodge et al. [4] propose a Fully convolutional network (FCN) for segmenting wall pixels, and Faster R-CNN framework to detect objects such as doors, sliding doors, kitchen, stoves and bathtubs. They use Google Vision API for text detection and character recognition. Liu et al. [11] propose a neural architecture based on ResNet-152 to first identify junction points in a given floor plan image, and then use integer programming to join the junctions to locate walls in the floor plan. This approach can only recognize layouts with rectangular rooms with walls of uniform thickness. Yamasaki et al. [19] apply Fully convolutional networks (FCN) to generate semantic segmentation for the apartment floor plan images. The approach simply recognizes pixels of different classes independently, thus ignoring the spatial relations among classes in the inference. Zeng et al. [20] propose a deep VGG encoder-decoder network to recognize elements in floor plan layouts. The method doesn't obtain vectorized results that users can utilize. Zhang et al. [21] propose direction-aware kernels and Generative Adversarial Networks (GAN) to recognize floor plan elements. The method fails to predict accurate semantic segmentation of basic elements such as walls, windows, rooms etc. and icons such as sinks, fireplaces, bathtubs etc. [21]. Most of these approaches require further post-processing to generate the final vector graphics using heuristics. Kalervo et al. [9] propose a large-scale floor plan data, Cubicasa5K, having 5000 samples comprising over 80 object categories. They apply a similar architecture as proposed by [11]. Lv et al. [13] propose an automatic framework for residential floor plan recognition and reconstruction that accurately recognizes the structure, type, and size of rooms. The approach only detects the room type and not icons. Unlike prior approaches, our method does not require any post-processing heuristics to restore the floor plan elements, including their geometry and semantics. Our approach provides an end-to-end trainable network to automatically parse floor plan elements including rooms, boundaries, and icons.

Floor Plan Datasets: Kalervo et al. [9] propose a large-scale floor plan data, Cubicasa5K, having 5000 samples comprising over 80 object categories. Liu et al. propose a dataset, R3D, of [12] 215 floor plans having a total of 1312 rooms, 6628 walls, 1923 doors, and 1268 windows. Liu et al. [11] provide 870 annotated floor plan images, named R2V having 11 room types and 8 icons. Lv et al. [13] provide a dataset, RFP, of 7,000 residential floor plan images with detailed vectorized and pixel annotations. We evaluate our approach on the Cubicasa5k dataset as this is one the largest dataset with diverse floor plans. Heras et al. [8] present a floor plan database, CVC-FP, consisting of 122 scanned floor plans. Sharma et al. [17] propose a dataset, ROBIN (Repository Of BuildIng plaNs), consisting of 510 floor plans having 3 categories. RFP [13] and R2V [11] datasets are not available online for us to evaluate our approach, due to license regulations.

3   Approach

We propose a multi-task deep attention-based network to recognize the room-boundary and room-type elements in CAD floor plans. Our network is based on U-Net architecture with attention module inspired from [14]. The model architecture is shown in Fig. 2. U-Nets are commonly used for image segmentation

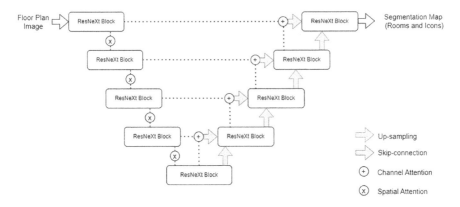

Fig. 2. The network follows a U-Net architecture consisting of encoder, decoder, and attention blocks. The ResNeXt blocks are used to capture complex features from the CAD floor plans. The left side of the U-Shape serves as an encoder and the right side as a decoder. Each block of the decoder receives the feature representation learned from the encoder and concatenates them with the output of the deconvolutional layer followed by the channel attention mechanism. The filtered feature representation after the attention mechanism is propagated through the skip connection.

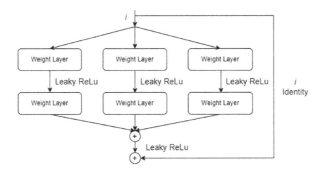

Fig. 3. A block of ResNext [18]

tasks because of their good performance and efficient use of GPU memory. Our model consists of two modules: encoder and decoder.

The encoder extracts features from the floor plan image and obtains a compact representation of these features through multiple levels. We use a ResNeXt block [18] in the encoder for extracting features from the floor plan images. The ResNeXt block allows the network to go deeper without affecting the performance. A ResNeXt repeats a building block that aggregates a set of transformations with the same topology. Compared to a ResNet, it exposes a new dimension, cardinality (the size of the set of transformations), as an essential factor in addition to the dimensions of depth and width [18]. The ResNeXt block is shown in Fig. 3. Down-sampling is performed by 2×2 max-pooling operation. During each down-sampling, the image size is reduced and the number of feature

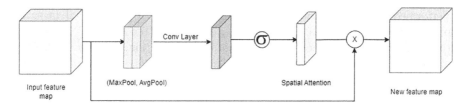

Fig. 4. Spatial Channel attention module to capture the inter-spatial relationship of features.

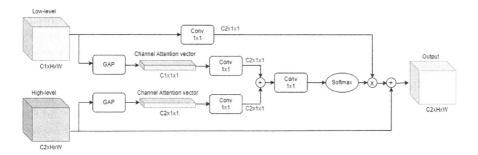

Fig. 5. Channel attention module to efficiently combine low-level feature maps with high-level feature maps.

channels is doubled. Between each ResNeXt block in the encoder, we apply spatial attention (as shown in Fig. 4) that focuses on the informative regions in the feature map. The spatial attention block utilizes the inter-spatial relationship of features. To compute the spatial attention, we first apply average-pooling and max-pooling operations along the channel axis to generate two spatial descriptors - F_s^{avg} and F_s^{max} respectively. We concatenate these two spatial descriptors to generate a single feature descriptor. Then, we apply a convolution layer on the concatenated feature descriptor followed by a sigmoid activation to generate a spatial attention map. We perform an element-wise multiplication between the input feature map and the spatial attention map to generate a new feature map focusing on spatial features. During the element-wise multiplication operation, the spatial attention values are broadcasted.

$$F' = F \otimes \sigma(f^{conv}(F_s^{avg}, F_s^{max})) \qquad (1)$$

where F is the input feature map, F_s^{avg} and F_s^{max} are the spatial descriptors generated by applying average-pooling and max-pooling operations on the input feature map, f^{conv} represents the convolution operation with the filter size of 4×4, and σ represents the activation function.

The decoder is used to up-sample the extracted feature map from the encoder to generate the segmentation image. The role of the decoder is to gradually restore the details and spatial dimensions of the image according to the image features and obtain the result of the image segmentation mask. Up-sampling is performed by bilinear interpolation. Finally, a 1×1 convolutional layer is applied to predict the class of each pixel, denoted as Conv(1×1, C), where C is the number of classes. For image semantic segmentation, C is set to 2. The decoder is structurally symmetrical with the encoder. The copy operation links the corresponding down-sampling and up-sampling feature maps.

The features obtained by the encoder contain less semantic information and are called low-level features, as they are computed early in the network. On the other hand, the features obtained by the decoder input the information obtained by the deep calculations of the network, and are called high-level features. The contextual transition structure between the encoder and the decoder plays a crucial role in the overall performance of the model. The decoder is used to upsample the extracted feature map to generate the segmentation image. The decoder recovers the position details by upsampling. However, upsampling leads to the blurring of edges and the loss of location details. Some existing works [14,15] adopt skip connections to concatenate the low-level features with the high-level features, which contributes to replenishing the position details. The role of skip connection is to directly connect the feature maps between the encoder and the decoder. However, due to the lack of semantic information in low-level features, it contains a lot of background information. This information may interfere with the segmentation of the target object. Pyramid pooling module (PPM) [6] and Atrous spatial pyramid pooling (ASPP) [2] propose to capture contextual information. PPM uses a multi-scale pooling operation to aggregate the input feature maps, and then re-fuse the features through convolution and up-sampling methods. ASPP performs multi-scale feature extraction on the input feature maps through dilated convolution operations with different dilated factors and then fuses the final multi-scale features. However, they fail to perform well in case of semantic segmentation. To effectively combine low-level feature maps with high-level feature maps, we use channel attention module to capture high-level semantic information and emphasize target features. The network is shown in Fig. 5. We perform Global average pooling to extract global context and semantic information. The global contextual information is transformed into an attentive vector that encodes the semantic dependencies, contributing to emphasizing the key features and filtering the background information. We apply 1×1 convolution with batch normalization on the vector to further capture semantic dependencies. We use softmax as the activation function to normalize the vector. The low-level feature maps are multiplied by the attentive vector to generate an attentive feature map. Finally, the attentive feature map is computed by adding it with the high-level feature map.

Training Objective: We apply multi-task loss as the training objective, which learns to predict semantic labels for pixels. The network outputs two segmentation maps. The first one is for segmenting background, rooms, and walls; and the second one is for segmenting the different icons and openings (windows and doors). The two segmentation tasks are trained using cross-entropy loss, commonly used for semantic segmentation tasks. Loss is given as

$$L_S = -\sum_{i=1}^{C} y_i.log(p_i) \tag{2}$$

where y_i is the label of the ith element in the floor plan, C is the number of floor plan elements and p_i is the prediction probability of the pixels of the $i-th$ element. L_S is a cross-entropy loss for the segmentation part and is composed of two cross-entropy terms - for room and icon segmentation tasks.

To compute the weights between different task losses, we leverage an approach from [10] that implicitly learns the relative weighting between the multi-task losses. It avoids complex, time-consuming and manual adjustment steps. The weights are implicitly learned through the homoscedastic uncertainty terms that are predicted as an extra output for each task.

Table 1. Performance of our approach along three metrics and comparison with baselines on Cubicasa5K dataset [9]

Approach	Types	Overall Accuracy		Mean Accuracy		Mean IoU	
		Validation	Test	Validation	Test	Validation	Test
UNet (ResNeXt)	Rooms	77.2	74.1	64.7	61.2	54.6	52.9
	Icons	71.4	68.1	59.5	56.4	47.1	44.8
Cubicasa5K	Rooms	84.5	82.7	72.3	69.8	61.0	57.5
	Icons	97.8	97.6	62.8	61.5	56.5	55.7
FPNet (ours)	Rooms	**91.3**	**89.4**	**80.4**	**78.1**	**73.4**	**70.5**
	Icons	**99.1**	**98.4**	**71.3**	**69.2**	**67.4**	**64.1**

4 Evaluation and Results

4.1 Dataset

We evaluate our approach using four CAD floor plan datasets - Cubicasa5K [9], Rent3D (R3D) [12], CVC-FP [8], and ROBIN [17].

Cubicasa5K: The Cubicasa5K dataset consists of 5,000 floor plans (with man-made annotations) that were collected and reviewed from a larger set of 15,000 mostly Finnish floor plan images [9]. The dataset is divided into three categories: high quality architectural, high quality, and colorful, containing 3,732, 992, and 276 floor plans respectively. We split the dataset into training, validation, and test sets consisting of 4,200, 400, and 400 floor plans respectively. The annotations are in SVG vector graphics format and contain the semantic and geometric annotations for all the floor plan elements.

R3D: The R3D dataset [12] consists of 215 floor plans having a total of 1,312 rooms, 6,628 walls, 1,923 doors, and 1,268 windows.

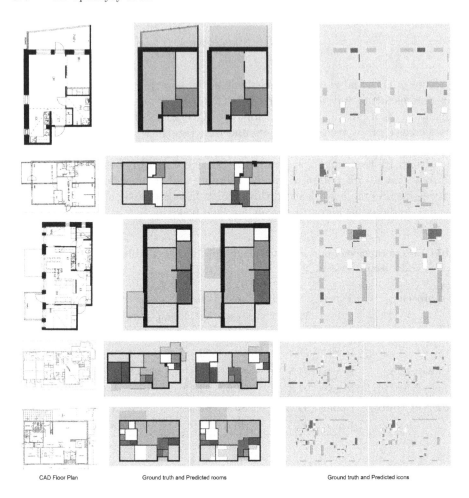

CAD Floor Plan   Ground truth and Predicted rooms   Ground truth and Predicted icons

Fig. 6. The predicted outputs from our network for rooms and icons along with their ground truths (Left image represents the ground truth and right image represents the output predicted from the network).

CVC-FP: The CVC-FP dataset consists of 122 scanned floor plan documents divided into 4 different subsets [8]. The dataset provides ground-truth information for the structural symbols: rooms, walls, doors, windows, parking doors, and room separations.

ROBIN: The ROBIN dataset contains 510 floor plans having 3 categories (3-rooms, 4-rooms, 5-rooms). The dataset only contains CAD drawings and doesn't provide ground-truth information for the structural symbols such as rooms, walls, doors, windows, etc. [17].

Table 2. Class-specific evaluation results for rooms (BG-Background, OD-Outdoor, W-Wall, KT-Kitchen, LR-Living room, BR-Bedroom, BH-Bath, HW-Hallway, RL-Railing, SG-Storage, GG-Garage)

Approaches	Metrics	Type	Rooms										
			BG	OD	W	KT	LR	BR	BH	HW	RL	SG	GG
UNet	IoU	Val.	80.5	54.1	57.1	65.4	60.9	74.3	61.2	57.2	24.3	45.2	26.2
		Test	77.6	51.3	54.2	61.2	58.3	69.8	57.8	54.2	21.8	41.5	23.3
	Acc.	Val.	85.4	62.1	77.2	74.8	72.3	80.8	71.5	66.2	25.1	49.8	31.5
		Test	81.2	60.3	75.1	71.5	69.7	76.4	69.6	63.9	22.8	45.5	27.8
Cubicasa5k	IoU	Val.	88.3	62.6	74.0	71.6	68.2	80.0	66.5	63.0	29.0	51.6	32.2
		Test	87.3	64.4	73.0	65.0	66.6	74.2	60.6	55.6	23.6	44.8	33.7
	Acc.	Val.	95.3	71.4	86.5	86.3	82.2	91.7	81.7	75.8	34.6	60.7	42.1
		Test	93.6	77.7	85.8	79.9	82.6	86.2	73.4	71.2	28.7	53.9	47.2
FPNet	IoU	Val.	91.2	75.3	81.4	83.2	77.6	87.2	78.4	76.9	67.4	69.8	62.4
		Test	90.3	73.2	80.6	81.8	76.1	85.3	77.1	74.1	64.1	65.5	61.5
	Acc.	Val.	98.2	81.8	92.2	91.8	89.4	95.4	88.7	86.2	69.9	72.4	69.8
		Test	97.5	79.3	91.5	91.2	88.9	93.8	85.4	83.4	66.5	69.2	68.1

4.2 Experimental Setup

We use Adam optimizer with an initial learning rate of 0.001. We train our model for 400 epochs with a batch size of 20. We train our model only on the Cubicasa5K dataset and perform the evaluation on all four datasets. Unlike Cubicasa5K, other datasets contain a very small number of floor plans.

4.3 Results

We evaluate our approach using three metrics [9]- (a) **Overall accuracy** indicating the proportion of correctly classified pixels, (b) **Mean accuracy** indicating the proportion of correctly classified pixels averaged over all the classes, and (c) **Mean intersection over union (IoU)** which indicates the area of the overlap between the predicted and ground truth pixels, averaged over all the classes.

Table 1 shows the performance of our approach with respect to these metrics on the Cubicasa5k dataset. We further report the class-specific IoUs and accuracies with respect to all room and icon classes in Table 2 and Table 3. In Table 4, we capture the performance of our approach on other floor plan datasets - R3D, CVC-FP, and ROBIN. Figure 6 shows the prediction from our network on the Cubicasa5K dataset for rooms and icons. Figures 7 and 8 shows the prediction on R3D and ROBIN dataset respectively.

Comparison with Baselines. We compare our approach with existing baselines (state-of-the-art) - UNet (ResNeXt as backbone) [15], TF2DeepFloorplan [20], DeepLabV3+ [3], Cubicasa5K [9], and PSPNet [22] along these three metrics. We observe a relative increase of 8% in Overall accuracy, 11.8% in Mean class accuracy, and 22.6% in Mean IoU for rooms (on the test set) for the Cubicasa5k

Table 3. Class-specific evaluation results for icons (EM-Empty, WI-Window, DR-Door, CT-Closer, EA-Electrical Appliance, TO-Toilet, SK-Sink, SB-Sauna bench, FP-Fire Place, BT-Bathtub, CH-Chimney)

Approaches	Metrics	Type	Icons										
			EM	WI	DR	CT	EA	TO	SK	SB	FP	BT	CH
UNet	IoU	Val.	88.5	59.4	49.6	59.1	60.5	58.1	46.7	64.3	31.2	11.3	10.1
		Test	85.1	56.5	47.2	57.3	56.8	56.2	43.1	61.9	29.1	10.2	8.5
	Acc.	Val.	87.1	64.4	54.2	69.1	67.5	63.8	59.6	72.2	37.5	12.7	8.2
		Test	84.8	61.6	53.1	67.3	65.1	61.1	57.4	69.5	35.3	10.7	7.6
Cubicasa5k	IoU	Val.	97.8	67.3	56.7	69.8	67.4	65.3	55.5	72.7	38.4	16.7	13.8
		Test	97.6	66.8	53.6	69.2	66.0	62.8	55.7	67.3	36.2	26.7	11.2
	Acc.	Val.	99.3	75.2	63.3	77.7	76.6	72.5	67.4	80.4	45.2	19.1	14.4
		Test	99.3	73.7	59.8	77.6	75.7	68.4	66.1	74.2	40.4	30.1	11.7
FPNet	IoU	Val.	**98.5**	**79.8**	**72.3**	**81.1**	**80.5**	**77.3**	**68.4**	**77.8**	**62.8**	**55.2**	**53.4**
		Test	**98.1**	**77.4**	**71.6**	**80.3**	**80.1**	**76.2**	**67.1**	**75.1**	**60.3**	**51.8**	**52.8**
	Acc.	Val.	**99.5**	**82.3**	**77.6**	**85.4**	**87.2**	**83.4**	**79.5**	**83.9**	**68.1**	**61.5**	**61.2**
		Test	**99.5**	**81.8**	**76.5**	**85.1**	**85.8**	**82.7**	**77.5**	**81.1**	**66.2**	**60.3**	**59.8**

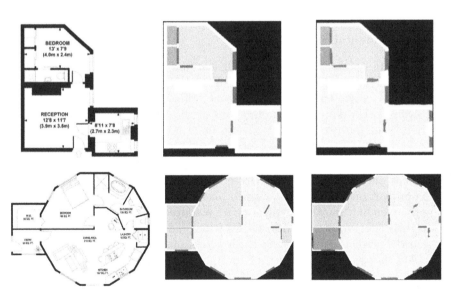

Fig. 7. The predicted outputs from our network for the R3D dataset (Left image represents the CAD floor plan as input, middle image represents the ground truth, and right image represents the output predicted from the network).

dataset. For icons, we observe a relative increase of 6.7% in Overall accuracy, 12.3% in Mean accuracy, and 30% in Mean IoU (on the test set). Our approach significantly outperforms the existing work across all the metrics when evaluated on multiple datasets (as shown in Table 1 and Table 4).

Table 4. Performance of our approach along three metrics and comparison with baselines on Rent3D (R3D) [20], CVC-FP [7], and ROBIN [17] datasets (W-Wall, DW-Door-and-window, BH-Bathroom & etc., LR-Living room & etc., H-Hall, BY-Balcony, CT-Closet)

Dataset	Approach	Acc.	Class Accuracy								IoU
			W	DW	BH	LR	BR	H	BY	CT	
R3D	TF2DeepFloorplan	0.9	0.98	0.83	0.78	0.93	0.79	0.68	0.49	0.54	0.66
	DeepLabV3+	0.83	0.93	0.60	0.57	0.90	0.40	0.44	0.0027	0.048	0.44
	PSPNet	0.81	0.91	0.54	0.50	0.89	0.40	0.23	0.11	0.086	0.41
	FPNet (ours)	**0.94**	**0.99**	**0.87**	**0.83**	**0.96**	**0.83**	**0.74**	**0.61**	**0.62**	**0.73**
CVC-FP	TF2DeepFloorplan	0.84	0.95	0.79	0.73	0.91	0.65	0.67	0.52	NA	0.62
	DeepLabV3+	0.81	0.89	0.61	0.54	0.87	0.47	0.49	0.35	NA	0.46
	PSPNet	0.74	0.86	0.57	0.42	0.79	0.41	0.39	0.33	NA	0.40
	FPNet (ours)	**0.91**	**0.97**	**0.9**	**0.84**	**0.95**	**0.74**	**0.79**	**0.67**	**NA**	**0.71**
ROBIN	TF2DeepFloorplan	0.92	0.93	0.86	0.82	0.93	0.85	NA	NA	NA	0.83
	DeepLabV3+	0.88	0.9	0.83	0.8	0.89	0.82	NA	NA	NA	0.81
	PSPNet	0.83	0.86	0.79	0.76	0.8	0.77	NA	NA	NA	0.78
	FPNet (ours)	**0.97**	**0.99**	**0.92**	**0.89**	**0.98**	**0.9**	**NA**	**NA**	**NA**	**0.87**

Table 5. Ablation study to analyze the performance of two modules - Channel Attention and Spatial Attention

Methods	Overall Accuracy	Mean Accuracy	Mean IoU
Setup1 (UNet with ResNeXt)	74.1	61.2	52.9
Setup2 (Setup1 + Spatial Attention)	84.2	72.6	62.7
Setup3 (Setup1 + Channel Attention)	86.3	75.2	66.3
Setup4 (FPNet)	**89.4**	**78.1**	**70.5**

Ablation Study. We conduct an ablation study to analyze the impact of different components within our network (FPNet). We consider three setups - (a) **Setup1**: We use ResNeXt block in the encoder and decoder unlike plain convolution in UNet [15], (b) **Setup2**: We follow a similar setup as the first one (Setup1) except we add Spatial Attention module, (c) **Setup3**: We follow a similar setup as the first one (Setup1) except we add Channel Attention module, and (d) **Setup4 (FPNet)**: We combine all the three setups (ResNeXt + Channel Attention + Spatial Attention).

Table 5 shows the ablation study results, where we evaluate different setups using the Cubicasa5K dataset [9]. We observe that our proposed network, FPNet, outperforms the other setups, indicating that the Spatial and Channel attention modules help in improving the floor plan recognition performance.

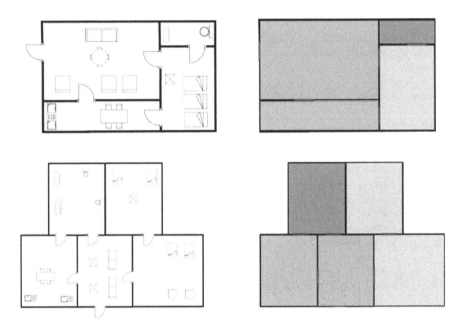

Fig. 8. The predicted outputs from our network for the ROBIN dataset (Left image represents the CAD floor plan as input and right image represents the output predicted from the network).

5   Conclusion

We propose a deep neural network, FPNet, for parsing and recognizing floor plan elements. We develop a multi-task deep attention network to recognize room-boundary and room-type elements in CAD floor plans. We evaluate our network on a large-scale floor plan dataset, Cubicasa5K, containing 5000 samples annotated into over 80 floor plan object categories. We compare our approach with the existing baselines and significantly outperform on all these metrics. Our model overcomes the limitation of previous models that need post-processing steps to generate the final vector graphics. As future work, we plan to extend the work to reconstruct 3D models of floor plans.

References

1. Ah-Soon, C., Tombre, K.: Variations on the analysis of architectural drawings. In: Proceedings of the Fourth International Conference on Document Analysis and Recognition, vol. 1, pp. 347–351. IEEE (1997)
2. Chen, L.C., Papandreou, G., Schroff, F., Adam, H.: Rethinking atrous convolution for semantic image segmentation. arXiv preprint arXiv:1706.05587 (2017)
3. Chen, L.C., Zhu, Y., Papandreou, G., Schroff, F., Adam, H.: Encoder-decoder with atrous separable convolution for semantic image segmentation. In: Proceedings of the European Conference on Computer Vision (ECCV), pp. 801–818 (2018)

4. Dodge, S., Xu, J., Stenger, B.: Parsing floor plan images. In: 2017 Fifteenth IAPR International Conference on Machine Vision Applications (MVA), pp. 358–361. IEEE (2017)

5. Dosch, P., Tombre, K., Ah-Soon, C., Masini, G.: A complete system for the analysis of architectural drawings. Int. J. Doc. Anal. Recognit. **3**(2), 102–116 (2000)

6. He, K., Zhang, X., Ren, S., Sun, J.: Spatial pyramid pooling in deep convolutional networks for visual recognition. IEEE Trans. Pattern Anal. Mach. Intell. **37**(9), 1904–1916 (2015)

7. de las Heras, L.P., Terrades, O.R., Robles, S., Sánchez, G.: CVC-FP and SGT: a new database for structural floor plan analysis and its groundtruthing tool. Int. J. Doc. Anal. Recognit. (IJDAR) **18**(1), 15–30 (2015)

8. de las Heras, L.P., Terrades, O., Robles, S., S'anchez, G.: CVC-FP and SGT: a new database for structural floor plan analysis and its groundtruthing tool. Int. J. Doc. Anal. Recognit. (2015)

9. Kalervo, A., Ylioinas, J., Häikiö, M., Karhu, A., Kannala, J.: CubiCasa5K: a dataset and an improved multi-task model for floorplan image analysis. In: Felsberg, M., Forssén, P.-E., Sintorn, I.-M., Unger, J. (eds.) SCIA 2019. LNCS, vol. 11482, pp. 28–40. Springer, Cham (2019). https://doi.org/10.1007/978-3-030-20205-7_3

10. Kendall, A., Gal, Y., Cipolla, R.: Multi-task learning using uncertainty to weigh losses for scene geometry and semantics. In: Proceedings of the IEEE Conference on Computer Vision and Pattern Recognition, pp. 7482–7491 (2018)

11. Liu, C., Wu, J., Kohli, P., Furukawa, Y.: Raster-to-vector: revisiting floorplan transformation. In: Proceedings of the IEEE International Conference on Computer Vision, pp. 2195–2203 (2017)

12. Liu, C., Schwing, A.G., Kundu, K., Urtasun, R., Fidler, S.: Rent3D: floor-plan priors for monocular layout estimation. In: Proceedings of the IEEE Conference on Computer Vision and Pattern Recognition, pp. 3413–3421 (2015)

13. Lv, X., Zhao, S., Yu, X., Zhao, B.: Residential floor plan recognition and reconstruction. In: Proceedings of the IEEE/CVF Conference on Computer Vision and Pattern Recognition, pp. 16717–16726 (2021)

14. Oktay, O., et al.: Attention u-net: learning where to look for the pancreas. arXiv preprint arXiv:1804.03999 (2018)

15. Ronneberger, O., Fischer, P., Brox, T.: U-Net: convolutional networks for biomedical image segmentation. In: Navab, N., Hornegger, J., Wells, W.M., Frangi, A.F. (eds.) MICCAI 2015. LNCS, vol. 9351, pp. 234–241. Springer, Cham (2015). https://doi.org/10.1007/978-3-319-24574-4_28

16. Ryall, K., Shieber, S., Marks, J., Mazer, M.: Semi-automatic delineation of regions in floor plans. In: Proceedings of 3rd International Conference on Document Analysis and Recognition, vol. 2, pp. 964–969. IEEE (1995)

17. Sharma, D., Gupta, N., Chattopadhyay, C., Mehta, S.: Daniel: a deep architecture for automatic analysis and retrieval of building floor plans. In: 2017 14th IAPR International Conference on Document Analysis and Recognition (ICDAR), vol. 1, pp. 420–425. IEEE (2017)

18. Xie, S., Girshick, R., Dollár, P., Tu, Z., He, K.: Aggregated residual transformations for deep neural networks. In: Proceedings of the IEEE Conference on Computer Vision and Pattern Recognition, pp. 1492–1500 (2017)

19. Yamasaki, T., Zhang, J., Takada, Y.: Apartment structure estimation using fully convolutional networks and graph model. In: Proceedings of the 2018 ACM Workshop on Multimedia for Real Estate Tech, pp. 1–6 (2018)

20. Zeng, Z., Li, X., Yu, Y.K., Fu, C.W.: Deep floor plan recognition using a multi-task network with room-boundary-guided attention. In: Proceedings of the IEEE/CVF International Conference on Computer Vision, pp. 9096–9104 (2019)
21. Zhang, Y., He, Y., Zhu, S., Di, X.: The direction-aware, learnable, additive kernels and the adversarial network for deep floor plan recognition. arXiv preprint arXiv:2001.11194 (2020)
22. Zhao, H., Shi, J., Qi, X., Wang, X., Jia, J.: Pyramid scene parsing network. In: Proceedings of the IEEE Conference on Computer Vision and Pattern Recognition, pp. 2881–2890 (2017)

Detection of Buried Complex Text. Case of Onomatopoeia in Comics Books

John Benson Louis(✉) and Jean-Christophe Burie

L3i Laboratory, SAIL Joint Laboratory La Rochelle Université,
17042 La Rochelle CEDEX 1, France
{john.louis,jean-christophe.burie}@univ-lr.fr

Abstract. Recent advances in scene text detection, boosted by deep neural networks, have revolutionized our ability to identify text in complex visual environments, including historical documents, newspapers, administrative records, and even text in the wild. However, traditional word-level bounding boxes struggle to capture irregular text shapes. Onomatopoeias in comic books, which epitomize those complex textual elements, are often mixed and buried with graphical components, that make fail conventional detection methods.

In this paper, we propose an innovative approach for text segmentation based on the Unet architecture and wisely integrating pre and post-processing. The method is specifically designed to accurately detect intricate text shapes and configurations. This cutting-edge technic offers exceptional accuracy, paving the way for innovative text detection applications in various environments.

The evaluations were carried out on the KABOOM-ONOMATO POEIA dataset and show the relevance of our method in comparison with methods of the literature, which makes it a promising tool in the field of scene text detection.

Keywords: Complex text detection · Character extraction · text and graphic analysis · Deep Learning · Onomatopoeia · Comics album

1 Introduction

Scene text detection has become a focal point of research in the computer vision domain due to its myriad applications in various fields. These applications include instant translation, image retrieval, and scene parsing, as well as the analysis of a wide range of documents, including newspapers, historical texts, and more recently, comic books. Comics are fine works of art with a diversity of text styles spread over images and speech balloons in various positions. The text of the dialogues is situated in the speech balloons [17] and is most of time quite easy to detect, then to extract, since the characters are horizontal and most of the time uniform in size. At the opposite, onomatopoeias are complex texts, with variable shapes and orientations, sometimes designed with an artistic

M. Coustaty and A. Fornés (Eds.): ICDAR 2023 Workshops, LNCS 14193, pp. 177–191, 2023.
https://doi.org/10.1007/978-3-031-41498-5_13

style. Buried in the graphical elements of the panels, they are more difficult to detect. The variability of their representation makes it difficult to design detection methods for this type of text. The growing interest in the field of comic book analysis highlights the need for robust algorithms and methods that can effectively address the challenges posed by this type of document. In recent years, the analysis of text, specifically within Franco-Belgian comics, manga or American comics, has become a subject of interest for researchers. This work aims to develop an efficient approach for detecting complex texts in comics, by tackling the challenges produced by the specific characteristics of the onomatopoeias.

The main contributions of this work include:

– The creation of a new annotated dataset focused on onomatopoeia at character-level. To the best of our knowledge, this is the first dataset on onomatopoeia in the context of Franco-Belgian comics, filling a significant gap in the available resources.
– A simple yet effective model based on the U-net architecture [18] that outperforms existing state-of-the-art methods for complex text detection in comics. This model and the proposed dataset will serve as solid baseline for future research in the field.
– A set of experiments for evaluating the relevance of the proposed approach compared to the methods in the literature (Fig. 1).

Fig. 1. Example of an onomatopoeia in a panel. The text overlaps the main characters and the graphic elements of the background.

The next section presents related works on text detection and comics analysis. Section 3 details the proposed approach. The KABOOM-ONOMATOPOEIA dataset is introduced in Sect. 4. The experiments and evaluation are given in Sect. 5. Finally, Sect. 6 presents the conclusions and future works.

2   Related Work

Text detection is a critical component of many applications in the field of document analysis or scene understanding. We present below some methods of the literature divided in three categories. The first one concerns text detection in comics books. The second one details methods based on region analysis. Finally, the third one presents related work based on text segmentation.

2.1   Text Detection in Comic Pages

Interesting work has been done around the text in the field of comics analysis. Some research has focused on the detection of speech bubbles [4,12,16]. These methods consider that the text is essentially present in the speech bubbles. This is true for the dialogues. However, it is important to highlight the complexity and nuances used by the authors to represent the text. Text is not always confined to speech bubbles, as evidenced by onomatopoeia.

2.2   Region Based Approaches

Various text detection approaches using box regression, adapted from well-known object detectors(Fast R-CNN [11], YOLO [15] etc.), have been proposed ranging from traditional sliding window approaches to more advanced deep learning-based methods. These methods involve localizing and classifying text regions within an image, then generating bounding boxes around the detected text instances. Texts, unlike general objects, often display irregular shapes and diverse aspect ratios. In TextBoxes [8], the authors modified convolutional kernels and anchor boxes to effectively capture a wide range of text shapes. East [22] and FOTS [10] were designed for text detection, at word-level, in unconstrained rotation environments.However, these methods present some difficulty in detecting long or distorted texts.

The Rotation-Sensitive Regression Detector (RSDD) [9] exploits rotation-invariant features by actively rotating convolutional filters. In the context of comic text detection, the works of Ogawa et al. [14] and Yanagisawa et al. [21] focused on detecting text region among multiple objects. Wei-Ta Chu and Chih-Chi Yu [3] also contributed to the field of text region using region proposal, classification, and regression appraoches.

Recently, Baek et al. proposed the CRAFT method [1] to detect unconstrained text in scene images. By examining individual characters and their affinities, they generate non-rigid word-level bounding boxes. However, methods that rely on rigid word-level bounding boxes may face limitations in accurately representing text regions with irregular shapes.

2.3   Segmentation Base Approaches

Some methods tackle text detection task as a pixel-wise classification problem. These approaches aim to label each pixel in the image as either text or non-text, resulting in a binary segmentation map. Methods such as Fully Convolutional Networks (FCN) [6] and U-Net [18] have been adapted for text segmentation tasks, showing their effectiveness in detecting text regions in challenging environments. More recent research has explored the combination of segmentation and region based approaches, like Mask RCNN, exploiting the strengths of both methodologies to improve overall text detection performance. The scientific community dedicated to document analysis has an extensive history of research in text segmentation [5] and image binarization [20, 23]. These studies have been conducted in various contexts, including historical manuscripts [13], cartographic materials [2], and handwritten text [19].

2.4   Conclusions

Most of the techniques mentioned earlier, which primarily focus on word-level text detection, exhibit certain limitations. These methods often experience difficulties with the variability in text sizes, orientations, and styles frequently encountered in comics. Consequently, they may require additional pre-processing or multi-scale processing strategies for efficient handling of diverse instances of text. Recognizing these weaknesses, in this paper, we introduce a character-level detector specifically designed to address the challenge of onomatopoeia detection, effectively overcoming the shortcomings of existing approaches.

3   Proposed Method

Segmentation techniques are particularly useful when the goal is to identify and extract specific objects, regions, or features within a complex image. Nowadays deep learning-based approaches are rather used for carrying out these kind of tasks because of their performance on various types of images. In this work, we propose an approach based on the Unet architecture. U-Net [18] has been shown to perform well for various segmentation tasks, especially biomedical image segmentation, due to its efficient and effective architecture that allow end-to-end training with relatively few training parameters. The architecture includes skip connections that allow the network to capture both high-level semantic information and low-level spatial information. Combined with an appropriate feature extractor, this architecture provides the ability to detect buried elements in comics since their detection requires a combination of both contextual and detailed information (font style, size, background features, etc.). Referring to onomatopoeias, those information can possibly be (*Text features, Font style and size, Text placement and orientation, Visual context etc..*)

3.1   Architecture

The proposed architecture is based on UNET [18] architecture and consists of an encoder-decoder structure used for semantic segmentation tasks, the encoder being built on the ResNet34 backbone.

The model takes an input image of dimension 512×512 pixels, passing through two original convolution layers, composed of double convolution blocks. These layers generate a feature map x_orig with 64 channels.

The model uses a pre-trained ResNet-34 as the backbone for the encoder path shown in Fig. 2. The ResNet-34 layers are split into different sections (down0, down1, down2, down3, down4). These layers successively downsample the input image, producing feature maps with an increasing number of channels and decreasing spatial dimensions.

The decoder path is composed of five up-sampling blocks (5 up convolution layer) as shown in Table 1. Each up-sampling block receives the output from the previous layer in the decoder path and the corresponding layer from the encoder path. The feature maps from the encoder and decoder are concatenated along the channel dimension and then passed through another double convolution block. This process is repeated in each up-sampling block, gradually increasing the spatial dimensions of the feature maps.

Finally, the output of the last upsampling block is passed through a 1×1 convolution layer (outc), which reduces the number of channels to the desired number of output classes for the segmentation.

Table 1. Proposed Unet Architecture

Layer	Type	Input Size	Output Size
x_orig0	inconv	(3, H, W)	(32, H, W)
x_orig1	double_conv	(32, H, W)	(64, H, W)
x1	ResNet down0	(3, H, W)	(64, H/2, W/2)
x2	ResNet down1	(64, H/2, W/2)	(64, H/4, W/4)
x3	ResNet down2	(64, H/4, W/4)	(128, H/8, W/8)
x4	ResNet down3	(128, H/8, W/8)	(256, H/16, W/16)
x5	ResNet down4	(256, H/16, W/16)	(512, H/32, W/32)
up1	up	(512, H/32, W/32)	(256, H/16, W/16)
up2	up	(256, H/16, W/16)	(128, H/8, W/8)
up3	up	(128, H/8, W/8)	(64, H/4, W/4)
up4	up	(64, H/4, W/4)	(64, H/2, W/2)
up5	up	(64, H/2, W/2)	(64, H, W)
outc	outconv	(64, H, W)	(n_classes, H, W)

Why Resnet34 as Backbone for the Proposed Model ?

ResNet-34 was chosen as the backbone for the Proposed Unet architecture due to a combination of factors, including its balance between model complexity and performance, as well as its widespread adoption in the computer vision community. Here are a few reasons for using ResNet-34 :

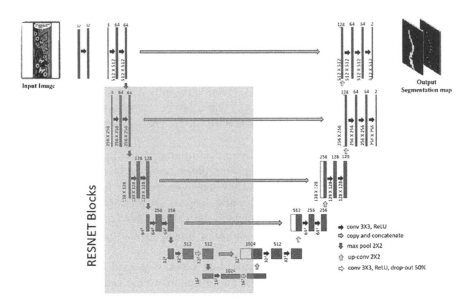

Fig. 2. Architecture of Proposed Unet

Balance Between Accuracy and Complexity: ResNet-34 strikes a good balance between model accuracy and complexity. It has 34 layers, allowing it to learn a rich set of features from the input data without being as computationally expensive as deeper ResNet variants, such as ResNet-50, ResNet-101, or ResNet-152. This balance is especially important for real-time applications or when training resources are limited.

Pretrained Models Availability: Pretrained ResNet-34 models are readily available, saving a significant amount of training time and resources. By using a pre-trained ResNet-34 as the backbone, the Proposed Unet Architecture architecture can leverage the features learned from large-scale datasets, such as ImageNet, and fine-tune them for the specific segmentation task. This transfer learning approach often results in faster convergence and better performance.

Compatibility with U-Net: ResNet-34's architecture is compatible with the U-Net design, as it provides a natural hierarchy of feature maps at different scales. This enables the seamless integration of skip connections between the encoder and decoder parts of the Proposed Unet architecture model, which helps retain spatial information and improves segmentation accuracy.

Note that other backbone networks could be used instead of ResNet-34, depending on the specific requirements and constraints of the problem. However, ResNet-34 serves as a solid starting point, offering in our case a good trade-off between model complexity for segmentation, performance, and ease of implementation.

3.2   Preprocessing Operations

Image preprocessing is a crucial step in deep learning pipelines, because it ensures that the input data is standardized, clean, and ready for efficient training. This section provides a detailed description of each function of the preprocessing pipeline and outlines their specific roles.

The objective is therefore to detect onomatopoeia in comics image, one of the specificities of this type of text is that it can be located on any position on the comic page, even askew on several panels. All input images are transformed into squares, so that the model can handle images at multiple scales. The following describes the different preprocessing tasks applied before the training process.

Square Image: To keep the shape of the elements intact in the image while having the input format required for our model, a function to transform the image into a square is used. The function is designed to resize and pad an input image to fit it into a square of the desired size while maintaining its original aspect ratio. This feature is essential when working with deep learning models that require input data to be in a consistent square format. The key mathematical operations in the $\mathtt{Square}$ function are scaling and padding. To maintain the original aspect ratio while resizing, the function calculates the new width (new_w) and height (new_h) using the following formulas:

$if\ w > h$

$$new_h = h \cdot \left(\frac{IMG\_SIZE}{w}\right)$$
$$new_w = IMG\_SIZE$$

$otherwise$

$$new_w = w \cdot \left(\frac{IMG\_SIZE}{h}\right)$$
$$new_w : IMG\_SIZE$$

where w and h are respectively the width and the height of the original image, IMG_SIZE corresponds to the size of the square image (here 512 px). The padding operation involves calculating the margin values for width and height by completing them with zero pixel values:

$$margin\_w : \left\lfloor \frac{IMG\_SIZE - new\_w}{2} \right\rfloor, margin\_h : \left\lfloor \frac{IMG\_SIZE - new\_h}{2} \right\rfloor$$

These margins are then used to center the resized image within the new square image array whose default pixel value is zero.

Coordinate System Modification: Our deep learning model requires input images to be of size IMG_SIZE (512 × 512 pixels). As the original image is resized, the coordinates to create the mask for the resized image must fit the new square image for the training process. This ensures that the model can accurately detect objects within the square input images without distortion or misalignment.

Mathematically, this process scales and translates the input coordinates based on the aspect ratio of the original image dimensions to ensure that the new coordinates are accurately mapped to the square image space. The new coordinates (x_s, y_s) are computed as follow:

if $w > h$

$$x_s = x + (new_w - w)/2$$
$$y_s = y \cdot (IMG\_SIZE/h)$$

otherwise

$$x_s = x \cdot (IMG\_SIZE/w)$$
$$y_s = y + (new_h - h)/2$$

where w and h are the width and the height of the original image, IMG_SIZE is the target size of the square image space, x and y are the original coordinates, new_w and new_h are the new width and height of the image after scaling,

Mask Creation: This operation generates a binary mask for an input image based on the target object labels, useful for tasks like image segmentation or object detection. The mask generate to train the proposed model returns a two-channel array "mask" with the same dimensions as the input square image. The first channel indicates the presence of an onomatopoeia character, while the second highlights the object centers (Centroid of each characters), as shown in Fig. 3.

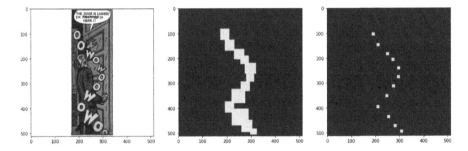

Fig. 3. Example of masks created from an onomatopoeia. (left) original square image, (middle) mask of each object: the bounding boxes of each character (right) mask of the center of each character.

Normalisation: A normalization is performed on the square image by adjusting its pixel values based on the local statistical properties of the image. This normalization is essential when working with deep learning models because it ensures that the input data has a consistent scale, which can improve the model training and generalization performance.

Why Normalized?
Normalization is a important preprocessing step when working with deep learning models. The primary reasons for normalizing input images are:

- Improvement of training stability: Normalizing the input data ensures that the model receives input values within a consistent range, reducing the risk of large input variations causing instability during training.

- Faster convergence: When the input data is normalized, the gradient updates during training tend to be more uniform, allowing the model to converge faster and more efficiently.
- Better generalization: Normalized input data helps the model to learn features that are invariant to the scale of the input, which can improve the model's ability to generalize to unseen data.

Loss Function: Selecting an appropriate loss function is a vital aspect when designing a machine learning model. We tested several loss function to train the proposed model on our dataset. We conducted the training of the proposed architecture with multiple loss for test before finally choosing the best one for the finale process. The following loss function are widely employed for segmentation tasks: **BCE(Binary cross entropy)**, **Dice coefficient(Or Sørensen-Dice index)**, **Logarithmic Dice Loss**, **focal loss**. We calculated our global loss function by combining BCE(Binary Cross Entropy) and a Dice loss where, the BCE loss measures the pixel-wise differences between the predicted and target masks, while the dice loss measures the overlap between the masks. By combining these two losses given by the formula:

$$loss = bce * bce_{weight} + dice * (1 - bce_{weight}) \qquad (1)$$

with a weighting factor (bce_{weight}), we can balance the importance of pixel-level differences and mask-level overlap in the overall loss. We conducted the training over 200 epochs, employing a maximum learning rate (lr) of 0.001. After experienced the combination of BCE with Dice Loss and Logarithmic Dice Loss given by -log(DiceLoss), we notice a better gradient descent using the Logarithmic loss, as shown in Fig. 4. Thus, this loss function has been selected for our experiments.

3.3   Post-processing Operations

Since onomatopoeia are objects often buried in images, our angle of approach consists in predicting mask which highlight the regions of interest on the processed images. However, the process does not stop at getting these output masks. A crucial post processing step is required to extract bounding boxes around the detected objects. Once the output mask of two dimensions (1: the region of the objects, 2: their center, is predicted, they are used to create a list of bounding boxes for each instance of object. In the initial step of extraction of the bounding boxes, some objects may have been missed. To improve the final result, a filter identifies these missed objects by analyzing the remaining uncovered areas in the predicted mask. This task checks regions with an area greater than 50% or a width greater than 1.5 times, respectively, the average area or the average width of the detected objects in the first step and add them to the list of the extracted bounding boxes.

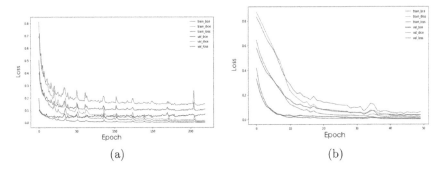

(a)   (b)

Fig. 4. Gradient descent of the loss function (a) Loss using Diceloss (b) Loss graph using -log(Diceloss)

Fig. 5. Images of the KABOOM-ONOMATOPOEIA dataset

4   Dataset Construction and Analysis

The images of the dataset come from the COMICBOOKFX website[1]. This website brings together a collection of approximately 1830 images. Each image is a panel extracted from a comics album including one or several onomatopoeia.

Qualitative Analysis: The onomatopoeia dataset includes panels extracted from comic book pages. In this dataset, onomatopoeic expressions display a diverse range of shape, size and orientation, even for instances of the same onomatopoeia, as shown at Fig. 5.

Annotations: Annotations have been carried out with VGG Image Annotator tool (VIA)[2]. The annotations were made at character level by defining a bounding box for each character of the onomatopoeia. A JSON-based data format is used to store the annotations.

---

[1] https://www.comicbookfx.com.
[2] https://www.robots.ox.ac.uk/~vgg/software/via/via.html.

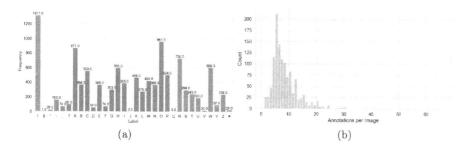

(a)   (b)

Fig. 6. Statistics of the dataset (a) number of occurrences for each class (character) (b) distribution of the number of annotations per panel.

Quantitative Analysis: Onomatopoeias are used to describe sounds, noise etc. However, onomatopoeias can vary considerably in terms of style, rotation, color and elastic effect on borders etc. depending on the corresponding events as can be seen in Fig. 5. Onomatopoeias consist of alphabetic characters, punctuation marks, and symbols. The annotation of the dataset has provides a set of bounding boxes around each element . Figure 6(a) gives an overview of the numbers of occurrences of each element in the dataset. Figure 6(b) presents the distribution of the number of annotations per panel. We have been meticulously annotated a total of 1,260 images, providing a solid foundation for conducting our experiments. The dataset includes a total of 10,732 annotated elements (*character of onomatopoeia*) divided into 32 distinct classes, providing a diverse range of examples from which the model can learn. The dataset is publicly available[3].

5   Experiments

In this section we present the experiments carried out to evaluate the performance of the proposed approach.

5.1   Evaluation Metrics

In order to evaluate the effectiveness of the proposed approach for onomatopoeia character detection, we use the F1-score. This score is calculated based on the precision and recall values obtained by comparing the detected bounding boxes to the ground truth data. Specifically, we train our models on onomatopoeia characters from the KABOOM-ONOMATOPOEIA dataset.

Using these metrics, we can calculate the precision and recall as follows:

$$Precision = \frac{TP}{TP + FP}$$

$$Recall = \frac{TP}{TP + FN}$$

---

[3] https://gitlab.univ-lr.fr/jcburie/kaboom_onomatopoeia_dataset/-/tree/DATASET.

The F1-score is then obtained from the harmonic mean of the precision and recall values using the following formula:

$$F1_{score} = \frac{2}{\frac{1}{Precision} + \frac{1}{Recall}}$$

By using the F1-score as an evaluation metric, we can assess the performance of the models in terms of their ability to detect onomatopoeia characters accurately and consistently.

5.2   Evaluation Protocol

The objective is to evaluate the performance of the proposed model compared to other models and approaches of the literature. The experiments are carried out in *2 phases*. In the **first phase**, The KABOOM-ONOMATOPOEIA dataset is divided in two parts: 80% which equals 1010 images, are used for the training and 20% a total of 250 images are kept as testing set. It consists in training the Craft [1] model, the mask RCNN and the proposed model with The training set and study their performances.Craft [1] and Mask R-CNN [7] have been chosen because they are known model which have proven themselves in object and character detection.

For the **second phase**, we proceed to a transformation of the training data provided to the models to have a greater diversity of data and to study the evolution of their performances.

First Experiment
For this step, every model have been trained using 80% of the KABOOM-ONOMATOPOEIA dataset. where 80% of the training, which is around 800 images for train and the other 200 images for validation .The Table 2 presents the results obtained for each model.

Table 2. Performance comparison of models in the first experiment

Model	Precision	Recall	F1-score
MASK-RCNN	0.512	0.408	0.4679
CRAFT	0.723	0.677	0.698
Proposed UNET	**0.92**	**0.902**	**0.91**

Second Experiment
It is imperative to ensure that the validation error decreases with the learning error. To do this, it is important that the training data can handle a set of samples as diverse as possible. Data augmentation is precisely an operation that proves to be suitable for ensuring the robustness of the models. We have generated 10000 new images by varying different parameters such as scale, brightness, contrast, saturation, hue, rotation, bringing the training set to a total of 10800 images.

The Table 3 presents the results obtained for the second experiment.

Table 3. Performance comparison of models in the second experiment

Model	Precision	Recall	F1-score
MASK-RCNN	0.572	0.463	0.52
CRAFT	0.77	0.712	0.745
Proposed UNET	**0.971**	**0.896**	**0.931**

Fig. 7. Qualitative results obtained by the different evaluated methods

5.3   Results Analysis

The first experiment shows in Table 2 that the proposed approach outperforms the results of the MASK-RCNN and CRAFT methods. Our method reaches respectively a precision, a recall and a F1-Score of 92%, 90.2% and 91%. In the second experiment where data augmentation is used, the Table 3 shows that the performance of each model is improved. However with our approach, the precision is improved of 5,5% and obtains a rate à 97,1%, while the F1 score reaches 93.1%. If the recall of our method always outperforms the other methods, the rate is a little bit lower (0,6%) than in the first experiments. We can conclude than the data augmentation improves the accuracy of the results.

To further analyze the performance of the models, a qualitative analysis of the detection results is provided in Fig. 7 for two images. In the first row, CRAFT detected only 3 characters out of 7, while MASK-RCNN performed better with 5 detections but some of them are incomplete. In contrast, the proposed UNET model demonstrated more accurate detection by detecting all characters. In the second raw, CRAFT almost detected the whole characters but two characters are associated with the same bounding box, indicating a segmentation error. MASK-RCNN, on the other hand, does not detect all characters, and one is partially detected. However, the proposed UNET model showed superior performance in accurately detecting each character and the barycenter of the bounding boxes.

The proposed model based on RESNET34 and UNET outperforms the CRAFT and MASK-RCNN approaches in character-level detection of ono-matopoeias in the KABOOM-ONOMATOPOEIA dataset. The UNET model consistently demonstrates higher precision, recall, and F1 scores in both experiments, as well as superior performance in qualitative analysis of individual images. The study shows that the UNET model is a promising approach for further research in onomatopoeia detection in comic book pages.

6   Conclusion and Perspectives

This paper presents a method for detecting onomatopoeias using a custom ResNet/UNET model. The proposed method is evaluated in two experiments using the KABOOM-ONOMATOPOEIA dataset. For both experiments, the results showed that the proposed UNET based architecture outperforms other methods of the literature.

The proposed method has demonstrated excellent performance in the detection of onomatopoeia at character-level, and the use of data augmentation techniques improves the robustness of the model.

In future work, we propose to incorporate other attributes such as character color composition, texture, distortion, and orientation for more accurate grouping of detected characters into their respective words. By considering these additional features, we aim to improve the performance of our model not only by detecting individual characters, but also by grouping them accurately to form complete onomatopoeic words.

Acknowledgements. This work is supported by the Research National Agency (ANR) in the framework of the LabCom program (ANR 17-LCV2-0006-01) and the Region Nouvelle Aquitaine for the ScanTrad project (2018-1R50116).

References

1. Baek, Y., Lee, B., Han, D., Yun, S., Lee, H.: Character region awareness for text detection. In: 2019 IEEE/CVF Conference on Computer Vision and Pattern Recognition (CVPR), pp. 9357–9366. IEEE. Long Beach, CA, USA (2019)
2. Biswas, S., Mandal, S., Das, A.K., Chanda, B.: Land map images binarization based on distance transform and adaptive threshold. In: 11th IAPR International Workshop on Document Analysis Systems. IEEE (2014)
3. Chu, W.T., Yu, C.C.: Text detection in manga by deep region proposal, classification, and regression. In: 2018 IEEE Visual Communications and Image Processing (VCIP), pp. 1–4 (2018)
4. Dubray, D., Laubrock, J.: Deep CNN-based speech balloon detection and segmentation for comic books. CoRR, abs/1902.08137 (2019)
5. Del Gobbo, J., Matuk Herrera, R.: Unconstrained text detection in manga: a new dataset and baseline. CoRR, abs/2009.04042 (2020)
6. He, D., et al.: Multi-scale FCN with cascaded instance aware segmentation for arbitrary oriented word spotting in the wild. In: 2017 IEEE Conference on Computer Vision and Pattern Recognition (CVPR), pp. 474–483 (2017)
7. He, K., Gkioxari, G., Dollár, P., Girshick, R.B.: Mask R-CNN. CoRR, abs/1703.06870 (2017)

8. Liao, M., Shi, B., Bai, X., Wang, X., Liu, W.: Textboxes: a fast text detector with a single deep neural network. In: AAAI (2017)
9. Liao, M., Zhu, Z., Shi, B., Xia, G.S., Bai, X.: Rotation-sensitive regression for oriented scene text detection. CoRR, abs/1803.05265 (2018)
10. Liu, X., Liang, D., Yan, S., Chen, D., Qiao, Y., Yan, J.: FOTS: fast oriented text spotting with a unified network. CoRR, abs/1801.01671 (2018)
11. Nagaoka, Y., Miyazaki, T., Sugaya, Y., Omachi, S.: Text detection by faster R-CNN with multiple region proposal networks. In: 14th IAPR Internayional Conference on Document Analysis and Recognition (ICDAR) (2017)
12. Nguyen, N.-V., Rigaud, C., Burie, J.-C.: Comic MTL: optimized multi-task learning for comic book image analysis. Int. J. Doc. Anal. Recogn. (IJDAR) **22**, 265–284 (2019)
13. Ntogas, N., Veintzas, D.: A binarization algorithm for historical manuscripts. In: International Conference on Intelligent Cloud Computing (2008)
14. Ogawa, T., Otsubo, A., Narita, R., Matsui, Y., Yamasaki, T., Aizawa, K.: Object detection for comics using manga109 annotations. CoRR, abs/1803.08670 (2018)
15. Redmon, J., Divvala, S.K., Girshick, R.B., Farhadi, A.: You only look once: Unified, real-time object detection. CoRR, abs/1506.02640 (2015)
16. Rigaud, C., Burie, J.-C., Ogier, J.-M.: Text-independent speech balloon segmentation for comics and manga. In: Lamiroy, B., Dueire Lins, R. (eds.) GREC 2015. LNCS, vol. 9657, pp. 133–147. Springer, Cham (2017). https://doi.org/10.1007/978-3-319-52159-6_10
17. Rigaud, C., Karatzas, D., Van de Weijer, J., Burie, J.C., Ogier, J.M.: Automatic text localisation in scanned comic books. In: Proceedings of the 8th International Conference on Computer Vision Theory and Applications (2013)
18. Ronneberger, O., Fischer, P., Brox, T.: arxiv:1505.04597
19. Solihin, Y., Leedham, C.G.: Integral ratio: a new class of global thresholding techniques for handwriting images. IEEE Trans. Pattern Anal. Mach. Intell. **21**(8), 761–768 (1999)
20. Wu, V., Manmatha, R., Riseman, E.M.: Finding text in images. In: Digital Library (1997)
21. Yanagisawa, H., Yamashita, T., Watanabe, H.: A study on object detection method from manga images using CNN. In: 2018 International Workshop on Advanced Image Technology (IWAIT), pp. 1–4 (2018)
22. Zhou, X., et al.: EAST: an efficient and accurate scene text detector. CoRR, abs/1704.03155 (2017)
23. Ioannis, P., Konstantinos, Z., Xenofon, K., Lazaros, T., Tanmoy, M., Isabelle M.-S.: ICDAR 2019 competition on document image binarization (DIBCO 2019). In: 2019 International Conference on Document Analysis and Recognition (ICDAR), pp. 1547–1556 (2017)

Text Extraction for Handwritten Circuit Diagram Images

Johannes Bayer[(✉)] [iD], Shabi Haider Turabi, and Andreas Dengel [iD]

German Research Center for Artificial Intelligence,
Trippstadter Str. 122, 67663 Kaiserslautern, Germany
{johannes.bayer,shabi.turabi,andreas.dengel}@dfki.de
https://www.dfki.de/web

Abstract. Paper-based handwritten electrical circuit diagrams still exist in educational scenarios and historical contexts. In order to check them or to derive their functional principles, they can be digitized for further analysis and simulation. This digitization effectively performs an electrical graph extraction and can be achieved by straight-forward instance segmentation, in which electrical symbols become the nodes and the interconnecting lines become the edges of the graph. For an accurate simulation however, the texts for describing the graph's items properties have to be extracted as well and associated accordingly.

The paper at hand describes a dataset as well as approaches based on optical character recognition, regular expressions and geometric matching. The source code of the described approach as well as the ground truth is integrated into a publicly available dataset and processing software.

Keywords: Schematic · OCR · Regular Expression

1 Introduction

Handwritten circuit diagrams are still being used today in situations where computer-aided design software is inappropriate like electrical engineering exams, sketching circuit concepts during initial design or examining historical sources. Understanding these documents automatically allows for subsequent applications like simulation or exam evaluation. Circuit diagrams mainly depict electrical graph structures, i.e. electrical symbols (nodes) along with their interconnections (edges). Additionally, text elements are used to describe electrical properties of individual elements.

While different technologies in the domain of machine learning and computer vision have already been applied to this domain, a publicly available and thus verifiable end-to-end solution incorporating text processing is still not addressed.

This paper describes a dataset as well as methods for processing texts from handwritten circuit diagram images during graph extraction by neural network-based character recognition, text classification based on regular expressions and geometric property assignment.

M. Coustaty and A. Fornés (Eds.): ICDAR 2023 Workshops, LNCS 14193, pp. 192–198, 2023.
https://doi.org/10.1007/978-3-031-41498-5_14

Both the described dataset extension[1] as well as the software in which the described text processing methods have been integrated[2] are publicly available.

2   Related Work

Most existing literature on circuit diagram processing focuses on the symbol and interconnection line extraction [2,4], but the neccesity of text processing is also acknowledged [1]. [5] considers texts in handwritten circuits for the special case of letters as terminal names. A more comprehensive approach exists for printed piping and instrumentation diagrams [6], however the training material has not been disclosed. [7] focuses on both handwritten circuit symbols and texts but neglects the graph structure and also does not disclose the training material.

2.1   Existing Dataset

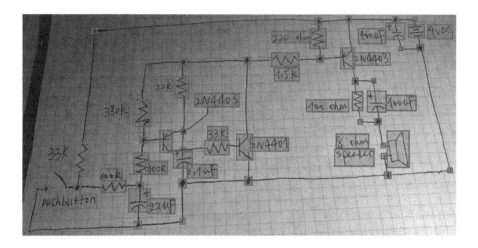

Fig. 1. CGHD Image Sample with Bounding Box Overlay

The CGHD Dataset [9] consists handwritten circuit images (see Fig. 1) along with bounding box annotations for electrical symbols as well as auxiliary structural elements and texts. The dataset has been extended by instance segmentation maps [3], including rotations for individual electrical symbols. In its current form, it contains 2.304 images and 193.850 bounding box annotations, from which 70.618 are of type text. The text annotations are intended to contain *semantic atoms*, i.e. single electrical values or component names, rather than having all texts related to a component in a single annotation (see Fig. 2).

---

[1] https://zenodo.org/record/7700896.
[2] https://gitlab.com/circuitgraph/.

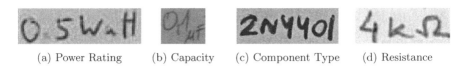

(a) Power Rating   (b) Capacity   (c) Component Type   (d) Resistance

Fig. 2. Text Snippets from CGHD

3   Methodology

3.1   Dataset Extension

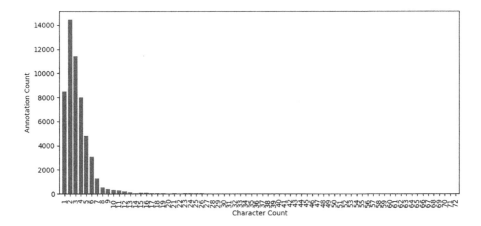

Fig. 3. Text Length Distribution

From the 70.618 text bounding box annotations, 54.156 have been annotated by strings manually, from which 3.902 are unique. Overall, the annotation strings contain 190.081 characters. Most texts have a length between one and seven characters while the longest string contains 72 characters (see Fig. 3).

Symbol Set. The symbol set contains both uppercase and lowercase Latin characters (including German umlauts), Arabic numerals (which are the most prevalent symbols) as well as mathematical operators and the Greek letters Ω and μ (see Fig. 4).

Rotation. While the vast majority of texts is only slightly rotated with respect to conventional reading direction, some texts appear deliberately rotated in the raw images. In order to aid the optical character recognition process, rotations are annotated in 90° steps for all text annotations.

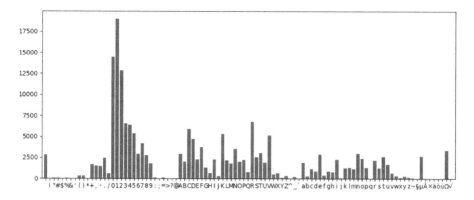

Fig. 4. Character Occurences within the Overall Dataset

Text Types. The annotated texts describe key electrical properties (e.g. resistance, capacity, inductivity), electrical specifications (e.g. maximum voltage), suggested device types or names of individual components as well as operation conditions (e.g. voltages or currents) of circuit parts or even describe the entire circuit (e.g. title).

3.2 Pipeline

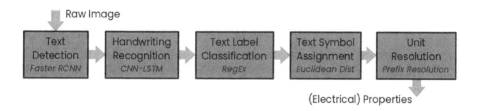

Fig. 5. Proposed Text Extraction Pipeline.

The proposed pipeline consists of the following steps (see also Fig. 5):

Text Detection. First of all, the list of text areas has to be identified from the raw image. As proposed in [9], this can be achieved by a Faster R-CNN [8], but generally this task is a object detection task. Since this entire text extraction pipeline is considered to take place in conjunction with an (electrical) graph extraction, the object detector can be shared to also detect other objects like (electrical) symbols or connections.

Handwriting Recognition. The extracted areas of interest form a list of line line text snippets for which an CNN and LSTM combination (trained using CTC) can be used to obtain text strings.

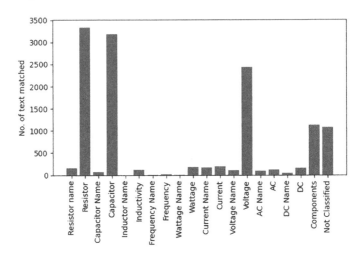

Fig. 6. Overall Dataset Text Class Distribution

Text Label Classification. Regular expressions are used to classify the text strings in categories (see Fig. 6).

Text Symbol Assignment. Each classified text string is assigned as property to an electrical symbol based on its euclidean distance to the closest symbol bounding box.

Unit Resolution. For electrical properties, the actual values have to be obtained from the strings by resolving the metric prefixes, stripping the optional units. For example the strings $2.2\,\mu F$, $2\,u2$ and $2200\,nF$ all encode for 0.0000022 Farads.

4   Evaluation

4.1   Text Label Classification

In order to validate the viability of the text string classification rules, the bounding box class distribution of the closest neighbors by euclidean center distance for individual text classes is considered (see Fig. 7). The vast majority of text class instances is closest to symbols of the their intended classes.

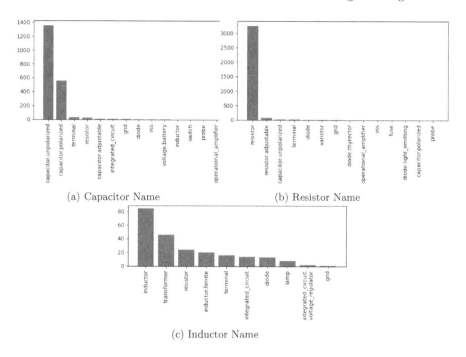

(a) Capacitor Name   (b) Resistor Name

(c) Inductor Name

Fig. 7. Distribution of Closest Neighbors by Class Bounding Box Annotation for selected Text Label Classes

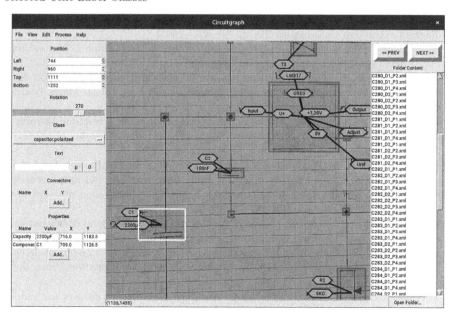

Fig. 8. Result of Property Assignment in the Circuitgraph UI

4.2   UI Integration

The text to property processing has been integrated into the publicly available Circuitgraph UI (see Fig. 8).

5   Future Work

Apart from an end-to-end performance evaluation, certain aspects need to be refined: Firstly, subscript markups (often occurring in component names) and negation lines are not yet covered by the annotations. Secondly, electrical values need to be homogenized (prefixed need to be resolved) for a uniform representation. Finally, a consistency check needs to be applied to verify the completeness of the graph.

Acknowledgements. This research was funded by the BMWE (Bundesministerium für Wirtschaft und Energie), project ecoKI, funding number: 03EN2047B. Furthermore, the authors like to thank all contributors to the dataset (circuit providers, drafters and annotators) as well as Muhammad Nabeel Asim for his support.

References

1. Bailey, D., Norman, A., Moretti, G., North, P.: Electronic Schematic Recognition. Massey University, Wellington, New Zealand (1995)
2. Barta, A., Vajk, I.: Document image analysis by probabilistic network and circuit diagram extraction. Informatica **29**(3), 291–301 (2005)
3. Bayer, J., Roy, A.K., Dengel, A.: Instance segmentation based graph extraction for handwritten circuit diagram images. arXiv preprint arXiv:2301.03155 (2023)
4. Lakshman Naika, R., Dinesh, R., Prabhanjan, S.: Handwritten electric circuit diagram recognition: an approach based on finite state machine. Int. J. Mach. Learn. Comput. **9**, 374–380 (2019)
5. Majeed, M.A., Almousa, T., Alsalman, M., Yosef, A.: Sketic: a machine learning-based digital circuit recognition platform. Turk. J. Electr. Eng. Comput. Sci. **28**(4), 2030–2045 (2020)
6. Mani, S., Haddad, M.A., Constantini, D., Douhard, W., Li, Q., Poirier, L.: Automatic digitization of engineering diagrams using deep learning and graph search. In: Proceedings of the IEEE/CVF Conference on Computer Vision and Pattern Recognition Workshops, pp. 176–177 (2020)
7. Rabbani, M., Khoshkangini, R., Nagendraswamy, H., Conti, M.: Hand drawn optical circuit recognition. Procedia Comput. Sci. **84**, 41–48 (2016)
8. Ren, S., He, K., Girshick, R., Sun, J.: Faster r-CNN: towards real-time object detection with region proposal networks. In: Advances in Neural Information Processing Systems, vol. 28 (2015)
9. Thoma, F., Bayer, J., Li, Y., Dengel, A.: A public ground-truth dataset for handwritten circuit diagram images. In: Barney Smith, E.H., Pal, U. (eds.) ICDAR 2021. LNCS, vol. 12916, pp. 20–27. Springer, Cham (2021). https://doi.org/10.1007/978-3-030-86198-8_2

Can Pre-trained Language Models Help in Understanding Handwritten Symbols?

Adarsh Tiwari, Sanket Biswas[✉][iD], and Josep Lladós[iD]

Computer Vision Center and Computer Science Department,
Universitat Autònoma de Barcelona, Cerdanyola del Vallès, Spain
adarshtd17@gmail.com, {sbiswas,josep}@cvc.uab.es

Abstract. The emergence of transformer models like BERT, GPT-2, GPT-3, RoBERTa, T5 for natural language understanding tasks has opened the floodgates towards solving a wide array of machine learning tasks in other modalities like images, audio, music, sketches and so on. These language models are domain-agnostic and as a result could be applied to 1-D sequences of any kind. However, the key challenge lies in bridging the modality gap so that they could generate strong features beneficial for out-of-domain tasks. This work focuses on leveraging the power of such pre-trained language models and discusses the challenges in predicting challenging handwritten symbols and alphabets.

Keywords: Symbol recognition · Sketch understanding · Large Language Models · Domain transfer · Graphical languages

1 Introduction

Human conceptual knowledge is integral to a wide range of abilities, such as perception, production, and reasoning [10]. This knowledge is characterized by its productivity and versatility, as individuals can utilize their internal models and mental representations to tackle new tasks without requiring extensive training [12]. Additionally, human conceptual knowledge is unique in the way it interacts with raw signals, as people can acquire new concepts directly from complex sensory data and identify familiar concepts within similarly complex stimuli. Pre-trained language models, like BERT [3] or GPT [2], have been designed primarily to understand and generate natural language text. While these models are powerful for text-related tasks, they are not inherently designed for understanding handwritten symbols directly.

Classical neurally-grounded models [5,10] have been help to perform more creative generative tasks to produce characters or graphic symbols which either imitate the exemplars from training set samples or showcased some interesting variations, making them easily identifiable from human perception. Symbolic programs like Bayesian Programming Language [10] provided a language for expressing the causal and compositional structure, by providing a dictionary of

M. Coustaty and A. Fornés (Eds.): ICDAR 2023 Workshops, LNCS 14193, pp. 199–211, 2023.
https://doi.org/10.1007/978-3-031-41498-5_15

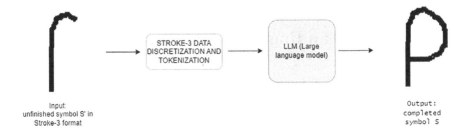

Input:
unfinished symbol S' in
Stroke-3 format

Output:
completed
symbol S

Fig. 1. Illustration of the Task: Given an uncompleted input symbol S', in stroke-3 format, a continuous data format, we discretize and tokenize this input so we can feed it into an LLM and reconstruct the entire handwritten symbol S predicting its missing parts.

simple sub-part primitives for generating handwritten character or symbolic concepts, and symbolic relations which define how to merge those sub-sequences into sequences (called strokes) and finally into a whole character or graphical symbol. We assume that pre-trained language models with multi-head self-attention [16] could provide an important additional prior for learning such kind of small subtasks or sub-concepts from online stroke information which could help to design a single task-agnostic or domain-adaptive neuro-symbolic generative system for understanding and generating new structured characters and symbols.

In summary, pre-trained Large Language Models (LLMs) cannot directly understand handwritten symbols but its inherent contextual knowledge transferability in zero-shot settings [2] could be exploited to process and analyze handwritten stroke sequences and generate conceptually meaningful characters as output. A small illustration of the aforementioned task is shown in Figure 1. Moreover, we add to the Graphics Recognition community a new confluence of research direction where the common interest of interpreting symbolic constructions follow a context-dependent language [14].The key investigations and insights deducted in this work can be divided into three-folds:

- Investigating the utility of pre-trained language models in graphic symbol recognition tasks and exploring the domain-shift problem between text-based data and stroke-3 data in graphic symbols
- Developing a task-general approach to discretize and tokenize stroke data to adapt to the language model input
- Comparing the performance of encoder-only and encoder-decoder based models on the graphic symbol recognition task

The rest of the paper is organized as follows. In Sect. 2 an overview of the related work is provided. Section 3 describes the proposed method for symbol completion based on LLM fine-tuning. The experimental evaluation is reported in Sect. 4. In Sect. 5 we analyze the obtained results and propose some continuation lines. Finally, Sect. 6 draws the conclusions.

2 Related Work

2.1 The Omniglot Challenge

The Omniglot benchmark challenge has been widely adopted in machine learning, as Lake *et al.* [11] reviewed on the current progress which stated that although there has been immense progress in the task of one-shot classification [6,17], there had been minor development in the task-general models which mimic the flexibility of human learners as addressed in significant works as in [5,10]. More recently, Fabi *et al.* [4] extended the Omniglot study for handwritten character generation and recognition by investigating how compositional structures are employed for faster concept-oriented learning, imitation, and understanding of handwritten styles. Souibgui *et al.* [15] further applied the Bayesian Program Language [10] principle to do a one-shot compositional generation for handwritten ciphered symbols and characters for developing a decent handwritten text recognition (HTR) system for low-resource manuscripts.

2.2 Large Language Models

Large Language Models (LLMs) have redefined the field of Artificial Intelligence (AI) through their ability to understand natural language information and generate human-like responses. These LLMs are trained on surprisingly simple objectives, like predicting the next token in a sequence of text given the preceding tokens, popularly known as Masked Language Modeling (MLM) or predicting the next sentence given the preceding sentence. The introduction of such pre-training objectives in NLP models like BERT [3] and RoBERTa [13] on sufficiently large and diverse corpora of text helped them generate coherent, contextual and human-like responses which could be used for a wide variety of tasks, such as creative content generation, machine translation, helping with coding tasks, and answering questions in a helpful and informative way. Later, the introduction of GPT series [2] of language models paved the way for zero-shot learning, meaning it can perform new tasks with little or no kind of fine-tuning. These autoregressive GPT architectures generated human-like responses to open-ended NLP tasks. More recently, Kojima *et al.* [9] further investigated the zero-shot reasoning capability of GPT models by introducing complex multi-step reasoning through step-by- step answer examples which helped them achieve state-of-the-art performances in neuro-symbolic reasoning.

In this work, we show the first direction towards addressing the capability of pre-trained LLMs towards understanding handwritten characters and graphic symbols by generating and reconstructing them with new conceptual variations. Inspired by related works as in autoregressive models like recurrent networks [7] for handwriting and generative modelling for sketches [1,8], we make an attempt to discover a new research direction towards utilizing the power of LLM's for producing creative handwritten symbols or characters.

3 Method

In this section we expand on the problem to solve, describing its basic notation. Furthermore, the section is devoted to describe the details of the proposed approach, and in particular the model and the learning objectives, together with some implementation details.

3.1 Problem Formulation

In order to frame the problem rigorously, we will establish a formal definition for the task at hand. We use the encoding of the handwritten input symbols proposed by [8] and called *stroke-3*. In this format, each sample is stored as list of strokes encoded each by a vector of three values: $[x, y, p]$, where x and y are the coordinate offsets, and p represents the status of the pen (on paper or on air). Our aim is to, given an incomplete handwritten symbol input S, represented in stroke-3 data format, predict the completion of the symbol and generate the completed graphic symbol S.

To achieve this, we formulate our approach in terms of a domain transfer architecture that leverages pre-trained LLMs (Large Language Models) that have been trained on a large corpus of text data. A symbol, consisting of a sequence of graphical primitives (strokes in our case), is an instance of a visual language. Based on this observation, our goal is to test the domain-shift capabilities of LLMs to capture stroke patterns from partially drawn symbols and generating coherent strokes to complete the symbol. Moreover since language models require to work with an alphabet of tokens, a dictionary learning-based approach D is used to discretize the continuous stroke data. Formally, we can represent our model for the completion of a graphic symbol as a function P_{LLM} defined by a specific LLM, which maps a discretized incomplete graphic symbol input $D(S')$ to the completed symbol, S:

$$S = P_{LLM}(D(S'); \theta_P) \tag{1}$$

where θ_P represents the parameters of the completion function P_{LLM} that are learned by our model.

The proposed approach presents three main challenges: (1) Since we are working with language models in this task, we need to take into account that these work on discrete tokens, so the continuous stroke-3 data needs to be discretized to be compatible with the language model input. (2) As we are working with stroke-3 type of data, and most language models are trained on text data, selecting the appropriate language model for the graphic symbol completion task is a serious challenge. We need a language model that can understand the structure and patterns of our data, which is different from the text data in which these language models are trained. (3) Stroke-3 data tokenization. Since stroke-3 is a continuous type of data, it does not have a natural tokenization strategy. The chosen approach to tokenize the data needs to be able to preserve the stroke continuity and at the same time ensure that it is compatible with the input of our language models.

Achieving a domain-shift task making use of a LLM trained on text data for a graphic symbol completion task is an important challenge. This work focuses on tackling the problem by exploring different language models for our task, while also attempting to address the challenges raised above.

3.2 Symbol Completion Approach

To build a model based on a pre-trained LLM (large language model) for our graphic symbol completion task, we will explain the proposed approach in two different parts: Training and Inference.

Training: The whole training process is illustrated in Fig. 2:   The training

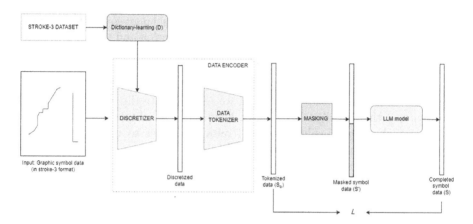

Fig. 2. Overview of the training process: The model consists of a Large Language Model (LLM) trained on Storke-3 data that previously has been discretized (dictionary-learning) and tokenized through an encoding process.

process needs two main inputs: the symbol we want to train with and a dictionary D. Our goal is to make the stroke-3 data, where each stroke in the graphic symbol is represented by a vector of three values: $[x, y, p]$ (x-coord, y-coord and pen_state), compatible with the inputs that the language model receives. Thus, it is crucial to discretize the data. This means that the continuous information needs to be transformed into discrete values. To achieve this, we make use of a dictionary-based learning strategy, more specifically, a K-means clustering. Through it, we define a dictionary D, generated by our whole stroke-3 dataset that maps similar stroke data points into clusters centers (obtained by the K-means clustering). This process allows the continuous data (in $[x, y, p]$ format) to be represented by a limited set of discrete values, which basically correspond to the center of each cluster. Formally, the dictionary D can be represented as follows:

$$D : \mathbb{R}^3 \rightarrow \{1, 2, \ldots, K\} \tag{2}$$

where $\mathbb{R}^3$ represents the continuous stroke-3 data and K is the number of clusters obtained by applying K-means clustering to the stroke data. The function D assigns to each stroke data (x, y, p) the index of the nearest cluster center, that is, $D(x, y, p) = k$ where k is the index of the cluster center that minimizes the Euclidean distance to the stroke data (x, y, p). The cluster centers themselves are represented by a dictionary of size K, which is learned during the training process.

Following the discretization step, the data is tokenized. Tokenization is necessary to convert the discrete data into a sequence of tokens, where each token corresponds to a specific word or symbol in the language model's vocabulary. We add some special tokens like *padding* tokens, *end of sequence* tokens, *start of sequence* tokens, *separator* tokens, and *mask* tokens, among others, to ensure that the data is correctly structured to be used in our language model.

After the encoding process completes the discretization and tokenization, the resulting data S_o is ready to be used as input of the language model. But, before doing so, we apply masking to some of the values, obtaining S'. This is done to evaluate the model's ability to reconstruct the masked values (the missing part of the data). By masking the data, we can evaluate the model's ability in reconstructing the complete symbol (S). Doing this, we obtain a reconstruction loss (L), which is used as part of our training process to update the model parameters and improve its performance.

Inference: During the inference time, the proposed model takes as input incomplete handwritten graphic symbol provided by the user, and returns a plausible completion for the given symbol. The completed symbol produced by our model can provide insights into the possible intentions and underlying concepts of the user, which can be valuable for various applications.

3.3   Model

The main objective of this work is to assess if a pre-trained language model can be transferred to hand drawn graphical symbols to predict the completion of partially written instances. We have experimented with two main variants of pre-trained Large Language Models (LLM) for the task. The first variant involved using encoder-type language models like BERT and RoBERTa, while the second variant was more focused on decoder-type language models, specifically we experimented with the GPT-2 model. To accomplish this, we will be utilize pre-trained models from *Hugging Face*[1], a leading provider of state-of-the-art natural language processing models.

---

[1] https://huggingface.co/.

Encoder Variant: Our first approaches are based on encoder-type language models. Our hypothesis was that due to the ability of these models to process data in a bi-directional manner, they can be specially good at capturing the context and meaning of our input data.

In the context of our graphic symbol completion task, the bi-directional nature of these encoder-type models allow them to capture the sequential dependencies and relationships between the different parts of the graphic symbol. This means that the model might be able to understand the relationship between different parts of the graphic symbol and use this knowledge to generate plausible completions, as it analyzes the symbols in both forward and backward directions.

Regarding the models used, we experimented with BERT as baseline and then we further moved on to RoBERTa, since it has shown to outperform BERT on tasks that require a deeper understanding of language semantics, patterns and context. This is especially useful in our case, since we are working with graphic symbol data, which is typically unstructured, with strokes of different lengths, orientations, etc. This makes the patterns and semantics of our data particularly complex.

Decoder Variant: The second variant of our approach was focused on decoder-type language models. Specifically, we experimented with the pre-trained GPT-2 model from Hugging Face, which is known for its strong performance on natural language generation tasks. Decoder-type models like GPT are commonly used for context generation based on a given prompt. This is achieved by training the model to predict the next token, known as Next Sentence Prediction (NSP) based on the previous tokens. Through this training process the model learns to generate coherent and plausible sequence based on the patterns and structure of the training data. Our hypothesis was that for our graphic symbol completion task we can use this ability to generate new content; the model can use as input the incomplete graphic symbols and generate its completions.

Moreover, since decoder-type models are trained to generate text in a coherent and structured way, they might also be able to learn how to generate graphic symbols that are consistent with the structure and composition of the original symbol. This means that the model can understand the sequential dependencies and relationships between the different parts of the handwritten symbol and use this knowledge to generate its completions.

4   Experimental Validation

In this section we report on the different experiments conducted to evaluate the performance of our approach and the different language models used. The aim is to assess the performance of both proposed models, namely the encoder-type and decoder-type, for hand drawn graphic symbol completion task. We want to verify whether our hypothesis regarding the bi-directional nature of encoder-type models and the patterns and structure learning capabilities of decoder-type models hold true in this context. Furthermore, these experiments will not only give us an idea of the performance of these models but also help us determine which approach works best for our task.

4.1   Dataset

We evaluate our strategy on the Omniglot dataset[2], which mainly contains sketches of handwritten symbols in stroke-3 data format. Each graphic symbol data is represented in form of a list of multiple strokes used to create the symbol. Moreover, each stroke is represented in the stroke-3 format, consisting of an x-coordinate, y-coordinate, and pen state (0 or 1). The entire dataset comprises 20,000 handwritten symbols for training and 1,000 handwritten symbols for validation.

4.2   Quantitative Results

To quantitatively evaluate the performance of the proposed approach we make use of the reconstruction loss L as our main evaluation metric, which measures the difference between the predicted graphic symbol and the original symbol. In this section, we will present the reconstruction losses obtained by the different language models used in our experiments, including BERT, RoBERTA, and GPT-2. This will allow us to assess the performance of each model and determine which one is the most effective for our graphic symbol completion task.

Looking at the reconstruction losses in Fig. 3, we can see clearly that the GPT-based model outperforms the two encoder-based models (BERT and RoBERTa). In terms of the reconstruction loss values, BERT achieved a training loss of 0.223968 and a validation loss of 0.466315, while RoBERTa had a training loss of 0.204341 and a validation loss of 0.463566. However, GPT performed the best with a training loss of 0.102047 and a validation loss of 0.117224.

Despite our initial hypothesis where we argued that since BERT and RoBERTa being bidirectional models might be a good choice for understanding the context and structure of our graphic symbols data, we observed that both models suffered from unstable training, with slow decreases in training loss and spikes in validation loss. This instability is indicative of significant overfitting and suggests that these models struggled to generalize to our stroke data.

---

[2] https://github.com/brendenlake/omniglot.

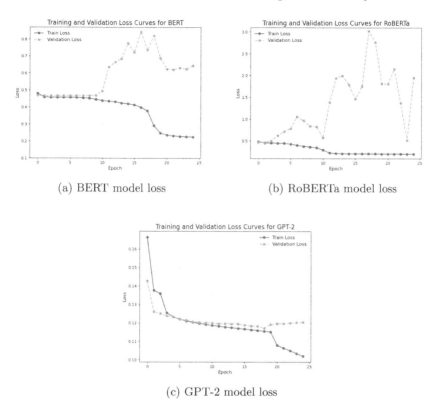

(a) BERT model loss

(b) RoBERTa model loss

(c) GPT-2 model loss

Fig. 3. Losses of the three models studied after 25 epochs.

In contrast, with the GPT model, we see a stable training in our reconstruction loss evaluation. Even though the GPT model also shows some signs of overfitting after several epochs, the overfitting is not as pronounced as what we observed with the BERT and RoBERTa models. Overall, analysing the losses we can see that the GPT model is learning more effectively than the other models. This observation supports our hypothesis that decoder-models such as GPT, which are designed for content generation, might work for our symbol completion tasks. The GPT model appears to have a better understanding of the sequential dependencies and relationships present in our graphic symbol data. Although our results are promising, further investigation is needed to fully understand the behavior of GPT and maybe other decoder-models when applied to graphic symbol tasks.

Table 1. Symbol completion results using our GPT-based model on five different examples.

Original symbol	Incomplete symbol	Predicted completion

4.3   Qualitative Results

In this section we show some qualitative results to demonstrate how our model reconstructs incomplete handwritten symbols. We will only focus on the best-performing model from our previous quantitative results, the GPT-based model. Through these examples, we aim to provide insights into the strengths and limitations of our approach, as well as demonstrate its potential for the task.

The qualitative results the we obtained using our GPT-based approach are shown in Table 1, in which we can observe five different examples. We display the original symbol, the incomplete symbol, and the symbol completion predicted by our model. The first four examples provide us some logical results, but still they are not very accurate and deviate from the original symbol. On the other hand the predicted symbol in the fifth example appears to be random and does not resemble our original symbol. As a result, we think that even though some results are quite promising, our model still needs to be improved in order to produce reliable symbol completion predictions.

Another interesting insight we can take from this results is that the performance of the model seems to be dependant heavily on the incomplete symbol it gets as input. In the first four examples where the models prediction appears to be quite coherent, we can see that the incomplete symbol provides more information about the original symbol. On the other hand, for the fifth symbol, the incomplete symbol is way more ambiguous and could belong to a wide range of symbols, which makes it more difficult for the model to predict the correct completion.

5   Discussion and Future Work

Having seen the results of our experiments with the different language models for our handwritten symbols completion task, we can observe some promising results for the GPT-based model, that showed the best performance among the tested models. Still, there is still much work to be done to improve its performance and effectiveness. Although the GPT-2 based model outperformed both BERT and RoBERTa based models, its overall effectiveness still falls short of our desired level of performance. Moreover, it is important to mention that the model overfits in few epochs and the completion of graphic symbols is often subpar. In this section we discuss several avenues for future research that may help to address these limitations and improve the performance of our approach.

Being overfitting one of the main issues we observed with the GPT-based model, we suspect this may be because of the limited size of the Omniglot dataset, which only contains only 20,000 handwritten graphic symbols for training. Given the huge size of GPT-2 model and its enormous number of parameters, it certainly requires more data to learn effectively and avoid overfitting. Another main issue that causes the poor performance of the model is the aggressive domain-shift from text-based data to stroke-3 data. Pre-trained LLMs like GPT are trained on huge amounts of text data, and our Omniglot dataset contains stroke-3 data, which is significantly different in nature. This domain-shift can cause the model to struggle with our task, and may also be contributing to the overfitting we observe.

To address this issue one possible approach could be to train a GPT-based model from scratch on stroke-3 data instead of text. One way to achieve this is by using the QuickDraw dataset [8], which consists of over 50 million hand-drawn sketches across 345 different categories in stroke-3 format. By training a GPT-based model on this large dataset and then fine-tuning it for our specific task of recognizing handwritten graphic symbols, we can potentially overcome the limitations of using a pre-trained language model that has been trained on text data. This approach can also help to mitigate the domain-shift issue, as QuickDraw data is already in the stroke-3 format.

6   Conclusion

In this work, we have studied the possibility to apply pre-trained Large Language Models (LLMs), which have been trained on large amounts of text data, to the task of recognizing handwritten graphic symbols in stroke-3 format. During the process we found several challenges, such as the continuous nature of our stroke data and the need for effective discretization and tokenization or the need to find a language model that can understand effectively the stroke data and it's patterns. Through our experiments, we saw that decoder-based models, particularly GPT-2, had a better performance in the task compared to encoder-based models like BERT or RoBERTa. These results suggest that decoder-based language models, such as GPT, may have potential for graphic symbol tasks despite having been trained on text data. This opens up opportunities for further investigation into the effectiveness of these models in this domain and their potential for advancing the field.

Still, although our study has shown promising results for using decoder language models for graphic symbol tasks, there is still much room for improvement. Future work includes training a GPT-based architecture from scratch on stroke-3 data using larger datasets to try to mitigate the domain-shift problem. With these efforts, we hope to further advance the field of recognizing graphic symbols with the use of large language models and enable new applications in various domains.

Acknowledgment. This work has been partially supported by the Spanish project PID2021-126808OB-I00, the Catalan project 2021 SGR 01559 and the PhD Scholarship from AGAUR (2021FIB-10010). The Computer Vision Center is part of the CERCA Program / Generalitat de Catalunya.

References

1. Bhunia, A.K., et al.: Pixelor: A competitive sketching AI agent. so you think you can sketch? ACM Trans. Graph. (TOG) **39**(6), 1–15 (2020)
2. Brown, T., et al.: Language models are few-shot learners. Adv. Neural Inf. Process. Syst. **33**, 1877–1901 (2020)

3. Devlin, J., Chang, M.W., Lee, K., Toutanova, K.: Bert: Pre-training of deep bidirectional transformers for language understanding. arXiv preprint arXiv:1810.04805 (2018)
4. Fabi, S., Otte, S., Scholz, F., Wührer, J., Karlbauer, M., Butz, M.V.: Extending the omniglot challenge: imitating handwriting styles on a new sequential dataset. IEEE Trans. Cogn. Dev. Syst. **15**, 896–903 (2022)
5. Feinman, R., Lake, B.M.: Learning task-general representations with generative neuro-symbolic modeling. arXiv preprint arXiv:2006.14448 (2020)
6. Finn, C., Abbeel, P., Levine, S.: Model-agnostic meta-learning for fast adaptation of deep networks. In: International Conference on Machine Learning, pp. 1126–1135. PMLR (2017)
7. Graves, A.: Generating sequences with recurrent neural networks. arXiv preprint arXiv:1308.0850 (2013)
8. Ha, D., Eck, D.: A neural representation of sketch drawings. In: ICLR 2018 (2018). https://openreview.net/pdf?id=Hy6GHpkCW
9. Kojima, T., Gu, S.S., Reid, M., Matsuo, Y., Iwasawa, Y.: Large language models are zero-shot reasoners. arXiv preprint arXiv:2205.11916 (2022)
10. Lake, B.M., Salakhutdinov, R., Tenenbaum, J.B.: Human-level concept learning through probabilistic program induction. Science **350**(6266), 1332–1338 (2015)
11. Lake, B.M., Salakhutdinov, R., Tenenbaum, J.B.: The omniglot challenge: a 3-year progress report. Current Opin. Behav. Sci. **29**, 97–104 (2019)
12. Lake, B.M., Ullman, T.D., Tenenbaum, J.B., Gershman, S.J.: Building machines that learn and think like people. Behav. Brain Sci. **40**, e253 (2017)
13. Liu, Y., et al.: Roberta: a robustly optimized Bert pretraining approach. arXiv preprint arXiv:1907.11692 (2019)
14. Lladós, J.: Two decades of GREC workshop series. conclusions of GREC2017. In: Fornés, A., Lamiroy, B. (eds.) GREC 2017. LNCS, vol. 11009, pp. 163–168. Springer, Cham (2018). https://doi.org/10.1007/978-3-030-02284-6_14
15. Souibgui, M.A., et al.: One-shot compositional data generation for low resource handwritten text recognition. In: Proceedings of the IEEE/CVF Winter Conference on Applications of Computer Vision, pp. 935–943 (2022)
16. Vaswani, A., et al.: Attention is all you need. In: Advances in Neural Information Processing Systems, vol. 30 (2017)
17. Vinyals, O., Blundell, C., Lillicrap, T., Wierstra, D., et al.: Matching networks for one shot learning. In: Advances in Neural Information Processing Systems, vol. 29 (2016)

CBDAR

CBDAR 2023 Preface

We are glad to welcome you to the proceedings of the 10th edition of the International Workshop on Camera Based Document Analysis and Recognition (CBDAR 2023). CBDAR 2023 buildt on the success of the previous nine editions in 2021 (Lausanne, Switzerland), 2019 (Sydney, Australia), 2017 (Kyoto, Japan), 2015 (Nancy, France), 2013 (Washington, DC, USA), 2011 (Beijing, China), 2009 (Barcelona, Spain), 2007 (Curitiba, Brazil) and 2005 (Seoul, Korea).

CBDAR aims to move away from the comfort zone of scanned paper documents and to investigate the innovative ways of capturing and processing both paper documents and other types of human created information in the world around us, using cameras.

Since the first edition of CBDAR in 2005, the research focus has shifted many times, but CBDAR's mission to provide a natural link between document image analysis and the wider computer vision community, by attracting cutting edge research on camera-based document image analysis, has remained very relevant.

In this 10th edition of CBDAR we received five initial submissions. However, two submissions were dropped due to insufficient material. Each submission was carefully reviewed by three expert reviewers. The Program Committee of the workshop comprised of 12 members. We would like to take this opportunity to thank all the reviewers for their meticulous reviewing efforts. Taking into account the recommendations of the reviewers, we selected 2 papers for presentation in the workshop.

The participation of attendees from both academia and industry has remained an essential aspect of CBDAR. The program of this edition was carefully crafted to appeal to both. Apart from the presentation of new scientific work, the workshop will have an invited talk.

We hope that the program of CBDAR 2023 attracted interest in the community and that the participants enjoyed the workshop, which was held along with ICDAR in San José, California, USA.

June 2023

Muhammad Muzzamil Luqman
Sheraz Ahmed
Muhamad Imran Malik

On Text Localization in End-to-End OCR-Free Document Understanding Transformer Without Text Localization Supervision

Geewook Kim[1]([⊠]), Shuhei Yokoo[2], Sukmin Seo[1], Atsuki Osanai[2], Yamato Okamoto[2], and Youngmin Baek[1]

[1] NAVER Cloud, Seongnam-si, Republic of Korea
gwkim.rsrch@gmail.com
[2] LINE, Tokyo, Japan

Abstract. This paper presents a simple yet effective approach for weakly supervised text localization in end-to-end visual document understanding (VDU) models. The traditional approach in VDU is to utilize off-the-shelf OCR engines in conjunction with natural language understanding models. However, to simplify the VDU pipeline and improve efficiency, recent research has focused on OCR-free document understanding transformers. These models have a limitation in that they do not provide the location of the text to the user. To alleviate it, we propose a simple yet effective method for text localization in OCR-free models that does not require any additional supervision, such as bounding box annotations. The method is based on properties of attention mechanism, and is able to output text areas with competitive high accuracy compared to other supervised methods. The proposed method can be easily applied to most existing OCR-free models, making it an attractive solution for practitioners in the field. We validate the method through experiments on document parsing benchmarks, and the results demonstrate its effectiveness in generalizing to various camera-captured document images, such as, receipts and business cards. The implementation will be available at https://github.com/clovaai/donut.

Keywords: Information Extraction · Text Localization · Visual Document Understanding · End-to-End Transformer

1 Introduction

Documents play a ubiquitous role in our lives and are used for a multitude of purposes. In recent years, the demand for an automated document processing has gained significant attention. With the advancement of Deep Learning and Transformer [31], there has recently been proposed OCR-free **Do**cument

---

G. Kim and S. Yokoo—Equal contribution.

© The Author(s), under exclusive license to Springer Nature Switzerland AG 2023
M. Coustaty and A. Fornés (Eds.): ICDAR 2023 Workshops, LNCS 14193, pp. 215–232, 2023.
https://doi.org/10.1007/978-3-031-41498-5_16

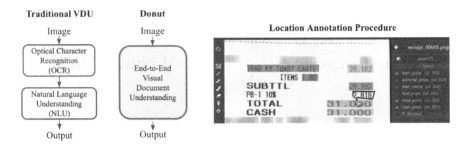

Fig. 1. Comparison of two VDU Pipelines and Associated Annotation Costs.
The Donut approach eliminates the need for pre-processing with an off-the-shelf OCR
(left) and also avoids the time-consuming task of text location annotation (right).

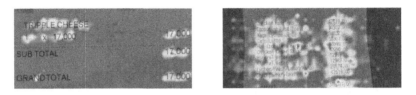

Fig. 2. Visualization of Attention Map in Donut. The figure shows the attention
map produced by the decoder module in Donut. The attention tends to be focused on
the text regions in the document image (left). However, there are instances where the
attention map appears blurred or dispersed (right).

Understanding Transformers [6,19,21] (denoted as **Donut** in this paper), which
enables various document processing tasks to be implemented by a single model
training, making it a highly efficient and effective tool in the field of Visual
Document Understanding (VDU).

In contrast to **Donut**, traditional VDU methods typically involve two separate components: Optical Character Recognition (OCR) [1,2] and a Natural
Language Understanding (NLU). In this approach, text information is extracted
from the input image, which is then processed in the NLU backbone, e.g.,
BERT [7]. However, this requires the prior use of OCR as a pre-processing
step, which can be unavailable or perform poorly. To achieve high performance
with OCR, practitioners must undergo the time-consuming and costly process of
training their own OCR model, including tasks such as annotating text bounding boxes (See Fig. 1, Right). These requirements make **Donut** a more efficient
and cost-effective option for document understanding.

Despite its huge advantages of low annotation costs and ease of model training, **Donut** has a critical disadvantage in that it does not provide the location
of the extracted text. While the absence of text location information may not
be an issue for certain applications, it can be a hindrance in others, e.g., masking private information from documents. The lack of text location information
makes it difficult to use **Donut** for automated document processing purposes
where text location is essential.

In this study, we introduce a new text localization algorithm for OCR-free Document Understanding Transformers, **Donut**. Our approach leverages the attention mechanism of the Transformer to identify text regions in document images. Figure 2 showcases an example of an attention map generated by **Donut** on an Indonesian receipt image [28], highlighting the high attention scores in the regions corresponding to the output text. The challenge, however, lies in converting these attention scores into accurate text bounding boxes or polygons, which we detail in Sect. 3. To assess the effectiveness of our proposed method, we conduct extensive experiments on two camera-captured document parsing benchmarks, and perform a comprehensive comparative analysis of the results. Additionally, we propose a new evaluation metric that considers both the information extraction and text localization capabilities of the model.

2   Related Work

2.1   Optical Character Recognition

Optical Character Recognition (OCR) has undergone substantial advancements in recent years, with a growing emphasis on the integration of deep learning models. OCR consists of two sub-tasks: text detection and text recognition. These sub-tasks are trained on large datasets, including synthetic images [10,17] and real images [18,27,29,33].

Text detection has seen advancements from early methods that utilized Convolutional Neural Networks (CNNs) to predict local text segments, followed by heuristics for merging them [12,38], to more recent techniques that incorporate region proposal and bounding box regression [23] and take advantage of the homogeneity and locality of text [2,30]. In text recognition, many OCR models follow a similar approach that combines various deep learning modules [1]. These models typically employ CNNs to encode the cropped text instance image into a feature space and then utilize a decoder to extract characters from the resulting features.

On the other hand, end-to-end models that perform both detection and recognition at the same time have been studied recently. Early models utilized region of interest (ROI) pooling or masking to extract relevant features from the regions detected by the detector [4,22,24,25]. However, these methods were found to limit the input features of the recognizer, leading to subpar performance. To mitigate this, Kim et al. [20] proposed another approach that eliminated ROI pooling, resulting in improved performance. The shift towards end-to-end models has brought about more efficient and streamlined approaches to OCR.

2.2   Visual Document Understanding

Visual Document Understanding (VDU) is a field of study that focuses on the automated extraction of meaningful information from visual documents, such as digital images, photos, and scanned documents. These documents may contain a variety of information such as text, tables, diagrams, and figures. The applications of VDU are broad and encompass document classification, information extraction, document summarization, and document-based question answering. With the recent advancements in deep learning and transformer models, there have been remarkable advancements in the field of VDU, leading to the development of more effective and efficient techniques for automated document processing [11,13,19,34,35].

OCR-Based VDU. Traditionally, Visual Document Understanding (VDU) has been approached through the integration of Optical Character Recognition (OCR) and Natural Language Understanding (NLU) models, where the output of OCR is utilized as input for the NLU model. For example, in the task of Document Parsing, Named Entity Recognition (NER) over OCR-extracted text is a commonly adopted approach [13]. The NER process enables the extraction of pre-defined entity field information from the document and its subsequent parsing into a structured form. These techniques have been widely researched and various models have been proposed and developed [11,13,34,35]. On the other hand, an effort to create a more efficient VDU pipeline has been made, proposing a method that does not rely on off-the-shelf OCR engines [19], which will be further explained in the following section.

OCR-Free VDU. In order to address the challenges posed by the traditional OCR-based Visual Document Understanding (VDU) pipeline, there has been proposed a new approach that leverages an end-to-end Transformer encoder-decoder architecture for conducting various VDU tasks without relying on OCR engines. This approach, as described in recent works such as [6,19], operates on raw document images and directly produces the desired form of information without the need for intermediate steps. It offers promising results by bypassing the limitations associated with OCR. However, a common issue with this approach is that it does not provide the location of the extracted text. One naïve approach to address this issue is to include the bounding box annotation in the output token sequence during model training. However, this approach sacrifices the low annotation cost advantage of the OCR-free VDU methods and introduces additional computational overhead due to the increased length of the target sequence to be predicted.

3   Text Localization in OCR-Free VDU

3.1   Preliminary: Document Understanding Transformer (Donut)

This section provides an overview of the recent advancements in OCR-free end-to-end document understanding transformers, as represented by the works of Kim et al. [19], Davis et al. [6], and Lee et al. [21]. These works, although differing in certain aspects of their model architectures, can be considered collectively as instances of a common framework, referred to as *Document Understanding Transformer* (**Donut**) in this paper.

The architecture of Donut is based on Transformer [31], comprised of two components: an encoder and a decoder. Unlike the original Transformer, Donut adopts a Vision Transformer [9] or its variants (e.g., Swin Transformer [26]) as the encoder module to process the raw input image $\mathbf{x} \in \mathbb{R}^{C \times H \times W}$. The encoder produces a set of visual embeddings, each representing a specific region of the document image, $\mathbf{Z} = \{\mathbf{z}_i | \mathbf{z}_i \in \mathbb{R}^d, 1 \leq i \leq n\}$, where d is the embedding dimension of the model, n is the number of image patches. If we set the patch size to $p \times p$, the number of patch is calculated as $n = H/p \times W/p$ (We assume the original Vision Transformer [9] for a simple explanation). The text decoder then references these visual embeddings through cross attention [31] to generate the final output token sequence $(\mathbf{y}_i)_{i=1}^m$, where $\mathbf{y}_i \in \mathbb{R}^d$ represents the vector representation of i-th output token (e.g., character, subword or word) in the vocabulary. The format of the output token sequence can be tailored to the specific task at hand. For example, Kim et al. [19] set the token sequence to be directly convertible to structured data in JSON format, which includes the target information to be extracted (See Fig. 3 for a detailed example).

In the decoder, *cross attention* mechanism [31] is used to dynamically weigh the contribution of different input elements to the computation of an output element. In Donut, cross attention can be interpreted as weighting the contribution of different image regions $\mathbf{Z} \in \mathbb{R}^{n \times d}$ to generate (predict) a corresponding output token $\mathbf{y}_{i+1} \in \mathbb{R}^{1 \times d}$. In details, most models adopt *multi-head cross attention* mechanism [31] which can be interpreted as an ensemble of *cross attention*. Formally, at the time step i, the multi-head cross attention mechanism computes attention maps $\mathbf{s}_i = \{\mathbf{s}_i^j | \mathbf{s}_i^j \in \mathbb{R}^{1 \times n}, 1 \leq j \leq h\}$ to predict the output token $\mathbf{y}_{i+1}$, where h is the number of heads. The computation can be expressed as follows,

$$\mathbf{s}_i = \{\mathbf{s}_i^i, \cdots, \mathbf{s}_i^h\}, \tag{1}$$

$$\mathbf{s}_i^j = \text{softmax}(\frac{\mathbf{q}_i \mathbf{K}^{j\top}}{\sqrt{d/h}}), \tag{2}$$

where $\mathbf{q}_i^j = \mathbf{y}_i \mathbf{W}_Q^j$ and $\mathbf{K}^j = \mathbf{Z} \mathbf{W}_K^j$ are i-th query and keys in the j-th attention head, respectively. The attention map is used to compute a weighted sum over the values, which are also another linear projections of the image region vectors $\mathbf{V}^j = \mathbf{Z} \mathbf{W}_V^j$. We assume $\mathbf{W}_Q^j \in \mathbb{R}^{d \times (d/h)}$, $\mathbf{W}_K^j \in \mathbb{R}^{d \times (d/h)}$ and $\mathbf{W}_V^j \in \mathbb{R}^{d \times (d/h)}$ for all $1 \leq j \leq h$ to simplify the notation. The final outputs $\{\mathbf{s}_i^j \mathbf{V}^j\}_j$ are concatenated and fed to a following module and used to predict the next time step token $\mathbf{y}_{i+1}$.

3.2 Proposed Method

Text localization in OCR-free Document Understanding Transformers (Donut) is challenging but important task. An naïve solution would be to incorporate text bounding box supervision into the target token sequence in the model training phase. While recent studies [5,37] have demonstrated the feasibility of extracting

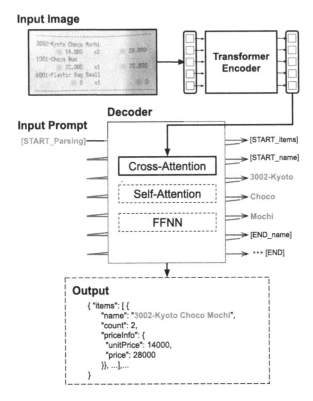

Fig. 3. Overview of Donut Pipeline. The encoder transforms a given document image into a set of image region embeddings, which are then fed into a cross-attention layer in the decoder. The decoder generates a sequence of tokens, using the encoded image region embeddings, which can be transformed into a structured form of the desired target information.

object or text locations from sequence generation models like Donut, this solution sacrifices the low annotation cost advantage that Donut provides and introduces additional computational overhead, which may negatively impact the model's inference speed due to the increased length of the target sequence to be predicted.

In this study, we aim to address the challenge of text localization in OCR-free Document Understanding Transformers (Donut) without introducing additional text localization supervision, such as text bounding boxes. Our objective is to enhance text localization capabilities while retaining the original benefits in training and inference methods of the model. To achieve this, we leverage the Cross Attention Map of Donut, which is discussed in detail in Sect. 3.2. This map provides valuable insights into the location of the desired text. However, we observed that the attention map is not always precise, and sometimes the region it highlights is scattered or blurred (See Fig. 2). To mitigate this issue, we conducted an in-depth analysis of the nature of attention and developed several techniques to effectively improve text localization.

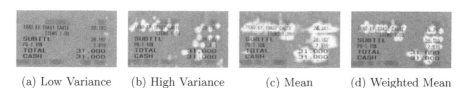

| (a) Low Variance | (b) High Variance | (c) Mean | (d) Weighted Mean |

Fig. 4. Visualization of Properties in Cross Attention Heads. (a) Attention head with low variance in attention scores appears to be less informative. (b) Conversely, Attention head with high variance seems to be more informative. (c) A simple average of the attention heads shows limited performance in text localization, whereas (d) a weighted mean proportional to the variance of the attention heads yields improved results.

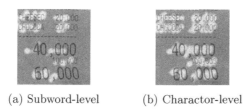

| (a) Subword-level | (b) Charactor-level |

Fig. 5. Visualization of Effects of Tokenization Level. (a) Subword-level tokenization divides words into fewer tokens, leading to biased attention towards the left side, as can be seen from the visualization of attention assigned to each token. (b) Character-level tokenization, on the other hand, resolves this issue by providing a more balanced distribution of attention.

Aggregation of Multi-head Cross Attention Map. The first approach to improve text localization involves the utilization of the multi-head cross attention mechanism in the Donut decoder [31]. As explained in Sect. 3.2, the decoder generates multiple attention maps, each of which exhibits a unique behavior. Our analysis revealed that the attention maps have varying levels of contribution to the localization task. Furthermore, we observed that the variance of the attention scores can serve as a useful metric in aggregating the attention maps. To this end, we proposed to aggregate the attention maps using a weighted mean, proportional to their variance. This approach leads to an improvement in the localization map compared to the simple average, as demonstrated in Fig. 4. The effectiveness of this technique will be quantitatively evaluated in our experiments and analyses (Sect. 4).

De-biased Attention Map via Character-Level Tokenization. Furthermore, we observed that the attention assigned to each character is influenced by the reading direction. As depicted in Fig. 5-(a), the attention tends to be biased towards the left due to the standard reading direction, which proceeds from upper left to lower right in a raster order. This issue is further compounded in subword-based decoders, where the smallest unit of text is represented subwords (i.e., frequent character n-grams). This results in a reduction in the number of

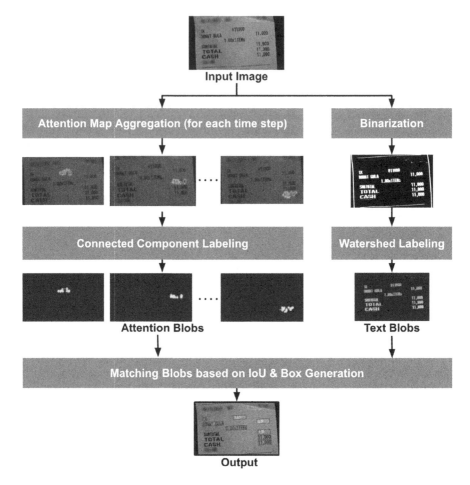

Fig. 6. Proposed Pipeline of Unsupervised Text Region Proposal and its Integration with Attention Map. Attention maps are transformed into attention blobs with Connected-component labeling [8] algorithm (i.e., identifying the connected active grid cells). The attention blobs are matched with text blobs achieved by the unsupervised region proposal module. We implement this with Binarization and Watershed-labeling [32] algorithms. This can be easily implemented with common computer vision libraries, e.g., OpenCV [16].

attention maps that can be utilized for text localization and even the actual visual length of each token can appear irregular, hindering accuracy. To address this challenge, we propose a simple solution by replacing the vocabulary of the text decoder with a set of characters (See Fig. 5). This idea, that an attention map at the character-level would be a better source of localization, was similarly explored in Xue et al. [36]. The impact of this modification will also be discussed in our experiments and analyses.

Integration with Unsupervised Text Region Proposals. Lastly, we introduce an unsupervised text region proposal module to augment the text localization capabilities of our approach. While attention provides valuable information, we acknowledge its limitations and therefore propose the addition of a module to extract *text-like* image regions with high recall but low precision. In line with our objective of achieving weakly supervised text localization, this module is implemented using basic computer vision algorithms without the need for additional text localization annotations. The proposed region proposal algorithm and its integration with the attention map are shown in Fig. 6. Binarization and watershed labeling are applied to produce a large number of text region proposals with a high recall rate. It is important to note that precision is not the primary focus, as the attention map serves as a guiding force in the text detection process. Our experimental results, discussed in a subsequent section, demonstrate the effectiveness of the proposed method in accurately localizing extracted text information.

4   Experiment and Analysis

In this section, we evaluate the effectiveness of the proposed method by integrating it into an existing OCR-free end-to-end document understanding model. Furthermore, we conduct an ablation study on the proposed method to determine the individual contribution of each technique described in Sect. 3.

4.1   Downstream Task and Datasets

To validate the effectiveness, following Kim et al. [19], we conduct a document parsing task [28]. Document parsing aims to extract structured information from raw input document images and is a core task for a wide range of real-world applications [13–15,19].

CORD (Public Benchmark Data). Consolidated Receipt Dataset [28] is a publicly available benchmark of receipt images. It comprises of 800 training images, 100 validation images, and 100 test images. The dataset contains a total of 30 unique fields, including information such as menu name, count, and total price. See Fig. 7 and 8 for examples.

Business Card (Private In-Service Data). This dataset is sourced from our active document parsing products[1] and consists of 20K training samples, 256 validation samples, and 70 testing samples of Japanese business cards. It encompasses 11 distinct fields, including the name, company name, address, among others. In accordance with stringent industrial policies regarding private industrial datasets derived from our active products, high-quality realistic samples are shown instead in Fig. 9.

The entire model pipeline, including the proposed unsupervised text region proposal algorithm, is tuned on the validation set and we report various evaluation scores on the test set.

---

[1] https://clova.ai/ocr.

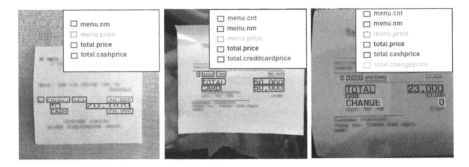

Fig. 7. Samples in CORD Dataset. The dataset comprises photographs of Indonesian receipts with texts, text boxes, and semantic category information.

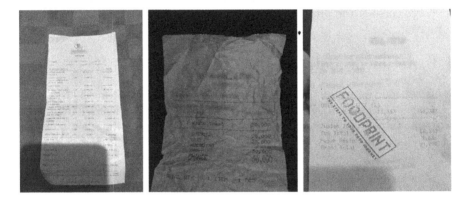

Fig. 8. Noises in CORD Dataset. The CORD dataset comprises various receipt documents captured using a real camera, resulting in samples that inherently exhibit diverse forms of noise, including light blurring, paper crinkling, and other artifacts.

4.2 Evaluation Metric

In this study, we present new evaluation metrics for document parsing systems that consider the accuracy of text localization, text recognition, and semantic category prediction. To the best of our knowledge, no existing work has comprehensively evaluated these aspects, thus calling for the development of new evaluation methods.

Entity Word Box Level Metrics. We start by incorporating commonly used OCR evaluation metrics, such as F1, $\text{CLEval}_{\text{Detection}}$ and $\text{CLEval}_{\text{End-to-End}}$, and apply them to all extracted entity word boxes. It is worth noting that, in contrast to traditional OCR evaluations, our approach only considers the entity word boxes, ignoring texts without semantic categories.

F1. F1 evaluation metric is comprised of two stages. Firstly, the model predictions are matched with the ground-truth (GT) bounding boxes. The match

Fig. 9. Synthetic Examples of Business Card Dataset. The dataset comprises photographs of Japanses business cards with texts, text boxes, and semantic category information.

is considered to be successful only when the center point of the prediction falls within the GT bounding box. Secondly, the recognized text within each matched bounding box is compared against the GT transcription. The predicted text is considered to contribute to the end-to-end accuracy only if it is identical to the GT.

CLEval$_{Detection}$ and CLEval$_{End-to-End}$. The CLEval evaluation metric, introduced in Baek et al. [3], is designed to evaluate the performance of text detection and recognition at the character-level granularity, rather than the conventional word-level or sentence-level evaluations. There are two variants of the CLEval metric, CLEval$_{Detection}$ and CLEval$_{End-to-End}$, that differ in whether they take into account text recognition accuracy or not. Unlike traditional metrics, CLEval metrics allow partial correctness by using character units. For example, if the ground truth is "reverside", partial scores are given to predictions that are split like "rever" and "side". Also, if the ground truth is "rever side" and the model predicts "reverside", you will receive a partial score, although there is a penalty. CLEval enables a more comprehensive evaluation of performance by differentiating between instances of entirely incorrect predictions and those that are only partially correct.

Entity Field Level Metrics. We notice that the aforementioned entity word box level metrics, while effective and easy to understand, do not fully account for the semantic categorization of the extracted entity word boxes. For instance, even if a menu name on a receipt is identified as a menu price, there is no penalty. To address this limitation, we propose modified versions of the aforementioned metrics, referred to as Field-level F1, Field-level CLEval$_{Detection}$, and Field-level CLEval$_{End-to-End}$. The key idea is to include an additional condition that evaluates not only the accuracy of text recognition but also the correctness of the predicted semantic category of the entity field. These metrics assess the performance of the document parsing system on the entity field text unit, by considering not only the recognized text, but also the accuracy of the predicted semantic field category.

Table 1. Experiment results on CORD dataset. The proposed **Free Donut**, which is a combination of the proposed method with the OCR-free VDU model of Kim et al. [19], achieved high performance in all metrics, especially in the text recognition and entity label prediction. Although it lagged behind the traditional OCR-based baseline in word-level text detection only, the efficacy of the proposed approach is evident in the overall performance evaluation. The advantage became more obvious in the field-level evaluation, which focuses on the semantic category of an entity field rather than just word box level comparison.

	Word-level			Field-level		
	$CLEval_{Det}$	$CLEval_{E2E}$	F1	$CLEval_{Det}$	$CLEval_{E2E}$	F1
LayoutLMv2 [34]	**97.8**	93.5	85.0	66.9	61.6	51.1
Free Donut (Proposed)	95.3	**94.7**	**88.6**	**80.5**	**77.6**	**64.9**
- Text region proposal	80.1	84.9	72.6	76.8	74.4	53.5
- Char-level tokenization	79.8	86.4	74.2	72.8	75.5	53.1
- Weighted aggregation	70.4	82.2	71.2	71.8	74.6	51.3

4.3   Training Setup and Baseline

We incorporate our method with the model developed by Kim et al. [19] and refer to it as the Bounding Box Annotation Free Document Understanding Transformer (**Free Donut**) in this study. Our document parsing model is achieved be starting from the same official pre-trained model weight (`donut-base`[2]). Our document parsing model is trained using NVIDIA V100 GPUs with a mini-batch size of 4 for CORD and 8 for Business Card, and the rest major hyperparameters, including the number of epochs and input image resolution, follow the setting of Kim et al. [19]. LayoutLMv2 [34] is used as a baseline model, which is one of the current state-of-the-art VDU models. In addition, CLOVA OCR API[3] is used for the baseline, as reported to achieve the best accuracy on the CORD dataset according to Kim et al. [19].

Table 2. Experiment results on Business Card dataset. The results indicate a consistent performance trend with those observed on the CORD dataset. This highlights the versatility of the proposed method in adapting to different document types and its potential for practical application in real-world scenarios.

	Word-level			Field-level		
	$CLEval_{Det}$	$CLEval_{E2E}$	F1	$CLEval_{Det}$	$CLEval_{E2E}$	F1
LayoutLMv2 [34]	**97.6**	**95.5**	66.5	**90.6**	79.0	52.8
Free Donut (Proposed)	93.6	93.9	**77.4**	84.9	**79.2**	**68.9**

---

[2] https://github.com/clovaai/donut.

[3] https://clova.ai/ocr.

4.4   Results

The results of the experiments, presented in Table 1 and Table 2, showcase the effectiveness of the proposed method. Despite the results of the word-level $\text{CLEval}_{\text{Det}}$ suggesting a slight lag behind traditional OCR-based methods in terms of text detection, the proposed method still manages to achieve competitive results. However, when evaluating the combined performance of text detection, recognition and entity recognition, the proposed method attains state-of-the-art performance when integrated with the OCR-free model. These findings indicate that the proposed method is a viable solution for practical applications where both information extraction and accurate text localization are crucial. The experiment outcomes demonstrate the feasibility of adopting an OCR-free model for scenarios requiring text localization.

In addition, we conduct ablation studies on CORD dataset. The results are presented in the second group in Table 1, offering valuable insight into the proposed set of techniques designed to enhance the performance of text localization.

Text Region Proposal. The removal of the unsupervised region proposal resulted in a significant decrease in performance. Nevertheless, the field-level metrics still surpassed the baseline performance even in the absence of the unsupervised region proposal.

Char-Level Tokenization. The comparison between character-level and subword-level tokenization demonstrated that subword-level tokenization mainly impacted the detection performance, yet, in some metrics, an increase in performance was observed. This can be attributed to the advantages of subword-level tokenization for language modeling, such as reducing the length of the sequence to be decoded, resulting in more efficient model training and inference. However, a large drop ($76.8 \rightarrow 72.8$) in detection performance at the field level and our qualitative analysis (see Fig. 5) suggest that character-level tokenization may be slightly more beneficial. Nonetheless, subword-level tokenization may still be a preferable choice in scenarios where text recognition performance is a critical concern.

Weighted Aggregation. Finally, the ablation study on the aggregation of multiple attention heads using variance revealed an overall performance gap, underscoring the importance of utilizing variance information.

5   Discussion

In this section, we delve into the limitations encountered while implementing the text localization feature within the OCR-free Document Understanding Transformer framework and discuss future directions for research. One of the key challenges is posed by the generative nature of the Donut model, which can lead to instances of false positive entity fields. These false positives pose a significant challenge for text localization, resulting in unnatural and difficult-to-interpret

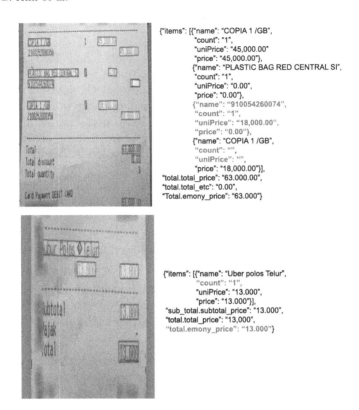

Fig. 10. Illustration of the Failure Cases of the Proposed Method. This figure showcases instances where the model fails to produce accurate results. In cases of underfitting, the Donut model may generate false positive entity fields, resulting in poor text location processing and a suboptimal user experience.

outputs, as demonstrated in Fig. 10. To effectively address this issue, improvements must first be made to the parsing performance of Donut. Further investigation into post-processing methods, such as correcting Donut's parsing output in instances where the text localization results are unnatural, is a potential avenue for future work.

On the other hand, our analysis has shown that the use of a subword tokenizer for text localization has some drawbacks. In particular, the attention map tends to be biased towards a particular text reading direction. To mitigate this, we propose the use of a character-level tokenizer, which leads to a more accurate, unbiased attention map. However, this approach also results in a longer output token sequence, which negatively impacts the overall efficiency of the model and reduces the performance of the language model. Given these trade-offs, further research is necessary to find an effective and efficient solution that balances the benefits of using a subword tokenizer with the importance of accurate text localization.

Another limitation of the proposed approach is related to the input image resolution (H, W). As described in Sect. 3.2, the size of the image patch embeddings $\mathbf{Z}$ is determined as $n = H/p \times W/p$, where (p, p) represents the patch size of the visual encoder. If n is not sufficiently large, the resolution of the attention map may be inadequate, resulting in a degradation of the quality of the generated attention-based region proposals. In our experiments, we adopted the resolution setting suggested by Kim et al. [19]. However, larger values of n come with increased computational cost, and further research is required to find a way to reduce n while preserving the accuracy of text localization.

Furthermore, the text-like region proposal module used in the OCR-free Document Understanding Transformer framework is another area for future work. While we utilized traditional computer vision preprocessing methods and tricks in our implementation, there are other options that could potentially provide improved results. For example, incorporating a text detector trained on synthetic scene texts or document images may yield better performance. Exploring these alternatives will be a valuable direction for future research.

6 Conclusion

In this work, we presented a simple yet effective attention-based text localization algorithm that can be integrated into OCR-free end-to-end document understanding transformer (Donut) models [6, 19, 21]. Our approach utilizes the cross attention mechanism of the Donut model to improve text localization without the need for additional annotations. Through a comprehensive analysis of the attention maps, we proposed a series of techniques to enhance the localization performance. The effectiveness of our method was demonstrated through experiments on two camera-captured document parsing benchmarks, and we proposed new evaluation metrics to assess both information extraction and text localization capabilities. We believe our work helps a significant step towards the development of effective OCR-free document understanding models.

References

1. Baek, J., Kim, G., Lee, J., Park, S., Han, D., Yun, S., et al.: What is wrong with scene text recognition model comparisons? dataset and model analysis. In: Proceedings of the IEEE/CVF International Conference on Computer Vision, pp. 4715–4723 (2019)
2. Baek, Y., Lee, B., Han, D., Yun, S., Lee, H.: Character region awareness for text detection. In: 2019 IEEE/CVF Conference on Computer Vision and Pattern Recognition (CVPR), pp. 9357–9366 (2019). https://doi.org/10.1109/CVPR.2019.00959
3. Baek, Y., et al.: CLEval: character-level evaluation for text detection and recognition tasks. In: Proceedings of the IEEE/CVF Conference on Computer Vision and Pattern Recognition Workshops, pp. 564–565 (2020)
4. Baek, Y.: Character region attention for text spotting. In: Vedaldi, A., Bischof, H., Brox, T., Frahm, J.-M. (eds.) ECCV 2020. LNCS, vol. 12374, pp. 504–521. Springer, Cham (2020). https://doi.org/10.1007/978-3-030-58526-6_30

5. Chen, T., Saxena, S., Li, L., Fleet, D.J., Hinton, G.: Pix2seq: a language modeling framework for object detection. In: International Conference on Learning Representations (2022). https://openreview.net/forum?id=e42KbIw6Wb

6. Davis, B., Morse, B., Price, B., Tensmeyer, C., Wigington, C., Morariu, V.: End-to-end document recognition and understanding with dessurt. In: Karlinsky, L., Michaeli, T., Nishino, K. (eds.) ECCV 2022. LNCS, vol. 13804, pp. 280–296. Springer, Cham (2023). https://doi.org/10.1007/978-3-031-25069-9_19

7. Devlin, J., Chang, M.W., Lee, K., Toutanova, K.: BERT: pre-training of deep bidirectional transformers for language understanding. In: Proceedings of the 2019 Conference of the North American Chapter of the Association for Computational Linguistics: Human Language Technologies, vol. 1 (Long and Short Papers), pp. 4171–4186. Association for Computational Linguistics, Minneapolis, Minnesota (2019). https://doi.org/10.18653/v1/N19-1423. https://aclanthology.org/N19-1423

8. Dillencourt, M.B., Samet, H., Tamminen, M.: A general approach to connected-component labeling for arbitrary image representations. J. ACM **39**(2), 253–280 (1992). https://doi.org/10.1145/128749.128750

9. Dosovitskiy, A., et al.: An image is worth 16x16 words: transformers for image recognition at scale. In: International Conference on Learning Representations (2021). https://openreview.net/forum?id=YicbFdNTTy

10. Gupta, A., Vedaldi, A., Zisserman, A.: Synthetic data for text localisation in natural images. In: Proceedings of the IEEE Conference on Computer Vision and Pattern Recognition (CVPR) (2016)

11. Hong, T., Kim, D., Ji, M., Hwang, W., Nam, D., Park, S.: Bros: A pre-trained language model focusing on text and layout for better key information extraction from documents. In: Proceedings of the AAAI Conference on Artificial Intelligence, vol. 36, no. (10), pp. 10767–10775 (2022). https://doi.org/10.1609/aaai.v36i10.21322. https://ojs.aaai.org/index.php/AAAI/article/view/21322

12. Huang, W., Qiao, Yu., Tang, X.: Robust scene text detection with convolution neural network induced MSER trees. In: Fleet, D., Pajdla, T., Schiele, B., Tuytelaars, T. (eds.) ECCV 2014. LNCS, vol. 8692, pp. 497–511. Springer, Cham (2014). https://doi.org/10.1007/978-3-319-10593-2_33

13. Hwang, W., et al.: Post-OCR parsing: building simple and robust parser via bio tagging. In: Workshop on Document Intelligence at NeurIPS 2019 (2019)

14. Hwang, W., Lee, H., Yim, J., Kim, G., Seo, M.: Cost-effective end-to-end information extraction for semi-structured document images. In: Proceedings of the 2021 Conference on Empirical Methods in Natural Language Processing, pp. 3375–3383. Association for Computational Linguistics, Punta Cana, Dominican Republic (2021). https://doi.org/10.18653/v1/2021.emnlp-main.271. https://aclanthology.org/2021.emnlp-main.271

15. Hwang, W., Yim, J., Park, S., Yang, S., Seo, M.: Spatial dependency parsing for semi-structured document information extraction. In: Findings of the Association for Computational Linguistics: ACL-IJCNLP 2021, pp. 330–343. Association for Computational Linguistics (2021). https://doi.org/10.18653/v1/2021.findings-acl.28. https://aclanthology.org/2021.findings-acl.28

16. Itseez: Open source computer vision library (2015). https://github.com/itseez/opencv

17. Jaderberg, M., Simonyan, K., Vedaldi, A., Zisserman, A.: Synthetic data and artificial neural networks for natural scene text recognition. In: Workshop on Deep Learning, NIPS (2014)

18. Karatzas, D., et al.: ICDAR 2015 competition on robust reading. In: 2015 13th International Conference on Document Analysis and Recognition (ICDAR), pp. 1156–1160 (2015). https://doi.org/10.1109/ICDAR.2015.7333942

19. Kim, G., et al.: OCR-free document understanding transformer. In: Avidan, S., Brostow, G., Cissé, M., Farinella, G.M., Hassner, T. (eds.) ECCV 2022. LNCS, vol. 13688, pp. 498–517. Springer, Cham (2022). https://doi.org/10.1007/978-3-031-19815-1_29

20. Kim, S., et al.: Deer: Detection-agnostic end-to-end recognizer for scene text spotting. arXiv preprint arXiv:2203.05122 (2022)

21. Lee, K., et al.: Pix2Struct: screenshot parsing as pretraining for visual language understanding (2023). https://openreview.net/forum?id=UERcQuXlwy

22. Liao, M., Pang, G., Huang, J., Hassner, T., Bai, X.: Mask TextSpotter v3: segmentation proposal network for robust scene text spotting. In: Vedaldi, A., Bischof, H., Brox, T., Frahm, J.-M. (eds.) ECCV 2020. LNCS, vol. 12356, pp. 706–722. Springer, Cham (2020). https://doi.org/10.1007/978-3-030-58621-8_41

23. Liao, M., Shi, B., Bai, X., Wang, X., Liu, W.: Textboxes: a fast text detector with a single deep neural network. In: Proceedings of the AAAI Conference on Artificial Intelligence, vol. 31, no. 1 (2017). https://doi.org/10.1609/aaai.v31i1.11196. https://ojs.aaai.org/index.php/AAAI/article/view/11196

24. Liu, X., Liang, D., Yan, S., Chen, D., Qiao, Y., Yan, J.: FOTS: fast oriented text spotting with a unified network. In: Proceedings of the IEEE Conference on Computer Vision and Pattern Recognition, pp. 5676–5685 (2018)

25. Liu, Y., Chen, H., Shen, C., He, T., Jin, L., Wang, L.: ABCNet: real-time scene text spotting with adaptive Bezier-curve network. In: Proceedings of the IEEE/CVF Conference on Computer Vision and Pattern Recognition, pp. 9809–9818 (2020)

26. Liu, Z., et al.: Swin transformer: hierarchical vision transformer using shifted windows. In: Proceedings of the IEEE/CVF International Conference on Computer Vision (ICCV), pp. 10012–10022 (2021)

27. Lucas, S., Panaretos, A., Sosa, L., Tang, A., Wong, S., Young, R.: ICDAR 2003 robust reading competitions. In: Seventh International Conference on Document Analysis and Recognition, 2003. Proceedings, pp. 682–687 (2003). https://doi.org/10.1109/ICDAR.2003.1227749

28. Park, S., et al.: CORD: a consolidated receipt dataset for post-OCR parsing. In: Workshop on Document Intelligence at NeurIPS 2019 (2019)

29. Phan, T.Q., Shivakumara, P., Tian, S., Tan, C.L.: Recognizing text with perspective distortion in natural scenes. In: Proceedings of the IEEE International Conference on Computer Vision (ICCV) (2013)

30. Tian, Z., Huang, W., He, T., He, P., Qiao, Yu.: Detecting text in natural image with connectionist text proposal network. In: Leibe, B., Matas, J., Sebe, N., Welling, M. (eds.) ECCV 2016. LNCS, vol. 9912, pp. 56–72. Springer, Cham (2016). https://doi.org/10.1007/978-3-319-46484-8_4

31. Vaswani, A., et al.: Attention is all you need. In: Guyon, I., et al. (eds.) Advances in Neural Information Processing Systems. vol. 30. Curran Associates, Inc. (2017). https://proceedings.neurips.cc/paper/2017/file/3f5ee243547dee91fbd053c1c4a845aa-Paper.pdf

32. Vincent, L., Soille, P.: Watersheds in digital spaces: an efficient algorithm based on immersion simulations. IEEE Trans. Pattern Anal. Mach. Intell. **13**(6), 583–598 (1991). https://doi.org/10.1109/34.87344

33. Wang, K., Babenko, B., Belongie, S.: End-to-end scene text recognition. In: Proceedings of the 2011 International Conference on Computer Vision, pp. 1457–1464.

ICCV 2011, IEEE Computer Society, USA (2011). https://doi.org/10.1109/ICCV.2011.6126402

34. Xu, Y., et al.: LayoutLMv2: Multi-modal pre-training for visually-rich document understanding. In: Proceedings of the 59th Annual Meeting of the Association for Computational Linguistics and the 11th International Joint Conference on Natural Language Processing, vol. 1 (Long Papers), pp. 2579–2591. Association for Computational Linguistics (2021). https://doi.org/10.18653/v1/2021.acl-long.201. https://aclanthology.org/2021.acl-long.201

35. Xu, Y., Li, M., Cui, L., Huang, S., Wei, F., Zhou, M.: LayoutLM: pre-training of text and layout for document image understanding. In: Proceedings of the 26th ACM SIGKDD International Conference on Knowledge Discovery & Data Mining, pp. 1192–1200 (2020)

36. Xue, C., Zhang, W., Hao, Y., Lu, S., Torr, P.H.S., Bai, S.: Language matters: a weakly supervised vision-language pre-training approach for scene text detection and spotting. In: Avidan, S., Brostow, G., Cissé, M., Farinella, G.M., Hassner, T. (eds.) ECCV 2022. LNCS, vol. 13688, pp. 284–302. Springer, Cham (2022). https://doi.org/10.1007/978-3-031-19815-1_17

37. Yang, Z., et al.: UniTAB: unifying text and box outputs for grounded vision-language modeling. In: Avidan, S., Brostow, G., Cissé, M., Farinella, G.M., Hassner, T. (eds.) ECCV 2022. LNCS, vol. 13696, pp. 521–539. Springer, Cham (2022)

38. Zhang, Z., Zhang, C., Shen, W., Yao, C., Liu, W., Bai, X.: Multi-oriented text detection with fully convolutional networks. In: 2016 IEEE Conference on Computer Vision and Pattern Recognition (CVPR), pp. 4159–4167 (2016). https://doi.org/10.1109/CVPR.2016.451

IndicSTR12: A Dataset for Indic Scene Text Recognition

Harsh Lunia[✉][ID], Ajoy Mondal[ID], and C. V. Jawahar[ID]

Centre for Vision Information Technology, International Institute of Information Technology, Hyderabad 500032, India
harsh.lunia@research.iiit.ac.in,
{ajoy.mondal,jawahar}@iiit.ac.in
http://cvit.iiit.ac.in/research/projects/cvit-projects/indicstr

Abstract. The importance of Scene Text Recognition (STR) in today's increasingly digital world cannot be overstated. Given the significance of STR, data-intensive deep learning approaches that auto-learn feature mappings have primarily driven the development of STR solutions. Several benchmark datasets and substantial work on deep learning models are available for Latin languages to meet this need. On more complex, syntactically and semantically, Indian languages spoken and read by 1.3 billion people, there is less work and datasets available. This paper aims to address the Indian space's lack of a comprehensive dataset by proposing the largest and most comprehensive real dataset - Indic-STR12 - and benchmarking STR performance on 12 major Indian languages (Assamese, Bengali, Odia, Marathi, Hindi, Kannada, Urdu, Telugu, Malayalam, Tamil, Gujarati, and Punjabi). A few works have addressed the same issue, but to the best of our knowledge, they focused on a small number of Indian languages. The size and complexity of the proposed dataset are comparable to those of existing Latin contemporaries, while its multilingualism will catalyse the development of robust text detection and recognition models. It was created specifically for a group of related languages with different scripts. The dataset contains over 27000 word-images gathered from various natural scenes, with over 1000 word-images for each language. Unlike previous datasets, the images cover a broader range of realistic conditions, including blur, illumination changes, occlusion, non-iconic texts, low resolution, perspective text etc. Along with the new dataset, we provide a high-performing baseline on three models: PARSeq (Latin SOTA), CRNN, and STARNet.

Keywords: Scene Text Recognition · Indian Languages · Synthetic Dataset · Photo-OCR · OCR · Multi-lingual · Indic Scripts · Real Dataset

1 Introduction

Language has enabled people world around to exchange and communicate. Different communities have recognized the importance of the same, which becomes evident by the diversity of languages across the globe. The textual representation of language, on the other hand, significantly broadens the scope of transfer. Semantically rich writing

M. Coustaty and A. Fornés (Eds.): ICDAR 2023 Workshops, LNCS 14193, pp. 233–250, 2023.
https://doi.org/10.1007/978-3-031-41498-5_17

found in the wild has powerful information that can substantially aid in understanding the surrounding environment in the modern era. Textual information found in the wild is used for various tasks, including image search, translation, transliteration, assistive technologies (particularly for the visually impaired), autonomous navigation, and so on. The issue of automatically reading text from photographs or frames of a natural environment is referred to as Scene Text Recognition (STR) or Photo-OCR. This problem is typically subdivided into two sub-problems: Scene text detection, which deals with locating text within a picture, and cropped word image recognition. Our work addresses the second sub-problem: recognizing the text in a clipped word image.

Traditionally, OCR has focused on reading printed or handwritten text in documents. However, as capturing devices such as mobile phones and video cameras have proliferated, scene text recognition has become critical and a problem whose solution holds promise for furthering the resolution of other downstream tasks. Although there has been considerable progress in STR, it has some unique issues, i) varying backgrounds in natural scenes, ii) varying script, font, layout, and style, and iii) text-related image flaws such as blurriness, occlusion, uneven illumination, etc. Researchers have attempted to address the aforementioned issue by amassing datasets specifically tailored to a given problem, each of which has some distinguishing features and represents a subset of challenges encountered in real-world situations. For example, [28] has perspective text, [40] has blurred text, and [29] has curved text.

Rather than designing and testing manually created features, nearly all current solutions rely on deep learning techniques to automate feature learning. Because of the data-intensive nature of these models, it has become standard practice to train the models on synthetically generated data that closely resemble real-world circumstances and test the trained models on difficult-to-obtain real datasets. STR solutions for Latin languages such as English have made significant progress. However, Latin STR having reached a certain level of maturity, has begun to train solely on available real datasets [3] to achieve nearly comparable performance compared to a mix of synthetic and real. Using an already available diverse set of public real datasets totaling almost 0.3 million image instances for English was a significant factor that led to similar results.

Because Indian language scripts are visually more complex, and their output space is much larger than English languages, not all Latin STR models can mimic performance in Indian STR solutions [24]. STR solutions have not progressed in the case of Indian languages, which are spoken by 18% of the world population, due to a lack of real datasets and models that are better equipped to handle the inherent complexities of the languages. Non-Latin languages have made less progress, and existing Latin STR models need to generalize better to different languages [8].

Contribution. Given the need for more data for non-Latin languages, particularly Indian languages, our work attempts to address the issue by contributing as follows:

1. We propose a real dataset (Fig. 1 (left)) for 12 Major Indian Languages, namely - Assamese, Bengali, Odia, Marathi, Hindi, Kannada, Urdu, Telugu, Malayalam, Tamil, Gujarati and Punjabi - wherein Malayalam, Telugu, Hindi, and Tamil word-images have been taken from [23] and [12]. Since the number of word instances proposed by [12] for Gujarati was less than 1000 work image instances, we augment

Fig. 1. Samples from IndicSTR12 Dataset: Real word-images (left); Synthetic word-images (right)

the proposed Gujarati instances to achieve numbers comparable to other languages in the proposed dataset.

2. We propose a synthetic dataset for all 13 Indian languages (Fig. 1 (right)), which will help the STR community progress on multi-lingual STR, in effect similar to SynthText [13] and MJSynth [16].

3. Finally, we compare the performance of three STR models - PARSeq [5], CRNN, and STARNet [20] - on all 12 Indian languages, some of which have no previous benchmark to compare with. The effectiveness of these models on the IndicSTR12 dataset and other publicly accessible datasets supports our dataset's claim that it is challenging (Table 6).

4. By simultaneously training on multiple languages' real datasets, we demonstrate how multi-lingual recognition models can aid models in learning better, even with sparse real data.

2   Related Works

Scene text recognition (STR) models use CNNs to encode image features. For decoding text out of the learnt image features in a segmentation freeway, it relies either on Connectionist Temporal Classification (CTC) [11] or encoder-decoder framework [37] combined with attention mechanism [4]. The CTC-based approaches [20,34] treat images as a sequence of vertical frames and combine prediction per frame based on a rule to generate the whole text. In contrast, the encoder-decoder framework [21] uses attention to align input and output sequences. There has been work on both CTC and attention-based models for STR. DTRN [15] is the first to use CRNN models, a combination of CNNs with RNNs stacked on them, to generate convolutional feature slices to be fed to RNNs. Using attention, [21] performs STR based on encoder-decoder model wherein the encoder is trained in binary constraints to reduce computation cost. Work on datasets of varying complexity has been done to promote research on STR for challenging scenarios as well. A few challenging ones in terms of occlusion, blur, small and multi orientation word-images are ICDAR 2015 [17], Total-Text [9], LSVT [35,36].

Indian Scene Text Recognition. The lack of annotated data is a hurdle, especially in the case of Indian languages, in realizing the success achieved in the case of Latin STR solutions. There has been attempt to address the data scarcity problem over the years in a scattered and very language-specific way for Indian languages. [7] is the first work to propose an Urdu dataset and benchmark STR performance on Urdu text. It contains 2,500 images, giving 14,100 word-images. The MLT-17 dataset [27] contains 18k scene images in multiple languages, including Bengali. Building on top of it, MLT-19 [26] contains 20k scene images in multiple languages, including Bengali and Hindi. To our knowledge, this is currently the only multilingual dataset, and it supports ten different languages. [23] trains a CRNN model on synthetic data for three Indian languages: Malayalam, Devanagari (Hindi), and Telugu. It also releases an IIIT-ILST dataset for mentioned three languages for testing, reporting a WRR of 42.9%, 57.2% and 73.4% in Hindi, Telugu, and Malayalam, respectively. [6] proposes a CNN and CTC-based method for script identification, text localization, and recognition. The model is trained and tested on MLT 17 dataset, achieving 34.20% WRR for Bengali. An OCR-on-the-go model [30] obtained a WRR of 51.01% on the IIIT-ILST Hindi dataset and a CRR of 35% on a multi-lingual dataset containing 1000 videos in English, Hindi, and Marathi. [12] explored transfer learning among Indian languages as an approach to increase WRR and proposed a dataset of natural scene images in Gujarati and Tamil to test the hypothesis further. It achieved a WRR gain of 6, 5 and 2% on the IIIT-ILST dataset and a WRR of 69.60% and 72.95% in Gujarati and Tamil respectively.

3   Datasets and Motivation

3.1   Synthetic Dataset

Table 1. Statistics of Synthetic Data

Language	Vocabulary Size	Mean, Std	Min, Max	Fonts
Gujarati	106,551	5.96 , 1.85	1 , 20	12
Urdu	234,331	6.39 , 2.20	1 , 42	255
Punjabi	181,254	6.45 , 2.18	1 , 31	141
Manipuri (Meitei)	66,222	6.92 , 2.49	1 , 29	24
Assamese	77,352	7.08 , 2.70	1 , 46	85
Odia	149,681	8.00 , 3.00	1 , 37	30
Bengali	449,986	8.53 , 3.00	1 , 38	85
Marathi	180,278	8.64 , 3.43	1 , 44	218
Hindi	319,982	8.76 , 3.20	1 , 50	218
Telugu	499,969	9.75 , 3.38	1 , 50	62
Tamil	399,999	10.75 , 3.64	1 , 35	158
Kannada	499,972	10.72 , 3.87	1 , 41	30
Malayalam	320,000	14.30 , 5.36	1 , 53	101

It has been an accepted practice in the community to train an STR model on a large synthetically generated dataset since [16] trained a model on 8 Million synthetically generated English word-images called MJSynth. This trend, as [2] points, is due to the high cost of annotating real data. Another synthetic dataset widely in use for English language is SynthText [13]. On the Indian language side, there have been some works like [12,23] which use synthetic datasets for a total of 6 Indian Languages. However, like real dataset scenario, a comprehensive synthetic dataset for all 12 major Indian languages is absent. We extend the previously referred work and propose a synthetic dataset for all 12 Major Indian languages by following the same procedure as [23]. The word images are rendered by randomly sampling words from a vocabulary of more than 100K words for each language (except for Assamese) and rendering them using freely available Unicode fonts Table 1. Each word image is first rendered in the foreground layer by varying font, size, and stroke thickness, color, kerning, rotation along the horizontal line, and skew. This is followed by the applying a random perspective projective transformation to the foreground layer and, consequently, a blending of the same with a random crop from a natural scene image taken from Places365 dataset [42]. Lastly, the foreground image is alpha composed with a background image which can either be a random crop from a natural scene image or one having a uniform color. The synthetic dataset proposed has more than 3 Million word images per language. For benchmarking STR performance, we have followed the same procedure as [12], using 2 Million word images for training the network and 0.5 Million for validation and testing.

3.2   Real Dataset

Table 2. Usage Statistics and General Information of Official Indian Languages not part of Indic-STR12 Dataset

Language	Script	Usage	Family
Bodo	Devanagari	1.4M	Sino-Tibetan
Kashmiri	Arabic & Devanagari	11M	Indo-Aryan
Dogri	Devanagari	2.6M	Indo-Aryan
Konkani	Devanagari	2.3M	Indo-Aryan
Maithali	Devanagari	34M	Indo-Aryan
Sindhi	Arabic & Devanagari	32M	Indo-Aryan
Santhali	Ol Chiki	7.6M	Austroasiatic
Nepali	Devanagari	25M	Indo-Aryan

According to the Census 2011 report on Indian languages [10], India has 22 major or scheduled languages with a significant volume of writing. All major Indian languages can be classified into four language families: Indo-Aryan, Dravidian, Sino-Tibetan, and Austro-Asian (listed in decreasing order of usage). Sanskrit, Bodo, Dogri, Kashmiri, Konkani, Maithili, Nepali, Santali, and Sindhi are among the languages not covered by the IndicSTR12 dataset Table 2. As a classical language, Sanskrit has a long history of heavily influencing all of the subcontinent's languages. It is now widely taught at the secondary level, but its use is limited to ceremonial and ritualistic purposes with no first-language speakers. Other languages that have been left out either have scripts that are similar to one of the included languages[1] or have minimal usage in the domain of scene text in natural settings[2]. According to the 2011 census report [10], the included languages cover 98% of the subcontinent's spoken language. IndicSTR12 is an extension of IIIT-ILST [23] and [12], which cover Telugu, Malayalam, Hindi, Gujarati, and Tamil, respectively. There has been no addition of images for any of the mentioned languages, except for Gujarati, which had less than 1000 word-images.

Fig. 2. IndicSTR12 Dataset: Font Variations for the same word - Gujarati or Gujarat

IndicSTR12 Curation Details. All the images have been crawled from Google Images using various keyword-based searches to cover all the daily avenues wherein Indic language text can be observed in natural settings. To mention some - Wall paintings, railway stations, Signboards, shop/temple/mosque/gurudwara name-boards, advertisement banners, political protests, house plates, etc. Because they have been crawled from a search engine, they come from various sources, offering a wide range of conditions under which images were captured. Curated images have blur, non-iconic/iconic text, low-resolution, occlusion, curved text, perspective projections due to non-frontal viewpoints, etc Figs. 3 and 2.

IndicSTR12 Annotation Details. All the words within the image are annotated with four-corner point annotation as done in [26] to capture both horizontal and curved word structures of scene texts. The annotators were encouraged to follow the reading direction and include as little background space as possible. To further ensure the quality of annotation, all the annotated data was reviewed by another entity to ensure proper removal/correction of empty or wrong labels. The reviewers were also tasked with further classifying each word instance into the three categories of Oriented Text, Low-resolution/Smaller Text, and Occluded Text. This will enable community members to assess which areas of a model's performance require special consideration. There are at least 1000 word images per language and their corresponding labels in Unicode. The dataset can also be used for the problem of script identification and scene text detection.

---

[1] Bodo, Dogri, Kashmiri, Konkani, Maithili, Nepali, and Sindhi.
[2] Santali.

3.3   Comparision with Existing Datasets

Table 3. Statistics of Various Public STR Real Dataset

Dataset	Word Images (train/test)	Language	Features	Tasks[a]
IIIT5K-Words [25]	2K/3K	English	Regular	R
SVT [40]	211/514	English	Regular, blur, low resolution	D,R
ICDAR2003 [22]	1157/1111	English	Regular	D,R
ICDAR2013 [18]	3564/1439	English	Regular, Stroke Labels	D,R
ICDAR2015 [17]	4468/2077	English	Irregular, Blur, Small	D,R
SVT Perspective [28]	0/639	English	Irregular, Perspective Text	R
MLT-19 [26]	89K/102K	Multi-Lingual[b]	Irregular	D,R
MTWI [14]	141K/148K	Chinese, English	Irregular	R
LSVT [35,36]	30K/20K	Chinese, English	Irregular, multi-oriented	D, R
MLT-17 [27]	85K/11K	Multi-Lingual[c]	Irregular	D,R
Urdu-Text [7]	14100	Urdu	Irregular, Noisy	D,R
CUTE80 [29]	0/288	English	Irregular, Perspective Text, Low Resolution	D,R
IndicSTR12 (Ours)[d]	20K/7K	Multi-Lingual[e]	Irregular, Low Resolution, Blur, Occlusion, Perspective Text	D,R

[a] 'D' Stands for Detection and 'R' for Recognition

[b] Arabic, Bangla, Chinese, Devanagari, English, French, German, Italian, Japanese, and Korean

[c] Arabic, Bangla, Chinese, English, French, German, Italian, Japanese, and Korean

[d] Extends IIIT-ILST [12,23]

[e] Assamese, Bengali, Odia, Marathi, Hindi, Kannada, Urdu, Telugu, Malayalam, Tamil, Gujarati and Punjabi

Real datasets were initially utilised for fine-tuning models trained on synthetic datasets and evaluating trained STR models because the majority of real data sets only contain thousands of word-images. There are many datasets available for English Languages that address a variety of difficulties, but the community has only now begun to consider training models with real data [3]. Broadly a real dataset can be seen as regular and irregular types. Regular datasets have most word-images that are iconic (frontal), horizontal and a small portion of distorted samples. In contrast, the majority of the word-images in irregular datasets are perspective text, low-resolution, and multi-oriented. The high variation in text instances makes these difficult for STR. For examples and more information, please refer to Table 3. Our Dataset has been curated with the goal of catering to both regular and irregular samples. Being extracted from the Google search engine via various keywords generally associated with scene and text, it tries to cover a wide range of natural scenarios, mostly seen in Indian Scene texts. The classification of word-images into categories was done to allow Indian STR solutions to assess their current standing on regular texts and, as the solutions advance, to provide a challenging subset to further refine the prediction models.

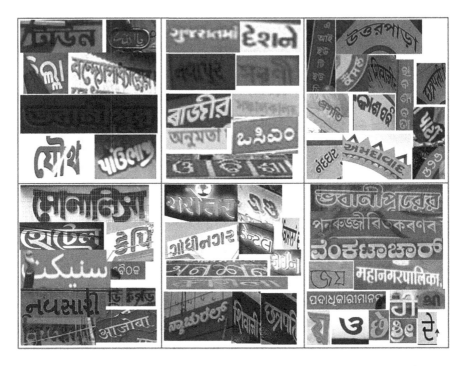

Fig. 3. IndicSTR12 Dataset Variations, clockwise from Top-Left: Illumination variation, Low Resolution, Multi-Oriented - Irregular Text, Variation in Text Length, Perspective Text, and Occluded.

4   Models

This section explains the models used to benchmark STR performance on 12 Indian languages in IndicSTR12 dataset. Three models were picked up to benchmark the STR performance - PARSeq [5] is the current state-of-the-art Latin STR model, CRNN [31] which has low accuracy than a lot of current models but is widely chosen by the community for practical usage because it's lightweight and fast. Another model called STAR-Net [20], extracts more robust features from word-images and performs an initial distortion correction, is also used for benchmarking. This model has been taken up to maintain consistency with the previous works on Indic STR [12,23].

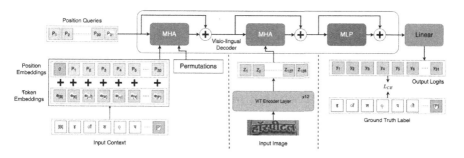

Fig. 4. PARSeq architecture. [B] and [P] begin the sequence and padding tokens. T = 30 or 30 distinct position tokens. L_{CE} corresponds to cross entropy loss.

PARSeq. Figure 4 PARSeq is a transformer-based model which is trained using Permutation Language Modeling (PLM). Multi-head Attention [39] is extensively used, MHA(q, k, v, m), where q, k, v and m refer to query, key, value and optional attention mask. The model follows an encoder-decoder architecture wherein the encoder stack has 12 encoder blocks while the decoder has only a single block.

The model uses 12 vision transformer encoder blocks each containing 1 self-attention MHA module. The image $x \in \mathbb{R}^{W \times H \times C}$ is tokenized evenly into $p_w \times p_h$ patches which are in-turn projected into d_{model} - dimensional tokens using an embedding matrix $W^p \in \mathbb{R}^{p_w p_h C \times d_{model}}$. Position embeddings are then added to tokens before sending them to the first ViT encoder block. All output tokens z are used as input to the decoder:

$$z = Enc(x) \in \mathbb{R}^{\frac{WH}{p_w p_h} \times d_{model}}$$

The Visio-lingual Decoder used is a pre-layerNorm transformer decoder with two MHAs. The first MHA requires position tokens, $p \in \mathbb{R}^{(T+1) \times d_{model}}$ (T being context length), context embeddings, $c \in \mathbb{R}^{(T+1) \times d_{model}}$, and attention mask, $m \in \mathbb{R}^{(T+1) \times (T+1)}$. The position token captures the target position to be predicted and decouples the context from the target position, enabling the model to learn from permutation language modeling. The attention masks vary at different use points. During training, they are based on permutations, while during inference it is a left-to-right look-ahead mask. Transformers process all tokens in parallel, therefore to enforce the condition of past tokens having no access to future ones, attention masks are used. PLM, which in theory requires the model to train on all $T!$ factorizations, in practice is achieved by using attention masks to enforce some subset K of $T!$ permutations.

$$h_c = p + MHA(p, c, c, m) \in \mathbb{R}^{(T+1) \times d_{model}}$$

The second MHA is used for image-position attention where no attention mask is used.

$$h_i = h_c + MHA(h_c, z, z \in \mathbb{R}^{(T+1) \times d_{model}})$$

The last decoder hidden state is used to get the output logits $\boldsymbol{y} = Linear(\boldsymbol{h}_{dec} \in \mathbb{R}^{(T+1)\times(S+1)}$ where S is the size of the character set and an addition of 1 is due to end of sequence token $[\boldsymbol{E}]$.

The decoder block can be represented by:

$$\boldsymbol{y} = Dec(\boldsymbol{z}, \boldsymbol{p}, \boldsymbol{c}, \boldsymbol{m}) \in \mathbb{R}^{(T+1)\times(S+1)}$$

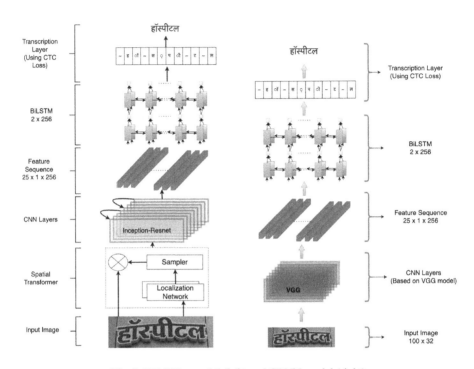

Fig. 5. STARNet model (left) and CRNN model (right)

CRNN. CRNN is a combination of CNN and RNN, as shown in Fig. 5 (right). The model primarily can be viewed as a combination of 3 components - (i) an encoder, here a standard VGG model [32], to extract features from word-image, (ii) a decoder consisting of RNN and lastly, (iii) a Connectionist Temporal Classification (CTC) layer which aligns decoded sequence to ground truth. The CNN encoder is made of 7 layers to extract the feature maps. For RNN, it uses a two-layer BiLSTM model, each with a hidden size of 256 units. During training, the CTC layer provides non-parameterized supervision to ensure that predictions match the ground truth. All of our experiments used a PyTorch implementation of the model by [31].

STARNet. STARNet in Fig. 5 (left), like CRNN, consists of three components: an encoder that is a CNN-based model, a RNN-based decoder, and a Connectionist Temporal Classification (CTC) layer to align the decoded sequence with the ground truths.

However, it differs from CRNN in two ways: it performs initial distortion correction using a spatial transformer, and its CNN is based on an inception ResNet architecture [38], which can extract more robust features required for STR.

5   Experiments

Each STR model is trained on 2 million and tested on 0.5 million synthetic word-images. To further adapt the model to real-world word-images, it is trained and tested on the proposed IndicSTR12. We use 75% of the word-images for training and 25% for testing in each language.

All PARSeq models are trained on dual-GPU platforms with Pytorch DDP for 20–33 epochs and 128 batch size. In conjunction with the 1cycle learning rate scheduler [33], the Adam optimizer [19] is used. As in the PARSeq model, we use K = 6 permutations with mirroring for PLM and an 8×4 patch size for ViTSTR. The maximum label length for the transformer-based PARSeq model is determined by the vocabulary used to create the synthetic dataset. We avoid using any data augmentation on synthetic datasets in accordance with community practise [3].

Using a spatial transformer, the STARNet model transforms a resized input image of 150×18 to 100×32. Both STARNet and CRNN encoders accept images of size 100×32, and the output feature maps are of size $23 \times 1 \times 256$. All CRNN and STAR-Net models are trained on 2 Million synthetic images on a batch size of 32 and with ADADELTA [41] optimizer for stochastic gradient descent. The number of epochs is fixed at 15. For each language, the models are tested on 0.5 million synthetic images.

We also run experiments on other Indic datasets, including MLT-17 [27] for Bengali (referred to as Bangla in MLT-17), MLT-19 [26] for Hindi, and Urdu-Text [7] for Urdu. For all other public real datasets, we finetune on the train split and test our models on the test split (if no test splits are available, the proposer's val split is used).

6   Result and Analysis

6.1   Benchmarking

In this section, we first list and compare the respective models' performance on a synthetic dataset, then on our proposed real dataset.

Performance on Synthetic Dataset. Table 4 shows the CRR and WRR for each of the 13 languages achieved by various models. According to the data, the PARSeq model [5] has clearly outperformed other models in terms of WRR and CRR in all 13 languages. This gain can be attributed to it being an attention-based model and using Permutation Language Modeling to further capture the context. Attention models have been shown to outperform CTC-based models in general by Latin solutions; the same holds true for the Indic language in the case of synthetic data points.

Table 4. Performance on Synthetic Data

Language	CRNN		STARNet		PARSeq	
	CRR	WRR	CRR	WRR	CRR	WRR
Kannada	83.44	48.69	90.71	66.13	**97.26**	**87.92**
Odiya	89.94	66.57	95.05	81.00	**98.46**	**93.27**
Punjabi	88.88	66.11	93.86	78.34	**97.08**	**87.89**
Urdu	76.83	39.90	86.51	58.35	**95.26**	**80.48**
Marathi	86.91	59.87	93.55	76.32	**99.08**	**95.82**
Assamese	88.86	67.90	94.76	82.93	**99.21**	**96.85**
Manipuri (Meitei)	89.83	67.57	94.71	80.74	**98.79**	**94.76**
Malayalam	85.48	48.42	93.23	69.28	**98.71**	**92.27**
Telugu	81.12	43.75	89.54	62.23	**96.31**	**83.4**
Hindi[a]	89.83	73.15	95.78	83.93	**99.13**	**95.61**
Bengali[a]	91.54	70.76	95.52	82.79	**98.39**	**92.56**
Tamil[a]	82.86	48.19	95.40	79.90	**97.88**	**90.31**
Gujarati[a]	94.43	81.85	97.80	91.40	**98.82**	**95.25**

[a] Values for CRNN and STARNet have been taken from Guna *et al.* [12] as the training parameters and synthetic data generator were same.

Performance on Real Dataset

IndicSTR12 Dataset: The CRR and WRR numbers for the three models on the Indic-STR12 dataset are listed in the Table 5. Because the dataset is an extension of [12] and IIIT-ILST [23], the Hindi and Tamil numbers have been directly quoted from their work. We conducted separate experiments for Malayalam and Telugu (also covered by the two papers) because the cited works used a larger number of synthetic data to achieve higher accuracies. This was done to make a more accurate comparison with other languages' performance and to accurately gauge the models' performance on real data. Furthermore, the existing data for Gujarati was less than the required minimum of 1000 word-images per language, so this extension also supplements the Gujarati dataset.

A careful examination of the WRR and CRR numbers reveals that the PARSeq model outperforms in almost all cases where the real dataset is large enough, say greater than 1500. In a few cases, PARSeq falls short of the other two due to a lack of word-image instances for the model to train on.

Other Public Dataset: The models' overall performance, as shown in Table 6, followed a similar pattern to that of the IndicSTR12 dataset. However, in contrast to trend seen for IndicSTR12, the performance of the STARNet models is comparable to that of PARSeq and somewhat better for MLT-19 Hindi and Urdu Text. Importantly, all models can be seen to achieve WRR substantially higher than in the case of the IndicSTR12 dataset if the number of word-images is similar. This trend is most pronounced for Urdu Text (refer Fig. 6, where 10% of the original dataset roughly equals the number

Table 5. Performance on IndicSTR12

Language	CRNN		STARNet		PARSeq		Word-images
	CRR	WRR	CRR	WRR	CRR	WRR	
Kannada	78.79	52.43	82.59	59.72	**88.64**	**63.57**	1074
Odiya	80.39	54.74	86.97	66.30	**89.13**	**71.30**	3650
Punjabi	83.15	68.85	84.93	62.5	**92.68**	**78.70**	3887
Urdu	63.68	26.7	74.60	41.48	**76.97**	**44.19**	1375
Marathi	70.79	50.96	83.73	58.65	**86.74**	**63.50**	1650
Assamese	59.25	43.02	80.97	51.83	**81.36**	**52.70**	2154
Malayalam	77.94	53.12	84.97	**70.09**	**90.10**	68.81	807
Telugu	78.07	58.12	85.52	63.44	**92.18**	**71.94**	1211
Hindi^a	**78.84**	46.56	78.72	**46.60**	76.01	45.14	1150
Bengali	59.86	48.21	80.26	57.70	**83.08**	**62.04**	3520
Tamil^a	75.05	59.06	**89.69**	**71.54**	87.56	67.35	2536
Gujarati (New)^b	52.22	23.05	**75.75**	41.80	74.49	**45.10**	1021
Gujarati	53.34	42.58	74.82	51.56	**85.02**	**60.61**	922

^{a} Values for CRNN and STARNet have been taken from Guna *et al.* [12] as the training parameters and real data were same.
^{b} Excluding [12] data

in IndicSTR12), MLT-19 Bengali, and MLT-17 Bengali. This further demonstrates that IndicSTR12 is more challenging due to all the irregular samples and the fact that most of the images in the MLT-17, MLT-19, and Urdu Text dataset are frontal captures or regular in form. In the case of the Urdu Text dataset, the models achieved nearly identical performance utilizing just half of the dataset.

Table 6. Performance on Other Public Datasets

Dataset	Word-images	CRNN		STARNet		PARSeq	
	Train/Test	CRR	WRR	CRR	WRR	CRR	WRR
MLT-17 (Bengali)	3237/713	79.98	55.30	85.24	65.73	**88.72**	**71.25**
MLT-19 (Bengali)	3935/-	82.80	59.51	89.46	71.25	**90.10**	**72.59**
MLT-19 (Hindi)	3931/-	86.48	67.90	91.00	**75.97**	**91.80**	75.91
Urdu -Text	12076/1480	93.33	82.45	97.91	**94.26**	**97.93**	93.92

Error Analysis. Some failure cases were found after analyzing the PARSeq model's ViT encoder attention maps for prediction mistakes (Fig. (7)) using the method proposed by [1]. The PARSeq model can recognize the textual region even in irregular word-image examples, but its predictions for the text are severely inaccurate for low-resolution images (Fig. (7a)) and only somewhat reliable for rotated or curved text

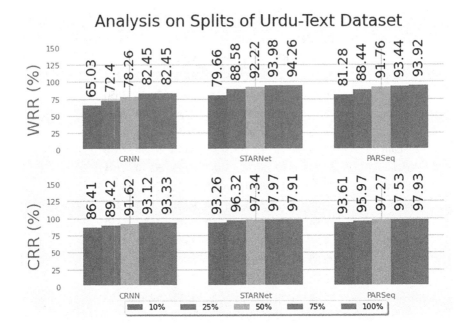

Fig. 6. Urdu Splits Analysis: Effect of training samples on STR models performance

(Fig. (7b)). Additionally, because lower and upper matra are not given adequate consideration, the model misses a lot of accurate predictions (Fig. (7d)). Another notable instance of failure is when distorted text or unusual fonts cause shadows or border thickness to be seen as a different character (Fig. (7e)). The model does reasonably well when dealing with typical iconic texts (Fig. (7f)) and does well overall when dealing with lengthy texts.

6.2   Multi-lingual Training

Table 7. Multi-Lingual Training

Lang.(s) Trained On	Lang. Tested On	PARSeq	
		CRR	WRR
Hindi	Hindi	76.01	45.14
Hindi-Gujarati	Hindi	**76.41**	**49.65**
Gujarati	Gujarati	74.49	45.51
Hindi-Gujarati	Gujarati	**75.68**	**45.70**

The community has investigated transfer learning for STR models [12] as well as multilingual models in the case of OCR [24] since there few real training examples available.

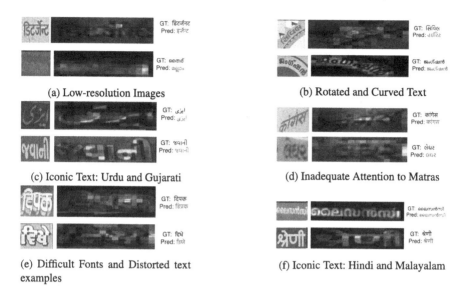

(a) Low-resolution Images

(b) Rotated and Curved Text

(c) Iconic Text: Urdu and Gujarati

(d) Inadequate Attention to Matras

(e) Difficult Fonts and Distorted text examples

(f) Iconic Text: Hindi and Malayalam

Fig. 7. Error Analysis using Attention Maps

It is warranted as Indic language groups share a syntactic and semantic commonality. We looked into the multi-lingual model perspective here for Indic Languages in STR to provide a demonstrative example. In order to have a fair comparison, since single language models are trained on 2M synthetic datasets and finetuned on their respective real images, we trained our multi-lingual model on the same number of synthetic images-1M Hindi and 1M Gujarati-and then finetuned on a combined Hindi-Gujarati real images dataset, to demonstrates that multi-lingual approach does indeed aid model in learning each individual language better. Results 7 indicate a 4.0% increase in WRR for Hindi and a 0.20% increase for Gujarati.

7   Conclusion

We assembled a real dataset for STR in Indian languages. IndicSTR12 is the most comprehensive dataset available, and it is a first in many languages. We generated 3 million synthetic word-images for all 13 languages in addition to real datasets for faster STR solution development for Indic languages. In addition to benchmarking STR performance on three models, this paper establishes the need for even more data to leverage the learning powers of SOTA Latin models. Because Indian scripts are more complex than Latin scripts, they necessitate a more comprehensive training resource. In the future, this large dataset in both the real and synthetic domains will aid the Indic Scene text community in developing solutions for Indic STR that are on par with Latin solutions.

Acknowledgement. This work is supported by MeitY, Government of India, through the NLTM-Bhashini project.

References

1. Abnar, S., Zuidema, W.: Quantifying attention flow in transformers. arXiv preprint arXiv:2005.00928 (2020)
2. Baek, J., et al.: What is wrong with scene text recognition model comparisons? dataset and model analysis. In: Proceedings of the IEEE/CVF International Conference on Computer Vision, pp. 4715–4723 (2019)
3. Baek, J., Matsui, Y., Aizawa, K.: What if we only use real datasets for scene text recognition? toward scene text recognition with fewer labels. In: Proceedings of the IEEE/CVF Conference on Computer Vision and Pattern Recognition, pp. 3113–3122 (2021)
4. Bahdanau, D., Cho, K., Bengio, Y.: Neural machine translation by jointly learning to align and translate. arXiv preprint arXiv:1409.0473 (2014)
5. Bautista, D., Atienza, R.: Scene text recognition with permuted autoregressive sequence models. In: Avidan, S., Brostow, G., Cissé, M., Farinella, G.M., Hassner, T. (eds.) ECCV 2022. LNCS, vol. 13688, pp. 178–196. Springer, Cham (2022). https://doi.org/10.1007/978-3-031-19815-1_11
6. Bušta, M., Patel, Y., Matas, J.: E2E-MLT - an unconstrained end-to-end method for multi-language scene text. In: Carneiro, G., You, S. (eds.) ACCV 2018. LNCS, vol. 11367, pp. 127–143. Springer, Cham (2019). https://doi.org/10.1007/978-3-030-21074-8_11
7. Chandio, A.A., Asikuzzaman, M., Pickering, M., Leghari, M.: Cursive-text: a comprehensive dataset for end-to-end Urdu text recognition in natural scene images. Data Brief **31**, 105749 (2020)
8. Chen, X., Jin, L., Zhu, Y., Luo, C., Wang, T.: Text recognition in the wild: a survey. ACM Comput. Surv. (CSUR) **54**(2), 1–35 (2021)
9. Ch'ng, C.K., Chan, C.S., Liu, C.L.: Total-text: toward orientation robustness in scene text detection. Int. J. Doc. Anal. Recogn. (IJDAR) **23**(1), 31–52 (2020)
10. GOI: Government Indian language report (2011). https://censusindia.gov.in/census.website/
11. Graves, A., Fernández, S., Gomez, F., Schmidhuber, J.: Connectionist temporal classification: labelling unsegmented sequence data with recurrent neural networks. In: Proceedings of the 23rd International Conference on Machine Learning, pp. 369–376 (2006)
12. Gunna, S., Saluja, R., Jawahar, C.V.: Transfer learning for scene text recognition in Indian languages. In: Barney Smith, E.H., Pal, U. (eds.) ICDAR 2021. LNCS, vol. 12916, pp. 182–197. Springer, Cham (2021). https://doi.org/10.1007/978-3-030-86198-8_14
13. Gupta, A., Vedaldi, A., Zisserman, A.: Synthetic data for text localisation in natural images. In: Proceedings of the IEEE Conference on Computer Vision and Pattern Recognition, pp. 2315–2324 (2016)
14. He, M., et al.: ICPR 2018 contest on robust reading for multi-type web images. In: 2018 24th International Conference on Pattern Recognition (ICPR), pp. 7–12. IEEE (2018)
15. He, P., Huang, W., Qiao, Y., Loy, C.C., Tang, X.: Reading scene text in deep convolutional sequences. In: Thirtieth AAAI conference on artificial intelligence (2016)
16. Jaderberg, M., Simonyan, K., Vedaldi, A., Zisserman, A.: Synthetic data and artificial neural networks for natural scene text recognition. arXiv preprint arXiv:1406.2227 (2014)
17. Karatzas, D., et al.: ICDAR 2015 competition on robust reading. In: 2015 13th International Conference on Document Analysis and Recognition (ICDAR), pp. 1156–1160. IEEE (2015)
18. Karatzas, D., et al.: ICDAR 2013 robust reading competition. In: 2013 12th International Conference on Document Analysis and Recognition, pp. 1484–1493. IEEE (2013)
19. Kingma, D.P., Ba, J.: Adam: a method for stochastic optimization. arXiv preprint arXiv:1412.6980 (2014)
20. Liu, W., Chen, C., Wong, K.Y.K., Su, Z., Han, J.: STAR-Net: a spatial attention residue network for scene text recognition. In: BMVC, vol. 2, p. 7 (2016)

21. Liu, Z., Li, Y., Ren, F., Goh, W.L., Yu, H.: SqueezedText: a real-time scene text recognition by binary convolutional encoder-decoder network. In: Proceedings of the AAAI Conference on Artificial Intelligence, vol. 32 (2018)

22. Lucas, S.M.: ICDAR 2003 robust reading competitions: entries, results, and future directions. IJDAR **7**, 105–122 (2005)

23. Mathew, M., Jain, M., Jawahar, C.: Benchmarking scene text recognition in Devanagari, Telugu and Malayalam. In: 2017 14th IAPR International Conference on Document Analysis and Recognition (ICDAR), vol. 07, pp. 42–46 (2017). https://doi.org/10.1109/ICDAR.2017.364

24. Mathew, M., Singh, A.K., Jawahar, C.: Multilingual OCR for Indic scripts. In: 2016 12th IAPR Workshop on Document Analysis Systems (DAS), pp. 186–191. IEEE (2016)

25. Mishra, A., Alahari, K., Jawahar, C.: Scene text recognition using higher order language priors. In: BMVC-British Machine Vision Conference. BMVA (2012)

26. Nayef, N., et al.: ICDAR 2019 robust reading challenge on multi-lingual scene text detection and recognition-RRC-MLT-2019. In: 2019 International Conference on Document Analysis and Recognition (ICDAR), pp. 1582–1587. IEEE (2019)

27. Nayef, N., et al.: ICDAR 2017 robust reading challenge on multi-lingual scene text detection and script identification-RRC-MLT. In: 2017 14th IAPR International Conference on Document Analysis and Recognition (ICDAR), vol. 1, pp. 1454–1459. IEEE (2017)

28. Phan, T.Q., Shivakumara, P., Tian, S., Tan, C.L.: Recognizing text with perspective distortion in natural scenes. In: Proceedings of the IEEE International Conference on Computer Vision, pp. 569–576 (2013)

29. Risnumawan, A., Shivakumara, P., Chan, C.S., Tan, C.L.: A robust arbitrary text detection system for natural scene images. Expert Syst. Appl. **41**(18), 8027–8048 (2014)

30. Saluja, R., Maheshwari, A., Ramakrishnan, G., Chaudhuri, P., Carman, M.: OCR on-the-go: robust end-to-end systems for reading license plates & street signs. In: 2019 International Conference on Document Analysis and Recognition (ICDAR), pp. 154–159. IEEE (2019)

31. Shi, B., Bai, X., Yao, C.: An end-to-end trainable neural network for image-based sequence recognition and its application to scene text recognition. IEEE Trans. Pattern Anal. Mach. Intell. **39**(11), 2298–2304 (2016)

32. Simonyan, K., Zisserman, A.: Very deep convolutional networks for large-scale image recognition. arXiv preprint arXiv:1409.1556 (2014)

33. Smith, L.N., Topin, N.: Super-convergence: Very fast training of neural networks using large learning rates. In: Artificial Intelligence and Machine Learning for Multi-domain Operations Applications, vol. 11006, pp. 369–386. SPIE (2019)

34. Su, B., Lu, S.: Accurate scene text recognition based on recurrent neural network. In: Cremers, D., Reid, I., Saito, H., Yang, M.-H. (eds.) ACCV 2014. LNCS, vol. 9003, pp. 35–48. Springer, Cham (2015). https://doi.org/10.1007/978-3-319-16865-4_3

35. Sun, Y., Liu, J., Liu, W., Han, J., Ding, E., Liu, J.: Chinese street view text: large-scale Chinese text reading with partially supervised learning. In: Proceedings of the IEEE/CVF International Conference on Computer Vision, pp. 9086–9095 (2019)

36. Sun, Y., et al.: ICDAR 2019 competition on large-scale street view text with partial labeling-RRC-LSVT. In: 2019 International Conference on Document Analysis and Recognition (ICDAR), pp. 1557–1562. IEEE (2019)

37. Sutskever, I., Vinyals, O., Le, Q.V.: Sequence to sequence learning with neural networks. In: Advances in Neural Information Processing Systems, vol. 27 (2014)

38. Szegedy, C., Ioffe, S., Vanhoucke, V., Alemi, A.: Inception-v4, inception-ResNet and the impact of residual connections on learning. In: Proceedings of the AAAI Conference on Artificial Intelligence, vol. 31 (2017)

39. Vaswani, A., et al.: Attention is all you need. In: Guyon, I., et al. (eds.) Advances in Neural Information Processing Systems, vol. 30. Curran Associates, Inc. (2017). https://proceedings.neurips.cc/paper/2017/file/3f5ee243547dee91fbd053c1c4a845aa-Paper.pdf

40. Wang, K., Babenko, B., Belongie, S.: End-to-end scene text recognition. In: 2011 International Conference on Computer Vision, pp. 1457–1464. IEEE (2011)

41. Zeiler, M.D.: ADADELTA: an adaptive learning rate method. arXiv preprint arXiv:1212.5701 (2012)

42. Zhou, B., Lapedriza, A., Khosla, A., Oliva, A., Torralba, A.: Places: a 10 million image database for scene recognition. IEEE Trans. Pattern Anal. Mach. Intell. **40**(6), 1452–1464 (2018). https://doi.org/10.1109/TPAMI.2017.2723009

IWCP 2023 Preface

Computational paleography is an emerging field investigating new computational approaches for analyzing ancient documents. Paleography, understood as the study of ancient writing systems (scripts and their components) as well as their material (characteristics of the physical inscribed objects), can benefit greatly from recent technological advances in computer vision and instrumental analytics. Computational paleography, being truly interdisciplinary, creates opportunities for experts from different research fields to meet, discuss, and exchange ideas. Collaborations between manuscript specialists in the humanities rarely overcome the chronological and geographical boundaries of each discipline. However, when it comes to applying optical, chemical or computational analysis, these boundaries are often no longer relevant. On the other hand, computer scientists are keen to confront their methodologies with actual research questions based on solid data. Natural scientists working either on the physical properties of the written artefacts or on the production of their digital "avatar" are the third link in this chain of knowledge. In many cases, only a collaboration between experts from different communities can yield significant results.

In this workshop, we aimed to bring together specialists of the different research fields analyzing handwritten scripts in ancient artefacts. It mainly targets computer scientists, natural scientists and humanists involved in the study of ancient scripts. By fostering discussion between communities, it facilitated future interdisciplinary collaborations that tackle actual research questions on ancient manuscripts.

The second edition was held in-person on August 24, 2023 in San Jose, in conjunction with ICDAR 2023. It was organized by Isabelle Marthot-Santaniello (Institute for Ancient Civilizations, Universität Basel, Switzerland) and Hussein Adnan Mohammed (Centre for the Study of Manuscript Cultures, Universität Hamburg, Germany). We had Dominique Stutzmann from IRHT-CNRS (Institut de Recherche et d'Histoire des Textes, French National Centre for Scientific Research), France as a keynote speaker. The Program Committee counted 14 members and was selected to reflect the interdisciplinary nature of the field.

For this second edition, we welcomed two kinds of contributions: full papers and abstracts. We received a total of 8 submissions. Each full paper was reviewed by two members of the program committee via EasyChair and five were accepted. A double-blind review was used for the full paper submissions. The two abstract submissions were evaluated and accepted by the organizers, and were published separately by the organizers in a dedicated website (https://www.csmc.uni-hamburg.de/iwcp2023.html).

June 2023   Isabelle Marthot-Santaniello
                                                                  Hussein Mohammed

Reconstruction of Broken Writing Strokes in Greek Papyri

Javaria Amin[1]($\boxtimes$), Imran Siddiqi[2]($\boxtimes$), and Momina Moetesum[3]($\boxtimes$)

[1] Bahria University, Islamabad, Pakistan
javaria.amin456@gmail.com
[2] Xynoptik Pty Limited, Melbourne, VIC, Australia
imran.siddiqi@xynoptik.com.au
[3] National University of Science and Technology, Islamabad, Pakistan
momina.moetesum@seecs.edu.pk

Abstract. In the recent years, there has been an increased trend to digitize the historical manuscripts. This, in addition to preservation of these valuable collections, also allows public access to the digitized versions thus providing opportunities for researchers in pattern classification to develop computerized techniques for various applications. A common pre-processing step in such applications is the restoration of missing or broken strokes and makes the subject of our current study. More specifically, we work on isolated Greek characters extracted from handwriting on papyrus and employ a denoising auto-encoder to reconstruct the missing parts of characters. Performance evaluation using multiple evaluation metrics reports promising results.

Keywords: Historical Manuscript · Degraded characters · Reconstruction · Denoising Autoencoders

1 Introduction

Over the past few decades, a substantial increase in the digitization of ancient documents has been observed which has opened many avenues for the pattern classification research community. Ancient documents are one of the key sources of information that can be used to infer valuable insights from the past [4]. Manuscripts have been found to be written on various materials like animal skin, stones, papyrus and palm leaves etc., each imposing its own unique challenges. Digitization of such rich collections of manuscripts, in addition to preservation, allows development and evaluation of various computerized solutions for problems like dating of manuscripts, identification of writers, writer retrieval and restoration of documents. Thanks to the recent advancements in artificial intelligence (machine learning in particular), many of these challenges have been substantially addressed.

In general, ancient manuscripts are highly degraded due to numerous factors such as the ageing of documents, discolouration of ink colour, back impressions,

© The Author(s), under exclusive license to Springer Nature Switzerland AG 2023
M. Coustaty and A. Fornés (Eds.): ICDAR 2023 Workshops, LNCS 14193, pp. 253–266, 2023.
https://doi.org/10.1007/978-3-031-41498-5_18

humidity, and fading of characters etc. [17]. As an example, Fig. 1 illustrates samples of handwriting on papyrus from the GRK-papyri dataset [11]. As it can be seen the documents are highly degraded and need a series of effective pre-processing steps before they can be further analyzed. A common problem in such documents is that of missing strokes i.e. parts of characters are lost over time. Reconstruction of these strokes can be a useful pre-processing step that can help in other tasks like transcription and retrieval etc. and, makes the subject of our current study.

Fig. 1. Sample images from the GRK-papyri dataset [11]

Handwriting is a collection of strokes. A stroke starts from the time when we start writing with a pen and ends when the pen is lifted. Every character has a smooth part with a specific radius, length, and curvature that corresponds to a stroke. Ancient manuscripts are present in distinct languages each having its own letters and characters. Although characters differ, few strokes are common to most of the languages. From the viewpoint of ancient manuscripts, writing strokes can be broken or missing hence increasing the complexity of other high-level tasks like scribe identification or transcription. Consequently, there is a need to investigate this problem and come up with effective solutions to complete the broken or missing strokes.

In this study, we present a technique to reconstruct the missing or broken strokes in handwriting using writing on papyrus as a use case. From a technical perspective, we employ a de-noising convolutional auto-encoder that is trained to complete the missing parts in character images. The key highlights of this study are outlined in the following.

- Reconstruction of missing or broken parts of characters through a de-noising auto-encoder.
- A rich experimental study on isolated Greek letters written on papyrus
- Qualitative as well as quantitative evaluation with promising findings.

We organize the contents of this paper as follows. In Sect. 2, we present a discussion on the related work. Section 3 introduces the proposed technique including data preparation and model architecture. Experimental study and the reported results are discussed in Sect. 4 while Sect. 5 concludes the paper with a summary of our key findings.

2 Literature Review

In the recent years, computer scientists and paleographers have shown a keen interest in the digitization and analysis of handwritten documents. Typical tasks that have been investigated on such documents include binarization, noise removal, identification of scribes, classification of writing styles, dating of manuscripts and so on.

Among various tasks, binarization of historical manuscripts has been thoroughly investigated to segment foreground and background. Both image processing and machine learning techniques have been employed for this purpose. Among one of the recent studies, Castellanos et al. [5] used selectional autoencoders (SAE) for document binarization. Experimental study was carried out on five different datasets including Dibco, Einsiedeln, Slazinnes, Phi and Palm and reported effective performance. Likewise, Rani et al. [15] employed a semi-adaptive thresholding technique to restore the deteriorated text in historical manuscripts. Evaluation on Palm and Dibco datasets revealed that while the method was able to restore text in most cases, the binarization was poor in case of thin strokes. In another work, Hanif et al. [6] carry out document restoration by removing the blind bleed-through employing Gaussian mixture models. The method effectively restores the bleed-through documents but fails to remove the other types of degradation present in the document.

Among other studies, Melnik et al. [10] employ a deep neural network to segment characters from historical documents. The network also completes the missing and broken parts. Likewise, Raha et al. [14] use a convolutional neural network for restoration of historical documents. The study was performed on a collection of 600 documents and the convolutional neural network was employed in the encoder-decoder architecture. The encoder was responsible for extracting the robust features in the image while the decoder produced the output image without noise. The proposed network reported a validation accuracy of 89%.

In other works on this subject, Wadhwani et al. [18] carry out restoration of historical documents using ConvNets which extract the foreground and background of an image. Experimental study on H-Dibco 2017 and H-Dibco 2018 datasets reported promising results. In another work, Nguyen et al. [13] proposed a technique to restore characters in degraded historical documents using generative adversarial networks (GANs). The study was carried out on Japanese Kanji characters and the method was able to remove the noisy background and also reconstruct the degraded parts of the characters. Likewise, Mosa and Nasruddin [12] restored broken characters using the genetic snake algorithm to fill the active contours by applying external and internal energies with divergence factors. In another work, Liu et al. [9] restore the damaged fonts of a historical document based on writing style. The authors employ conditional GANs to repair the fonts. Experiments were carried out on the FontDataset containing Chinese characters written in the same style and high values of PSNR and SSIM are reported.

Among other methods, Sober et al. [16] employ computer-aided techniques for restoring the character strokes present in historical handwritten document images. Reconstruction of strokes is carried out using the energy functional minimization framework and gradient descent. In another study, Huang et al. [8] perform character reconstruction based on strokes using a weighted quadratic Bezier curve. The proposed method is also able to prevent any adversarial attack in recognizing the characters. The method relies on extracting strokes and then using the subsequent strokes to construct a clean image of a character without any noise.

Arvind et al. [2] present an interesting study on restoration of handwritten strokes. The authors first detect the diagonals and the 'U' strokes by calculating the angles and distances. Subsequently, Bresenham's line detection algorithm is employed that determines the two points making a straight line hence, completing the stroke. In another work, Assael et al. [3] attempt to restore the damaged inscriptions that were found to be written on the stones, metals and pottery. The authors make use of a deep neural network and named their network architecture as 'ithaca'. In addition to restoring the words of damaged inscription, the proposed model was able to find the geographical region and the date of the inscription as well. In other similar studies, Ho et al. [7] and Arun et al. [1] demonstrate the effectiveness of character restoration as a preprcessing step prior to recognition that significantly improves the recognizer performance.

The above discussion reveals that while the earlier research endeavors primarily focused on image enhancements techniques to restore degraded documents, a recent trend is the investigation of machine learning-based solutions which, in most cases, have resulted in better performance. Such methods, naturally, assume the availability of labeled training data so that appropriate ML models can be trained. In our techniques, we have also chosen to employ a learning-based method as presented in the next section.

3   Proposed Method

This section presents the technical details of the proposed method for reconstructing the missing or broken writing strokes. Figure 2 shows the overall pipeline of the method that starts with a data preparation step where noisy images (with missing strokes) are generated. The restoration relies on de-noising auto-encoder which receives the noisy image as input and the ground truth image as target. Once reconstructed, we compare the produced image with the ground truth and quantify the performance.

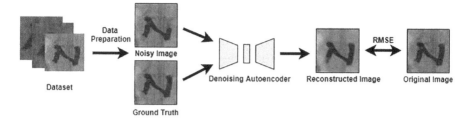

Fig. 2. Overview of key processing steps in the system

3.1 Dataset

The current study employs a collection of isolated handwritten Greek characters on Iliad papyri. The dataset consists of numerous manuscripts containing Iliad texts from the 3rd century BC to the 7th century AD. Each character is annotated by D-Scribes using Research Environment for Ancient Documents (READ) software. The total number of character images is 9,182, while the distribution of images in different character classes is illustrated in Fig. 3. Sample character images from the employed dataset are presented in Fig. 4.

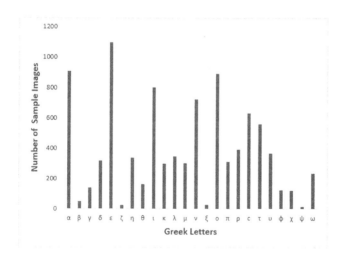

Fig. 3. Distribution of images in different character classes

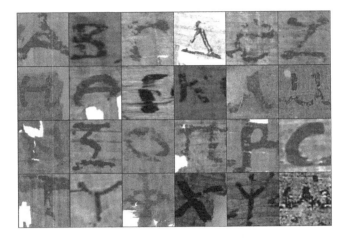

Fig. 4. Sample images from the dataset

3.2 Data Preparation

Since we employ a machine learning-based solution, labeled training data is needed to train the model(s). More specifically, for the de-noising auto-encoder, we need the noisy (broken characters) and the ground truth image of a character. If we directly pick images with broken characters from the dataset, the ground truth information is not known. Consequently, we generate the training data by creating pairs of ground truth images and those with missing strokes by producing the later through a series of processing steps. Though the trained model is evaluated on actually broken character images as well, having ground truth images also allow to report quantitative performance which otherwise would not be possible.

As a first step, we convert all images to gray scale and all subsequent processing is carried out on these single channel images. To generate the noisy images, we carry out the following steps. These steps can be correlated with the illustration in Fig. 5.

- A copy of the character image is binarized (and inverted) using the deep learning-based binarization method DP-LinkNet [19]. It must be noted that binarization is applied to a copy of the image, only to generate the noisy image. The completion of missing parts is carried out on gray scale images.
- A separate binary image of the same size as the original image is generated with a rectangular mask (of white pixels) randomly positioned within the image.
- Bit-wise intersection is applied to the two binary images (character image and the mask) producing the parts of characters that will be removed.

– In the gray scale image of the character, the pixels values of the intersected region of the two binary images are replaced with the values of the background neighbouring pixels. This creates a noisy image with broken characters corresponding to the input (ground truth) image.

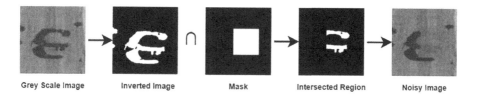

Grey Scale Image   Inverted Image   Mask   Intersected Region   Noisy Image

Fig. 5. Key steps in generating noisy images

Figure 6 presents pairs of images with the noisy characters and the corresponding ground truth images. It can be observed that the generated images highly resemble the characters seen in originally degraded images thus making the training pairs resemble the real world scenarios.

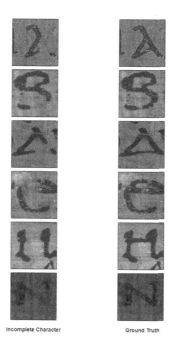

Incomplete Character   Ground Truth

Fig. 6. Pairs of images showing noisy and ground truth characters

3.3 Completion with a De-noising Auto-encoder

For completion of broken characters, we train a de-noising auto-encoder. The encoder and decoder parts are made up of convolutional layers with down sampling (in the encoder) and up sampling (in the decoder) layers. The model is fed with images with broken characters as input and the ground truth images as target. Although de-noising auto-encoders are typically employed for noise removal, we exploit the key idea of the encoder learning robust latent representation and reproducing the clean image to complete missing parts of characters. While the encoder compresses the input image through a series of convolutional and down sampling layers, the decoder takes the encoded representation and maps it back to the higher-dimensional representation to reconstruct the image.

The architecture of the employed auto-encoder is presented in Fig. 7. Input character images (resized to $64 \times 64 \times 1$) are fed to the encoder that comprises two convolutional layers with depths of 32 and 64 respectively. Each convolutional layer is followed by max pooling (down sampling) layer (2×2) and batch normalization layer to speed up the training process. Finally, the encoding layer has a depth of 128. Likewise, the decoder consists of two convolutional layers each followed by up sampling (through transposed convolutions). The output layer is $64 \times 64 \times 1$ representing the reconstructed image. A layer-wise summary of the model including output volume shape and number of parameters in each layer is presented in Table 1. Model hyper-parameters are optimized on the validation part of the dataset while more details are presented in the next section.

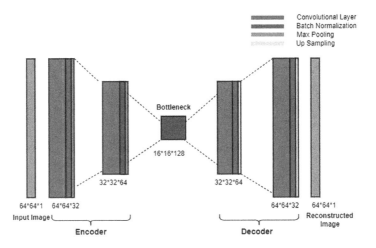

Fig. 7. Architecture of the de-noising auto-encoder for completing the broken characters

Table 1. Layer-wise volume shape and number of parameters in the model

Layer (Type)	Output Shape	Parameters
Input Layer	(64,64,1)	0
Conv2D	(64,64,32)	320
Batch Normalization	(64,64,32)	128
Conv2D	(64,64,32)	9,248
Batch Normalization	(64,64,32)	128
Max pooling	(32,32,32)	0
Conv2D	(32,32,64)	18,496
Batch Normalization	(32,32,64)	256
Conv2D	(32,32,64)	36,928
Batch normalization	(32,32,64)	256
Max pooling	(16,16,64)	0
Conv2D	(16,16,128)	73,856
Batch Normalization	(16,16,128)	512
Conv2D	(16,16,128)	147,584
Batch Normalization	(16,16,128)	512
Conv2D	(16,16,64)	73,792
Batch Normalization	(16,16,64)	256
Conv2D	(16,16,64)	36,928
Batch Normalization	(16,16,64)	256
Up sampling	(32,32,64)	0
Conv2D	(32,32,32)	18,464
Batch Normalization	(32,32,32)	128
Conv2D	(32,32,32)	9,248
Batch Normalization	(32,32,32)	128
Up sampling	(64,64,32)	0
Conv2D	(64,64,1)	289
Total Parameters : 427,713		

4   Experiments and Results

This section presents the details of the experimental protocol, evaluation metrics and the reported results along with the accompanying discussion.

4.1   Experimental Protocol

As discussed previously, we employ a total of 9,182 images of Greek letters from the Iliad-Papyri dataset. Out of these, 70% of the images constitute the training set, 20% the validation set and 10% of images make up the test set. The distribution of images into these subsets is summarized in Table 2.

Table 2. Distribution of Iliad-Papyri images into training, validation and test sets

Dataset	Iliad-papyri
Total Images	9,182
Training set	6,612
Validation set	1,836
Test set	734

The model is trained using binary cross entropy as loss function and 'Adam' as an optimizer with a learning rate of 0.001. All layers employ ReLu as the activation function with the exception of last layer that employs a sigmoid activation function (each pixel value of output image is in the range 0 to 1). A mask size of 30×30 is employed to generate the broken characters. A summary of the key hyper-parameters to train the model is presented in Table 3.

Table 3. Hyper-Parameter values to train the auto-encoder

Hyper-parameter values	
Epochs	100
Batch size	32
Activation function	Relu
Loss function	Binary Cross entropy
Optimizer	Adam
Learning Rate	0.001

4.2   Evaluation Metrics

For quantitative evaluation of the method, we employ a number of standard evaluation metrics. These include root mean square error (RMSE), peak signal to noise ratio (PSNR) and structural similarity index method (SSIM), comparing how good the reconstructed image is given the ground truth reference image. For completeness, each of these metrics are defined in the following.

– **RMSE** measures the pixel-wise difference between the reconstructed image and the actual image. Lower the value of RMSE, better is the reconstruction.

$$RMSE = \sqrt{\frac{1}{N} \sum_{i=1}^{N} (\hat{y}_i - y_i)^2} \qquad (1)$$

N is the total number of pixels in the image while $\hat{y}_i$ and y_i represent the generated and the ground truth values of pixel i respectively.

- **PSNR** is typically defined via the mean squared error(MSE) as follows.

$$PSNR = 20 \log_{10} \left(\frac{MAX}{\sqrt{MSE}} \right) \tag{2}$$

Here, MAX is the maximum possible pixel value of the image, higher the PSNR, better is the reconstruction.

- **SSIM** is another standard metric that is used to measure the similarity between two images. If the value is closer to 1, it implies that the two images being compared are highly similar, while values closer to 0 represent the dissimilarity between the two images.

$$\text{SSIM}(x, y) = \frac{(2\mu_x\mu_y + c_1)(2\sigma_{xy} + c_2)}{(\mu_x^2 + \mu_y^2 + c_1)(\sigma_x^2 + \sigma_y^2 + c_2)} \tag{3}$$

In the above equation, x and y are the original and the reconstructed images. μ_x and μ_y are the respective mean pixel intensities, σ_x and σ_y represent the respective standard deviations whereas σ_{xy} is the covariance between two image.

4.3   Results and Discussion

We first present the performance of our baseline model that comprises two layers each in the encoder and the decoder parts. The model is trained on the pairs of images in the training set with a batch size of 32 (with early stopping). The corresponding evolution of loss on training and validation sets is presented in Fig. 8. Quantitative results using the previously introduced evaluation metrics are presented as a function of the number of layers in the model and are summarized in Table 4. It can be seen from the reported results that all three evaluation metrics exhibit a similar trend once the number of model layers is changed. The best performance is reported with two layers reading an RMSE value of 0.04, a PSNR of 28.24 while a SSIM value of 0.90. These results are indeed very promising and validate the effectiveness of the de-noising auto-encoders for character completion problem in the challenging use case of historical manuscripts.

Table 4. Quantitative results as a function of number of layers in the model

No. of layers	RMSE	PSNR	SSIM
2	0.04	28.24	0.90
3	0.05	26.61	0.85
4	0.06	24.50	0.74

Since the missing parts in originally degraded images can be of varying sizes, we also carry out an experiment by varying the mask size used to generate the noisy images. We vary the mask size from 20×20 to 60×60, train the model for each and report the results on the test set. Greater the mask size, greater is the missing part in the character, making its completion more challenging. These results are summarized (in terms of PSNR) in Fig. 9 where it can be seen that the PSNR is more or less stable across varying mask sizes hence validating the robustness of the completion technique to different sizes of missing parts.

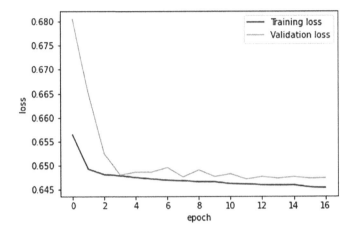

Fig. 8. Evolution of training and validation loss as a function of training epochs

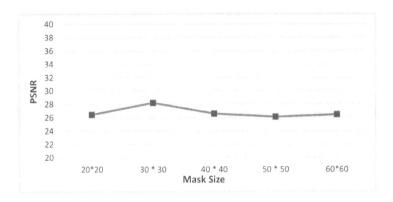

Fig. 9. PSNR values for varying sizes of mask to generate the noisy images

In addition to quantitative measures, to provide deeper insights, we also present a visual comparison of the original and the reconstructed images. Figure 10 illustrates sample triplets of images including the original image, the generated noisy image and the model-reconstructed image where it can be observed that in most cases, the produced images are very similar to the original ones. In few cases, though the output images have a blurred-effect, the character is correctly completed.

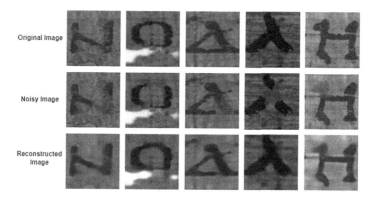

Fig. 10. Original, generated (noisy) and model reconstructed sample images

5 Conclusion

We presented a machine learning-based technique to complete the missing or broken parts of characters using Greek letters on papyri as a use case. The reconstruction relies on a convolutional de-noising auto-encoder that is trained on pairs of images (those with broken characters and the corresponding ground truth images). For model training and quantitative evaluation, the noisy images are generated using a sequence of image processing steps. Experimental study reveals the effectiveness of auto-encoders in completing the missing parts of characters on a challenging collection of images. The technique can be employed primarily as a pre-processing step for other high-level problems like transcription or scribe identification.

The current study relies on isolated character images as units of processing. In our further research on this subject, we plan to investigate how the technique can be extended to complete pages of handwriting. Furthermore, since each scribe has a unique writing style, it would also be interesting to employ writer-specific completion models to fill the missing parts in their specific handwriting. Generative models can also be investigated for this and other similar problems.

References

1. Arun, E., Vinith, J., Pattar, C., George, K.: Improving Kannada OCR using a stroke-based approach. In: TENCON 2019–2019 IEEE Region 10 Conference (TENCON), pp. 1611–1615 (2019). https://doi.org/10.1109/TENCON.2019.8929606
2. Arvind, K., Kumar, J., Ramakrishnan, A.: Line removal and restoration of handwritten strokes. In: International Conference on Computational Intelligence and Multimedia Applications (ICCIMA 2007), vol. 3, pp. 208–214. IEEE (2007)
3. Assael, Y.: Restoring and attributing ancient texts using deep neural networks. Nature **603**(7900), 280–283 (2022)

4. Baird, H.S., Govindaraju, V., Lopresti, D.P.: Document analysis systems for digital libraries: challenges and opportunities. In: Marinai, S., Dengel, A.R. (eds.) DAS 2004. LNCS, vol. 3163, pp. 1–16. Springer, Heidelberg (2004). https://doi.org/10. 1007/978-3-540-28640-0_1
5. Castellanos, F.J., Gallego, A.J., Calvo-Zaragoza, J.: Unsupervised neural domain adaptation for document image binarization. Pattern Recognit. **119**, 108099 (2021)
6. Hanif, M., et al.: Blind bleed-through removal in color ancient manuscripts. Multimed. Tools Appl. **82**, 1–15 (2022)
7. Ho, L.T., Tran, S.T., Dinh, D.: Nom document background removal using generative adversarial network. In: 2021 IEEE International Conference on Signal and Image Processing Applications (ICSIPA), pp. 100–104. IEEE (2021)
8. Huang, Z., Heng, W., Tao, Y., Zhou, S.: Stroke-based character reconstruction. arXiv preprint arXiv:1806.08990 (2018)
9. Liu, R., et al.: SCCGAN: style and characters inpainting based on CGAN. Mobile Netw. Appl. **26**, 3–12 (2021)
10. Melnik, G., Yekutieli, Y., Sharf, A.: Deep segmentation of corrupted glyphs. ACM J. Comput. Cult. Heritage (JOCCH) **15**(1), 1–24 (2022)
11. Mohammed, H., Marthot-Santaniello, I., Märgner, V.: GRK-Papyri: a dataset of Greek handwriting on papyri for the task of writer identification. In: 2019 International Conference on Document Analysis and Recognition (ICDAR), pp. 726–731. IEEE (2019)
12. Mosa, Q.O., Nasrudin, M.F.: Broken character image restoration using genetic snake algorithm: deep concavity problem. J. Comput. Sci. **12**(2), 81–87 (2016)
13. Nguyen, K.C., Nguyen, C.T., Hotta, S., Nakagawa, M.: A character attention generative adversarial network for degraded historical document restoration. In: 2019 International Conference on Document Analysis and Recognition (ICDAR), pp. 420–425. IEEE (2019)
14. Raha, P., Chanda, B.: Restoration of historical document images using convolutional neural networks. In: 2019 IEEE Region 10 Symposium (TENSYMP), pp. 56–61. IEEE (2019)
15. Rani, N.S., Nair, B.B., Chandrajith, M., Kumar, G.H., Fortuny, J.: Restoration of deteriorated text sections in ancient document images using a tri-level semi-adaptive thresholding technique. Automatika **63**(2), 378–398 (2022)
16. Sober, B., Levin, D.: Computer aided restoration of handwritten character strokes. Comput. Aided Des. **89**, 12–24 (2017)
17. Sulaiman, A., Omar, K., Nasrudin, M.F.: Degraded historical document binarization: a review on issues, challenges, techniques, and future directions. J. Imaging **5**(4), 48 (2019)
18. Wadhwani, M., Kundu, D., Chakraborty, D., Chanda, B.: Text extraction and restoration of old handwritten documents. In: Mukhopadhyay, J., Sreedevi, I., Chanda, B., Chaudhury, S., Namboodiri, V.P. (eds.) Digital Techniques for Heritage Presentation and Preservation, pp. 109–132. Springer, Cham (2021). https:// doi.org/10.1007/978-3-030-57907-4_6
19. Xiong, W., Jia, X., Yang, D., Ai, M., Li, L., Wang, S.: DP-LinkNet: a convolutional network for historical document image binarization. KSII Trans. Internet Inf. Syst. (TIIS) **15**(5), 1778–1797 (2021)

Collaborative Annotation and Computational Analysis of Hieratic

Julius Tabin[1] (iD), Mark-Jan Nederhof[2(✉)] (iD), and Christian Casey[3] (iD)

[1] Department of Organismic and Evolutionary Biology, Harvard University,
Cambridge, MA 02138, USA
jtabin@g.harvard.edu
[2] School of Computer Science, University of St Andrews, St Andrews KY16 9SX, UK
markjan.nederhof@googlemail.com
[3] Department of History and Cultural Studies, Freie Universität Berlin,
14195 Berlin, Germany
christiancasey86@gmail.com

Abstract. We introduce a new open-source web application for the collaborative annotation and study of hieratic texts, in the form of facsimiles and/or photographs. Functionality includes the automatic classification of occurrences of hieratic signs in Unicode, aided by image processing and optical character recognition techniques. Relying on various forms of dimensionality reduction, the tool also allows visualization of differences in sign shapes between texts, periods, genres, and geographical areas. This is motivated by recent work that demonstrated the value of dimensionality reduction for such purposes as identifying scribal hands. The interactive user interface can also be used for teaching students to read hieratic.

Keywords: Hieratic · OCR · Dimensionality reduction · Visualization

1 Introduction

There are currently a number of databases that gather photographs or facsimiles of texts, with the aim of facilitating palaeographic research, that is, investigations into the shapes of glyphs and the possible conclusions that can be drawn from these shapes about dating and geographical areas [25], scribal hands [17], writing materials [18], and evolution of scripts [20]. In most cases, there is a substantial gap between the breadth of the collected data and the computational analysis of the shapes. Bridging this gap typically involves manual selection of a few tokens that are deemed to be of interest. This limits the kinds of computational analysis that can be done, as much of the variation is lost in the selection.

One example of a project that does exhaustive labeling of glyphs is *the Demotic Palaeographical Database Project*.[1] However, it merely excises areas around glyphs, without representing the shapes of the glyphs explicitly. Similarly,

---

[1] http://www.demotischdemotisch.de/.

© The Author(s), under exclusive license to Springer Nature Switzerland AG 2023
M. Coustaty and A. Fornés (Eds.): ICDAR 2023 Workshops, LNCS 14193, pp. 267–283, 2023.
https://doi.org/10.1007/978-3-031-41498-5_19

DigiPal[2] and *HebrewPal*[3] merely encode rectangles around selected occurrences of letters in photographs of texts, and moreover do not exhaustively annotate any texts. The *ICDAR2023 Competition on Detection and Recognition of Greek Letters on Papyri*[4] comprised an exhaustive annotation of many dozens of texts in the MS COCO format [9], and encoded rectangles around occurrences of letters.

In contrast to all of the above, the *Paläographie des Hieratischen and der Kursivhieroglyphen*[5] collects the shapes of (selected) glyphs, extracted from facsimiles. However, at least at present, the project does not exhaustively annotate any texts.

In the case of alphabetical scripts, there is little variation in letter shapes in any given text, and the added value of exhaustive annotation, as opposed to selection of a few instances, has limited value to palaeography. However, Ancient Egyptian has many more signs, which each occur with smaller relative frequency and with greater variation. This suggests that exhaustive annotation holds greater potential for Egyptological studies than in the case of alphabetical scripts. This holds particularly well for the hieratic script, which is notable for its ligatures [4,21], where multiple glyphs are joined together, further extending the number of unique signs to be categorized. Hieratic lends itself relatively well to a digital representation in grayscale and to image processing techniques. However, with off-the-shelf tools, exhaustive annotation of the shapes of hieratic glyphs requires a prohibitive amount of manual labor.

The first contribution of this paper is a demonstration that exhaustive annotation of hieratic facsimiles is feasible with our customized tool, called *Isut*, which is implemented as a web application.[6] After explaining the structure of the data in Sect. 2, Sect. 3 describes the user interface for annotating a text, and Sect. 4 presents several ways to check annotations for accuracy. Section 5 summarizes the applied technology.

The second contribution, addressed in Sect. 6, is to demonstrate that exhaustively annotated texts allow analyses that would be impossible with sparsely annotated texts where only a very limited number of tokens per type would be collected. This builds on [23], which explored the use of UMAP dimensionality reduction for making inferences about scribal hands, for automatic discovery of distinct hieratic types of the same underlying hieroglyphic sign, and for identifying distinct signs whose graphical realizations can be easily confused. Section 7 then provides an example of specific hypotheses that can be tested against the available data.

Lastly, Sect. 8 presents different scenarios of use for collaborative annotation and addresses potential applications.

---

[2] http://www.digipal.eu/.

[3] https://www.hebrewpalaeography.com/.

[4] https://lme.tf.fau.de/competitions/2023-competition-on-detection-and-recognition-of-greek-letters-on-papyri/.

[5] https://aku-pal.uni-mainz.de/.

[6] From , *sš (n) jswt*, 'writings of the ancient ancestors'.

2   Data

The data of Isut consists of two parts: a traditional database and a directory structure of image files. There is one database record for each text, consisting of:

- (unique index,)
- name,
- creator (of the facsimiles),
- provenance (if known),
- period (typically a dynasty or range of dynasties),
- genre,
- notes (such as copyright information),
- edit history (list of usernames and dates),
- zero or more pages.

For each page, the record contains:

- (unique index,)
- type of the main image of the page (generally 'facsimile'),
- zero or more alternative images (e.g. photographs),
- zero or more lines (or columns).

For each line (or column), the record contains:

- (unique index,)
- line number,
- polygon around the line (in terms of coordinates in the main image of the page),
- direction of writing (either 'horizontal' for a line of text or 'vertical' for a column),
- zero or more glyphs.

For each glyph, the record contains:

- (unique index,)
- Unicode string identifying a hieroglyphic sign or a combination of signs,
- coordinate of the top-left corner of the bounding box, in terms of pixel positions in the main image of the page, as well as the width and height measured in the number of pixels.

A Unicode string may include control characters.[7] For example, ⟨glyph⟩, which is the normalized representation of hieratic ligature ⟨glyph⟩, can be encoded as ⟨glyphs⟩. The limitations of normalized hieroglyphs for encoding hieratic signs have been

---

[7] Unicode 15, introduced in September 2022, brings the total number of control characters for Ancient Egyptian to 17, in combination with a number of modifying and other special characters [5].

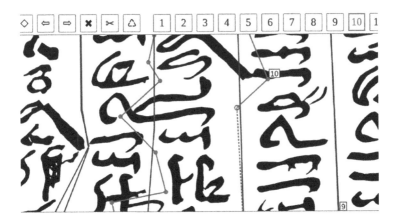

Fig. 1. Polygons delimiting columns of text, which may overlap due to 'tails' of signs protruding into neighboring columns.

acknowledged in the literature, and more fine-grained encodings have been proposed that were designed specifically for hieratic [6,13]. Such encodings can differentiate between hieratic forms that correspond to the same hieroglyphic sign. Our motivation for a more traditional and coarse-grained encoding is that it offers the best prospects for stability in the medium term, and allows us to explore how different hieratic forms can be discovered with the help of computational techniques.

The image files are arranged into a directory structure, with file and directory names identified by the unique indices referred to above. For each line, there is a directory gathering all grayscale image files of glyphs in that line. These images have a transparent background, so they can be overlaid on top of the facsimile or photograph. For each page, there is a directory containing the facsimile and possibly a photograph or other additional images, and subdirectories for the individual lines. For each text, there is a directory that has one subdirectory for each page.

3   Annotation

Exhaustive annotation of texts with occurrences of hieratic signs is facilitated by a number of customized graphical user interfaces. The first step in the annotation of a page is to delineate lines (or columns) of text. This is primarily done by a series of mouse clicks on the image of the facsimile, to form a polygon. The polygon is intended to encompass all glyphs belonging to a line, but may include parts of glyphs belonging to neighboring lines. It is quite frequent in hieratic that a 'tail' of a sign in one line protrudes into a neighboring line, as exemplified in Fig. 1.

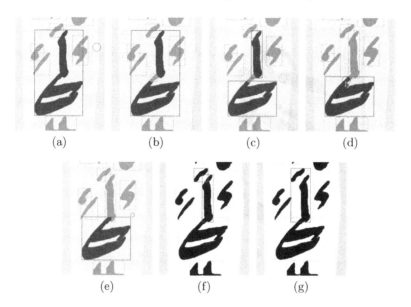

(a)   (b)   (c)   (d)

(e)   (f)   (g)

Fig. 2. (a) A blob (within the solid red rectangle) contains parts of two signs. (b) Pixels connecting the two signs are erased, leaving the upper sign intact. (c) The selected blob is split into two new blobs. (d-e) The erased pixels from the lower glyph are restored. (f) Two blobs are selected (in solid red boxes). (g) These are then merged into one glyph. (Color figure online)

Each polygon around a line represents an excised part of a page, which is presented in a second type of user interface. In the most typical case, the next step in the annotation is to invoke the program's ability to automatically identify all "blobs" in the line. A *blob*, also called a *connected component*, is a maximal collection of black (or, in general, dark) pixels that are connected to one another. Blobs that are smaller than a user-defined threshold are ignored.

Initially, the candidate glyphs in a line are the automatically obtained blobs. However, hieratic signs may consist of several blobs and one blob may consist of several hieratic signs. In the latter case, the blob may be a generally recognized ligature or it may be the result of an accidental touching or crossing of glyphs, even between neighboring lines, as exemplified earlier. To deal with these cases, a number of glyphs may be selected and merged into one glyph or a glyph may be split into several by first erasing pixels from a blob by an 'eraser' tool, effectively 'cutting up' a blob into multiple blobs. The separated glyphs are then 'repaired' by filling the pixels back in that were previously erased, by a 'paint' tool that only paints pixels that are dark in the facsimile. Figure 2 illustrates this process.

Another tool can be used to fill in patches of light pixels surrounded by dark pixels. This can help annotate facsimiles where glyphs in red ink are suggested by drawing only the outlines. The tool then effectively fills in the space between the outlines.

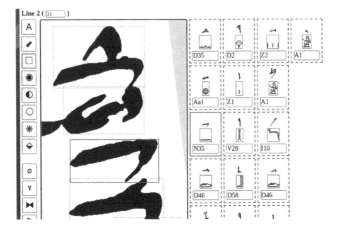

Fig. 3. Glyphs (left) with their labels (right)

In the final stage of annotation, the candidate glyphs are classified, that is, they are labeled with a Unicode string of normalized hieroglyphs, possibly including control characters. First, each glyph is given a label automatically, which can then be manually corrected by entering the correct name or through the HieroJax graphical Unicode editor.[8] The view at the end of this process is exemplified by Fig. 3.

Automatic classification of glyphs is implemented as follows. Glyphs are scaled to 16 by 16 bilevel grids and the set of such grids is subjected to principal component analysis (PCA). For each glyph, a vector consisting of the first 40 principal components is stored in a table, together with its ground-truth label. For a new glyph to be classified, a new vector is calculated in the same way, which is then compared to the vectors in the table, in terms of Euclidean distance. The label belonging to the nearest vector is returned. The reason we have opted for PCA is that it is very fast and runs on any platform without dedicated hardware, such as GPUs. Moreover, PCA does not require large amounts of training data, which can be a bottleneck for hieratic, as there can be relatively many types and few tokens per type. A single sign occurrence seen before may suffice to recognize a new occurrence of the same sign.

To be able to report an objective measure of accuracy, we have tested our classifier for individual signs. We excluded ligatures because opinions vary when a pair (or triple, ...) of overlapping or touching hieratic signs should be seen as a ligature. This left 7893 tokens and 325 types. These were randomly shuffled. We then set aside a quarter (1973 tokens) for testing and applied PCA on the remainder (5920 tokens) to build the table. For each test token we then determined the type of the nearest token in the table, in terms of the Euclidean distance, which resulted in an accuracy of 73%. (In over 90% of the cases, the 5

---

[8] https://github.com/nederhof/hierojax.

types with the nearest tokens include the correct type; this is over 92% for 10 types.)

Our 73% accuracy may seem lower than the 85% accuracy reported by [3] for a 8k dataset. However, that dataset had only 39 types, an order of magnitude below our 325 types, and it seems unlikely that their experimental conditions were representative. We found that the 39 most common types only account for 77.5% of the tokens.

There is a stark contrast with alphabetical scripts for which many tokens for relatively few types are available for training. This allows effective parameter estimation of neural networks, which can for some applications reach almost perfect accuracy [1]. Somewhat more similar to hieratic may be Chinese handwriting, for which accuracies above 90% have been reported [26]. However, this required two orders of magnitude more training data than was available for our experiments. A fair comparison is therefore hard to make.

The annotation process has been extensively tested by the first two authors. Facsimiles of two texts (the Shipwrecked Sailor and the Eloquent Peasant B1) were previously prepared by the first author as part of the work described in [23], and were annotated by the second author. The remaining facsimiles were taken from [15] and were annotated by the first author. Labeling of glyphs was done on the basis of published hieroglyphic transcriptions. We found that the time needed for annotation can vary depending on the degree of overlap of glyphs, as the procedure of "cutting up and repairing", as discussed earlier, is the most time-consuming part of annotation. If almost no glyphs overlap, annotation of a single line typically takes on the order of two minutes. If there is more overlap, then it can take up to 10–15 min per line. We hope to do systematic user studies at a later stage.

4   Validation

In the annotation of large numbers of facsimiles, it is inevitable that mistakes are made. The application offers a number of views of the data that help identify and correct such mistakes. First, all occurrences of glyphs with the same label (normalized transcription as hieroglyphs) can be listed and viewed next to each other. The glyphs can also be overlaid on a slightly bigger square excised from the facsimile, to see the glyphs in context; see Fig. 4.

Second, dimensionality reduction (cf. Sect. 6) can reveal outliers in shapes, which may indicate errors.

Lastly, a view of an annotated facsimile is provided that shows glyphs in different colors, such that no two glyphs can have the same color if their bounding boxes overlap or are close to one another, as illustrated by Fig. 5. A distinct color results where two glyphs overlap, unless the mouse hovers over a glyph, in which case that glyph's color supersedes the colors of the other glyph(s), as shown by Fig. 6. In this way, one can see at a glance whether any ink has remained unaccounted for, or if ink has been attributed to the wrong glyph. In addition, the labels of the glyphs are shown at a fixed distance from the glyphs themselves.

Fig. 4. Listing of glyphs for ♀ (D2), with surrounding context.

5   Implementation

Isut is a Node.js web application, and has been tested on Linux and macOS. The database technology is MongoDB. Source code and data are available on GitHub.[9]

Dimensionality reduction and OCR were implemented in Python, running on the server. This is motivated by the existence of standard Python packages that offer the required functionality and are highly optimized for speed. The latency caused by communication between client and server for the purposes of dimensionality reduction and OCR is not likely to be a hindrance.

In contrast, functionality to annotate pages and lines is done completely client-side. For example, the user interface to add points to a polygon that delineates a line is implemented in HTML canvas and JavaScript responding to mouse clicks and mouse movements. Also, recognition and manipulation of blobs are implemented in client-side JavaScript. In this way, web pages remain responsive even in the case of slow internet connections. Modified annotations are asynchronously sent to the server every few seconds, to minimize the risk that work is lost if there is any disruption to connectivity.

6   Analysis

Dimensionality reduction turns high dimensional data into a form that is more amenable to analysis and visual representation, with the potential to gain new insights. Isut applies dimensionality reduction on representations of glyphs as vectors of length $16 * 16$. Such a vector is obtained by first rescaling a glyph to a grid of 16 by 16 pixels, and then converting the image to bilevel.

---

[9] https://github.com/nederhof/isut.

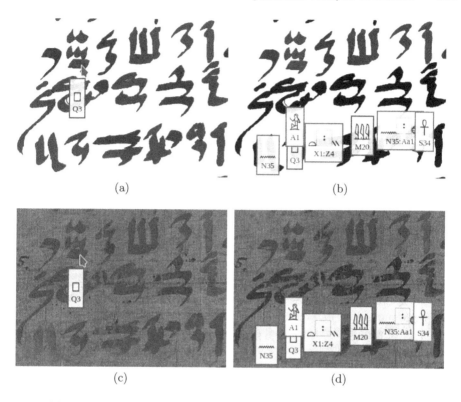

(a)   (b)

(c)   (d)

Fig. 5. (a) In one view, different glyphs are distinguished by color, and by hovering the mouse over a glyph, its label is shown. (b) In another view, labels of all glyphs of a selected line are shown. (c–d) The glyphs can also be superimposed on a photograph.

The following methods are currently available:

- principal component analysis (PCA) [8]
- t-distributed stochastic neighbor embedding (t-SNE) [10]
- multidimensional scaling (MDS) [12]
- Isomap [24]
- spectral embedding [2]
- locally linear embedding [22]
- uniform manifold approximation and projection (UMAP) [11]

With each of these methods, the user can opt for reduction to 1, 2, or 3 dimensions. The default is PCA to 2 dimensions, which often gives the best results, exemplified by Fig. 7. This simple example deals with one sign represented in two texts, and, for each token, a circle appears with a color that depends on the text. In general, one can include any number of signs and any number of texts, and can group texts together into periods, genres, or geographical areas, to, for example, obtain one distinct color for each period. By hovering over a circle, the shape of the glyph is shown; by simultaneously pressing the space bar, the

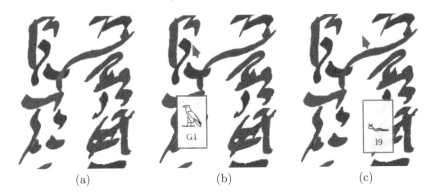

Fig. 6. (a) Overlap of glyphs is shown in a color distinct from the colors of the individual glyphs. (b–c) By hovering the mouse over a glyph, it is raised over neighboring glyphs.

shape is added to one of the corners of the graph, as illustrated here. Clicking on a circle opens a page showing the facsimile, with the selected occurrence of the glyph marked by its bounding box.

When viewing the output from dimensionality reduction to 3 dimensions, the user can use the mouse to rotate the model and inspect it from different angles. Points closer to the 'camera' appear larger, as shown in Fig. 8. Dimensionality reduction to 1 dimension is illustrated by Fig. 9.

7   Case Study

Isut's aforementioned dimensionality reduction techniques have utility beyond the overall visualization of the data and its many didactic functions. Substantial cross-textual insights have previously been uncovered using dimensionality reduction [23], and Isut builds on this by making its repertoire of techniques quickly accessible and efficient. In this way, a researcher can search the data to make meaningful inferences about texts or illuminate trends in sign development and evolution.

For example, when dimensionality reduction is used to plot the data for sign ⌑ (D21), the majority of the examples of the sign from the later time periods (Dynasties 15 and 18) cluster together at one end of the graph, indicating similarity (Fig. 10). The 12th Dynasty glyphs that cluster with them are those from Texte aus Hatnub, which can be disregarded here, given that they were carved, rather than written. There is partial overlap, with some tokens from Dynasties 15 and 18 clustering with the tokens from Dynasties 12 and 13, but, nevertheless, the trend is clear. Once a hypothesis has been generated from observation of the data (e.g., the form of sign D21 changed over time, with a distinct form dominating the corpus in the later dynasties), Isut's mouse-over

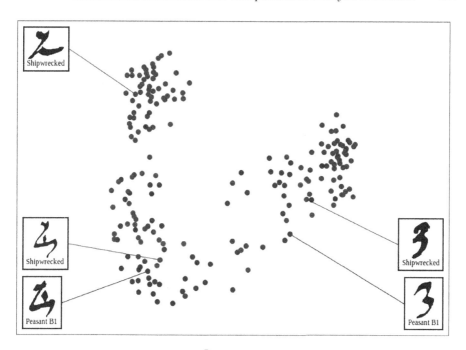

Fig. 7. PCA identifies three types of 𓅓 (G17) in the combination of the Shipwrecked Sailor and the Eloquent Peasant (manuscript B1). One of the types is found in Shipwrecked and Peasant B1 (bottom left). Another is found in Shipwrecked and Peasant B1 (bottom right). A third type is almost solely found in Shipwrecked (top left).

function, allowing individual tokens to be seen relative to their position on the graph, provides the potential to visualize these trends with unprecedented ease.

The selected tokens in Fig. 10 are representative of the tokens around them. From such inspection, one can see that the later form of D21, shown here using individual signs from Papyrus Rhind and Westcar, is much shorter than the Dynasty 12/13 version, in both width and tail. In the earlier dynasties, there was already a smaller D21 by width, presented here using a glyph from Peasant B1, but, as can be seen, this is morphologically distinct from the later version of the sign, particularly with respect to tail length. This is a promising avenue for further research, since knowledge of the evolution of such a common and integral glyph can further our understanding of the physical development of the language as a whole. While this is not a new area of research (indeed, Möller [14] was categorizing the development of hieratic characters over 100 years ago), never before have researchers been able to view and compare hieratic data on this scale.

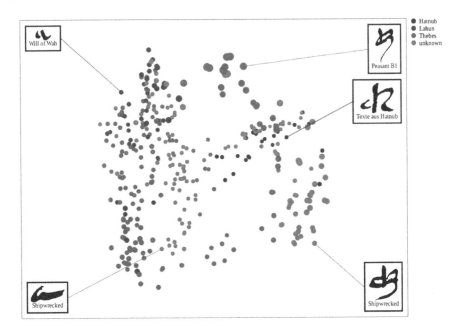

Fig. 8. Dimensionality reduction with Isomap for 𝕪 (A1) in 3 dimensions, per provenance.

Using this same functionality, Isut also allows hypotheses to be tested. For instance, if sign D21 truly changed form over time, one might predict that the often nearly indistinguishable sign ⌒ (X1) would have undergone a similar transformation; by form alone, the two hieratic signs are often indistinguishable to humans, as well as to OCR methods. With a single button press, a user can generate a corresponding dimensionality reduction graph for X1 in seconds (Fig. 11). As expected, the X1 signs from Dynasties 15 and 18 largely group together, apart from the earlier dynasty material. As with D21, hovering over individual points reveals the trend: once again, there is a truncation of the tail of the sign. Thus, D21 and X1 changed in largely the same way, adding support for the idea that there was a distinct historical shift in the writing of these signs. By checking Isut's graphs for other signs, one can confirm that the later dynasty material does not always cluster together apart from the rest of the data (data not shown). This demonstrates that some signs, such as D21 and X1, were subject to greater change than others in the time between Dynasties 12 and 18. Investigations into these varying evolutionary rates of sign morphology is an important area for future work, which can be facilitated using this application.

Of course, these results will only become clearer as more data is added to Isut. Currently, the uploaded 18th Dynasty texts are restricted to Papyrus Ebers and Papyrus Westcar and the only 15th Dynasty text is the Rhind Papyrus. While the above D21/X1 trend is visually obvious given the graphical output, the nuances of the morphological shift cannot be fully ascertained without a greater amount

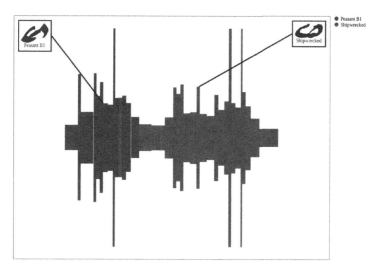

Fig. 9. PCA to one dimension for ⊜ (Aa1), rendered in a two-dimensional diagram. Tokens are positioned from left to right according to the one dimension, but each token is represented by a rectangle with a fixed area, such that the rectangles of neighboring tokens touch but do not overlap. Thereby, clusters of tall rectangles suggest tokens that are close together. This graphic shows that the shapes of Aa1 in the Eloquent Peasant (B1) can be almost perfectly separated from those in the Shipwrecked Sailor.

of data. For instance, it is conceivable that a new text from a later dynasty could be added to Isut's corpus containing D21/X1 signs with long tails, changing the overall interpretation from "D21/X1 were shortened over time" to "D21/X1 became more variable over time". The value of this tool, beyond its accessibility and power for large-scale research, lies in its adaptability, continually growing with more information and allowing an ever-expanding number of increasingly significant comparisons to be made.

8   Deployment

The current implementation can be deployed centrally or locally. Central deployment requires a publicly accessible web server. One or more accounts can be created for 'editors'. An editor can create accounts of 'contributors' or other editors and assign preliminary passwords. Both editors and contributors can add and modify texts and annotations. Local deployment can dispense with user administration and credentials.

Texts and their annotations can also be easily exchanged. From one instance of the application, a text consisting of its database record and image files can

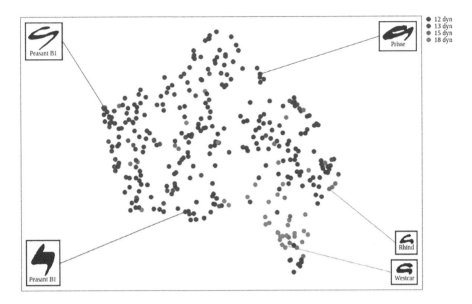

Fig. 10. UMAP graph for ⌢ (D21), by period.

be downloaded into a single zip file. This zip file can be uploaded to another instance of the application. This allows for distributed creation of annotated texts, which, if desired, can be merged centrally, possibly after validation by moderators.

At present, material for teaching hieratic mainly consists of printed books. This includes [14,15], which regrettably do not cover more recently discovered papyri. Other material is more up to date [16], but is in most cases still only available in printed form. There are good grounds to develop teaching material for hieratic in digital form [19], which can be more easily expanded and corrected on the basis of fresh evidence. Isut creates new opportunities for teaching hieratic, as new manuscripts can be added with relative ease, and the organization of the data allows development of tools for training and assessing hieratic reading skills.

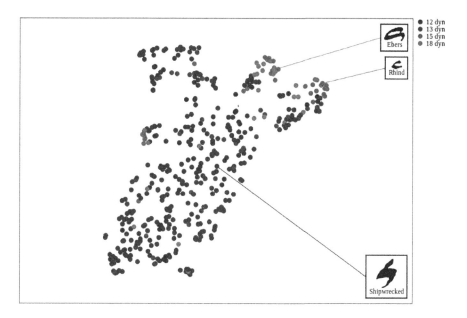

Fig. 11. UMAP graph for ⌒ (X1), by period.

9   Conclusions

We have introduced Isut, a tool to exhaustively annotate hieratic texts. The resulting data creates new potential for analysis and teaching of hieratic. As the data set expands, the power of analyses will grow and new insights will be able to be gained. Thus, the focus in the immediate future will be on expanding the collection of data, which currently comprises 691 distinct signs, and 8880 sign occurrences altogether, from thirteen texts from the 12th to the 18th Dynasty, covering three geographical areas and seven genres.

References

1. Ali, S., Abdulrazzaq, M.: A comprehensive overview of handwritten recognition techniques: a survey. J. Comput. Sci. **19**, 569–587 (2023). https://doi.org/10.3844/jcssp.2023.569.587
2. Bengio, Y., Delalleau, O., Le Roux, N., Paiement, J.F., Vincent, P., Ouimet, M.: Learning eigenfunctions links spectral embedding and kernel PCA. Neural Comput. **16**, 2197–2219 (2004). https://doi.org/10.1162/0899766041732396
3. Bermeitinger, B., Gülden, S., Konrad, T.: How to compute a shape: optical character recognition of hieratic. In: Gracia Zamacona, C., Ortiz-García, J. (eds.) Handbook of Digital Egyptology: Texts, chap. 7, pp. 121–138. Editorial Universidad de Alcalá (2021). https://doi.org/10.25358/openscience-6757

4. Gasse, A.: Les ligatures dans les textes hiératiques du Nouvel Empire (à partir des ostraca) - entre pragmatisme et maniérisme. In: Gülden et al. [7], pp. 111–133. https://doi.org/10.25358/openscience-403

5. Glass, A., Grotenhuis, J., Nederhof, M.J., Polis, S., Rosmorduc, S., Werning, D.: Additional control characters for Ancient Egyptian hieroglyphic texts (2021). https://www.unicode.org/L2/L2021/21248-egyptian-controls.pdf

6. Gülden, S.: Paläographien und Hieratogramme — digitale Herausforderungen. In: Gülden et al. [7], pp. 83–109. https://doi.org/10.25358/openscience-390

7. Gülden, S., van der Moezel, K., Verhoeven, U. (eds.): Ägyptologische "Binsen"-Weisheiten III. Franz Steiner Verlag (2018)

8. Jolliffe, I.: Principal Component Analysis, 2nd edn. Springer, New York (2002). https://doi.org/10.1007/b98835

9. Lin, T.Y., et al.: Microsoft COCO: common objects in context. In: Fleet, D., Pajdla, T., Schiele, B., Tuytelaars, T. (eds.) Computer Vision - ECCV 2014. Lecture Notes in Computer Science, vol. 8693, Zurich, Switzerland, pp. 740–755 (2014). https://doi.org/10.1007/978-3-319-10602-1_48

10. van der Maaten, L., Hinton, G.: Visualizing data using t-SNE. J. Mach. Learn. Res. **9**, 2579–2605 (2008)

11. McInnes, L., Healy, J., Melville, J.: UMAP: uniform manifold approximation and projection for dimension reduction (2020). https://arxiv.org/abs/1802.03426

12. Mead, A.: Review of the development of multidimensional scaling methods. J. Roy. Stat. Soc. Ser. D **41**, 27–39 (1992). https://doi.org/10.2307/2348634

13. van der Moezel, K.: On signs, lists and standardisation. In: Gülden et al. [7], pp. 51–81. https://doi.org/10.25358/openscience-391

14. Möller, G.: Hieratische Paläographie. Otto Zeller, Osnabrück (1909–1936). Three volumes

15. Möller, G.: Hieratische Lesestücke für den akademischen Gebrauch. J.C. Hinrichs (1927–1935). Three volumes

16. Möschen, S.: Lernen leicht gemacht? Arbeiten an einer Chrestomathie des Hieratischen. In: Gülden et al. [7], pp. 319–327. https://doi.org/10.25358/openscience-405

17. Papaodysseus, C., Rousopoulos, P., Arabadjis, D., Panopoulou, F., Panagopoulos, M.: Handwriting automatic classification: application to ancient Greek inscriptions. In: 2010 International Conference on Autonomous and Intelligent Systems, pp. 1–6. Povoa de Varzim, Portugal (2010). https://doi.org/10.1109/AIS.2010.5547045

18. Piquette, K., Whitehouse, R.: Material Practice: Substance. Surface and Medium. Ubiquity Press, London (2013). https://doi.org/10.5334/bai

19. Polis, S.: Methods, tools, and perspectives of hieratic palaeography. In: Davies, V., Laboury, D. (eds.) Oxford Handbook of Egyptian Epigraphy and Paleography, chap. IV.3, pp. 550–565. Oxford University Press (2020). https://doi.org/10.1093/oxfordhb/9780190604653.013.39, Online edition

20. Rajan, V.: How handwriting evolves: an initial quantitative analysis of the development of Indic scripts. In: 17th Biennial Conference of the International Graphonomics Society. Pointe-à-Pitre, Guadeloupe (2015)

21. Regulski, I.: Writing Habits as Identity Marker: On Sign Formation in Papyrus Gardiner II. In: Gülden et al. [7], pp. 235–265. https://doi.org/10.25358/openscience-395

22. Roweis, S., Saul, L.: Nonlinear dimensionality reduction by locally linear embedding. Science **290**, 2323–2326 (2000). https://doi.org/10.1126/science.290.5500.2323

23. Tabin, J.: From Papyrus to Pixels: Optical Character Recognition Applied to Ancient Egyptian Hieratic. B.A. research project, University of Chicago (2022). https://doi.org/10.6082/uchicago.3695

24. Tenenbaum, J., de Silva, V., Langford, J.: A global geometric framework for nonlinear dimensionality reduction. Science **290**, 2319–2323 (2000). https://doi.org/10.1126/science.290.5500.2319

25. Verhoeven, U.: Stand und Aufgaben der Erforschung des Hieratischen und der Kursivhieroglyphen. In: Verhoeven, U. (ed.) Ägyptologische "Binsen"-Weisheiten I-II, pp. 23–63. Franz Steiner Verlag (2013). https://doi.org/10.11588/propylaeumdok.00003321

26. Wang, Y., Yang, Y., Chen, H., Zheng, H., Chang, H.: End-to-end handwritten Chinese paragraph text recognition using residual attention networks. Intell. Autom. Soft Comput. **34**, 371–388 (2022). https://doi.org/10.32604/iasc.2022.027146

Efficient Annotation of Medieval Charters

Anguelos Nicolaou[1]([✉])[iD], Daniel Luger[1][iD], Franziska Decker[1],
Nicolas Renet[2][iD], Vincent Christlein[2][iD], and Georg Vogeler[1][iD]

[1] Center for Information Modeling (ZIM), University of Graz,
Elisabethstraße 59/III, 8010 Graz, Austria
`anguelos.nicolaou@gmail.com`,
`{daniel.luger,franziska.decker,georg.vogeler}@uni-graz.at`
[2] Pattern Recognition Lab, Friedrich-Alexander-Universität Erlangen-Nürnberg,
91058 Erlangen, Germany
`vincent.christlein@fau.de`

Abstract. Diplomatics, the analysis of medieval charters, is a major field of research in which paleography is applied. Annotating data, if performed by laymen, needs validation and correction by experts. In this paper, we propose an effective and efficient annotation approach for charter segmentation, essentially reducing it to object detection. This approach allows for a much more efficient use of the paleographer's time and produces results that can compete and even outperform pixel-level segmentation in some use cases. Further experiments shed light on how to design a class ontology in order to make the best use of annotators' time and effort. Exploiting the presence of calibration cards in the image, we further annotate the data with the physical length in pixels and train regression neural networks to predict it from image patches.

Keywords: Diplomatics · Paleography · Object detection · Resolution

1 Introduction

1.1 Diplomatics

Diplomatics is the study of diplomatic charters, medieval documents of a legal nature having a highly regular form. While diplomatics is mostly focused on the content of the charters, many visual aspects are subject of investigation.

Sigillography (the study of diplomatic seals), forgery detection, document identification, dating, etc. are all strongly dependent on analysing visual attributes of the documents making image analysis tools indispensable. Although diplomatics data has been part of larger datasets of historical document images, pure image analysis on charters has not been that common. Standing out among recent studies, Boroş et al. [1] demonstrates that Named Entity Recognition directly on images works better than with a two stage pipeline and the work by Leipert et al. [8] which addresses the problem of class imbalance in the context of image segmentation. In Fig. 1 two typical charters, one verso and one recto, can be seen.

M. Coustaty and A. Fornés (Eds.): ICDAR 2023 Workshops, LNCS 14193, pp. 284–295, 2023.
https://doi.org/10.1007/978-3-031-41498-5_20

Fig. 1. Left: a typical charter front (recto); right: a typical charter back-side on fine parchment (verso)

1.2   Monasterium.net

Our work is motivated by the need to analyse online archives, in particular monasterium.net, which constitutes the largest collection of continental European archives. It is focused on central Europe, containing 170 archives in total from Austria, Czech Republic, Estonia, Germany, Hungary, Italy, North Macedonia, Poland, Romania, Serbia, Slovenia, and Slovakia. The data in such archives is highly inhomogeneous in nature and many details about the acquisition process are unavailable and can only be guessed at. Thus, there is a need to perform automatic image analysis in order to obtain a proper overview of the quality of such archives.

1.3   Challenges

At the same time, in traditional methods of diplomatics, much of the scientist's attention and effort is given to each specific charter. In the late medieval period, charters grow to the hundreds of thousands, much more than a scholar can embrace. This makes the need of automatic analysis assistance unavoidable. Among humanists such automation is known by the term *distant reading*, it is used to signify automatic text analysis instead of analysing the original documents; in contrast with the manual analysis of each particular document. this notion also extends to visual analysis known as *distant seeing*. Distant reading and seeing, as opposed to typical image analysis, is practically impossible to scrutinise experimentally and the extent to which such methodologies are falsifiable is a nuanced discussion. These epistemological challenges mandate a very good understanding of the underlying data and biases in the employed methods. From an engineering perspective, we considered that the most precious resource is the diplomatist's time and most decisions were made by trying to minimize the scholar's time needed for annotation and then trying to maximize the use of the spent time.

The main contributions of this paper are as follows:

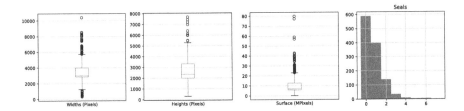

Fig. 2. Distributions of image cardinalities in the 1000-charters dataset, and a histogram of occurring seals per image

1. We introduce a highly efficient data annotation tool called FRAT.[1]
2. We present an object detection diplomatics dataset and report experiments using it.[2]
3. We present a dataset with manually annotated image resolution estimates and present a model that learns to infer them.[3]

2 Layout Analysis

2.1 1000 Charter Dataset

In order to make sure we gain a proper understanding of the data available in `monasterium.net`, we randomly chose 1000 charters from it. The heterogeneous nature of the dataset is illustrated by Fig. 2, in which we plot the distribution of width, height, and surface in pixels and mega-pixels, respectively. Additionally, the number of diplomatic seals occurring in each charter image is given.

The dataset not only aims at understanding the quality and nature of the diplomatics data in monasterium.net but also at supporting the creation of pipelines that will offer distant viewing and distant reading. Specifically, we model the layout analysis problem as an object detection problem. As our goal is to maximize the efficiency of diplomatic scholars, we devised an annotation strategy that would require as little time as possible per charter. Diplomatic charters are considered highly formulaic [3], commonly containing a single dominant text-block as can be seen in Fig. 1. Since the images do not seem to need any dewarping, we modeled all layout annotation tasks as defining two-point rectangles, i.e., a rectangle extending from the top left corner to the bottom right corner. Each rectangle belongs to one of 11 specific classes. While these

---

[1] Software available under an open-source licence at https://github.com/anguelos/frat and in pypi https://pypi.org/project/frat/.

[2] The code has been forked from the original YOLOv5 repository, the specific scripts used for the experiments are in the bin folder https://github.com/anguelos/yolov5/tree/master/bin.

[3] The code used to run the resolution regression experiments, including links to the datasets, is available in a github repository https://github.com/anguelos/resolution_regressor.

classes are from the point of view of annotation flat and independent, they imply a hierarchical structure:

- *"No Class"*: Reserved, should never occur normally.
- *"Ignore"*: Used for possible detections that should be tolerated but not required when doing performance evaluation. Might make sense to exclude this class of objects when training a model, e. g., weights placed on charters in order to hold it in place for photography.
- *"Img:CalibrationCard"*: Calibration cards are usually the 21 cm Kodak ones but not always; they usually occur on both sides of the charter.
- *"Img:Seal"*: Many charters contain one or more seals. Seals are of particular interest for sigillographers, i.e. a small sub-community of paleographers focused on the study of seals.
- *"Img:WritableArea"*: This class represents the background material on which a charter was written. It consists typically of parchment but paper is also common in late medieval times. The following classes can only reasonably occur inside this class.
- *"Wr:OldText"*: This class represents the typically single text block of the charter. This rectangle should aim to contain the whole tenor, if possible only the tenor of the charter.
- *"Wr:OldNote"*: This refers to pieces of text put there during the charter's creation/early life that are not part of the tenor.
- *"Wr:NewText"*: This refers to pieces of text added at later times, typically by archivists. From a legal point of view it should be considered irrelevant.
- *"Wr:NewOther"*: This refers to items added to the charter after its creation that are not purely text, e. g., stamps and seal imprints.
- *"WrO:Ornament"*: This refers to items put on the charter at creation time for principally aesthetic purpose.
- *"WrO:Fold"*: This class is meant to indicate a typical charter element, the plica. A fold that is not strictly a plica but is of significant size is still considered a fold.

As the class names imply: calibration cards, seals, and writable areas can be perceived as children of the root element "image", and all other classes are children of the writable area. This structural interpretation is suggested but not enforced in any way.

2.2   Object Detection

As an object detector we employed an off-the shelf YOLOv5 [6], which was only modified to include fractal-based augmentations, that are especially well suited for document images [11]. The small model variant was chosen, which consists of 7.2M parameters at half precision and it was trained on images resized to 1024×1024 pixels. The dataset contains 1184 annotated images, which were split into 741 images for the train set and 440 images for the validation set.

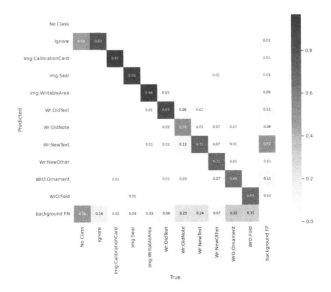

Fig. 3. Confusion Matrix for all classes at 50% IoU

The mean average precision (mAP) for all classes at 50% Intersection over Union (IoU) obtained is 73.21%, while at 95% IoU it is 52.72%. These numbers demonstrate very good results, for some classes, the results are even nearly flawless. The confusion matrix is shown in Fig. 3. We can see that classes *Img:Seal*, *Img:WriteableArea*, and *Img:CalibrationCard* all score at 94% precision or higher. These classes are the ones for which annotators only need little interpretation. Class *"Wr:OldText"* is also performing fairly well with an accuracy of 87%, especially if one takes into account that 8% of missed objects are detected as *"Wr:OldNote"*, which indeed can easily be confused. This class is also quite important as it is the one that detects text blocks, which can subsequently be segmented into text lines so that they can be passed to a handwritten text recognition (HTR) process.

Figure 4 shows the precision recall curves of each class along with their mAP at 50% IoU. It can be observed that the mAP scores are significantly higher than the precision values reported in the confusion matrix.

2.3   Annotations with FRAT

We introduce a lightweight annotation tool for document image annotation. The tool can annotate rectangles by just two-points with an arbitrary set of classes. Other than a class, each rectangle can also have an associated Unicode transcription as well as a comment. The tool is called Fast Rectangle Annotation Tool (FRAT). It is web-based and written in Python and Javascript and therefore runs on Linux and Mac OsX natively, and on Windows using Docker. FRAT is available in pypi and can be easily deployed with pip. In order to accommodate this very

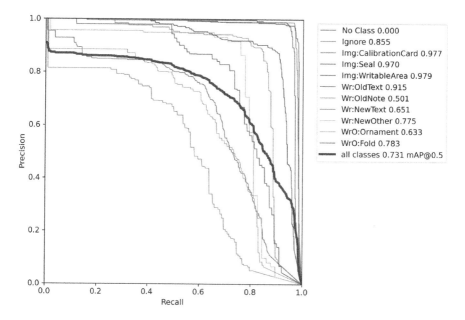

Fig. 4. Precision Recall curves for all classes at 50% IoU

generic data model, the tool has its own JSON-based file format reflecting these fundamental assumptions and nothing more. The User Interface (UI) is aimed at being extremely lightweight and fast, to a certain extent at the expense of precision. It is inspired by gaming interfaces, where all buttons are accessible as single keys on the keyboard, see Fig. 5. Set operations such as rectangle union and subtraction allow the annotator to figure out efficient use patterns that best suit the domain of data employed.

A user-overridable JSON configuration file defines all user preferences as well as the class labels to be employed. All that is needed in order to begin a new project is to define the class labels in the config file. Class labels are considered by FRAT to be flat although they might be perceived to form a hierarchy For example, if a class is *textline* and another class is *word*, FRAT will ignore the fact that a textline consists of words but simple heuristics could be made to realise this hierarchical relation, but this hierarchical relation could be realized through fairly straightforwrd post-processing heuristics, that would create textlines out of word bounding boxes. Custom exporting to PAGE-XML [12] for text-styled textlines is demonstrated in the code. Since object detection fits best, the scenario FRAT was built around, generic exporting to the MS-COCO dataset format [9]. YOLOv5 [7] is also available but no transcriptions are exported.

From a user interface perspective, FRAT features two modes of image annotation: The first mode, "Label", works like a typical generic object annotation tool, with an extra option of transcribing or commenting each annotated object.

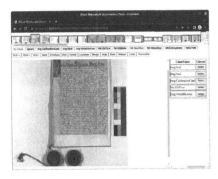 

Fig. 5. FRAT's "Label" (left) and "Transcription" (right) modes.

Defining an object is done by a single drag along its diagonal. Single-key short-cuts allow for editing, manipulating, and navigating through all defined objects. The second mode, "Transcribe", is aimed at transcribing text and reviewing object classes. In this mode a list of all defined objects, their classes, their transcriptions, and their comments is displayed in a scrollable list. This mode is most effective when annotating transcriptions of textlines or words, since allowing to transcribe something out of its context might make the groundtruthing more realistic not allowing annotators to use high-level semantic cues from the nearby text. FRAT has also been used to annotate data including class and transcriptions in [13].

3   Resolution Prediction

Given that many images of diplomatic charters contain calibration cards, it is possible to infer an image resolution approximation. This is quite useful as it allows for quality analysis of large archives but would also allow to associate visual objects of the charter seal, plica, etc. with physical world sizes. As is often the case, the data we are dealing with has been acquired in a non-standardized way and any information about the acquisition process is considered irrecoverable. Under the assumption that during acquisition the charter, the calibration card, and the seals lie on the same plane and that local depth deviations are negligible, we can use the marked distances on the calibration cards to get an estimate for the width and height of any pixel that lies on that plane. Assuming a view angle ϕ, and assuming that calibration cards are predominantly on the horizontal edge of the cameras field of view, the estimated pixel resolution lies in between 1 and $\cos(\frac{\phi}{2})$ of the estimated pixel resolution. In the case of a 50 mm focal length, this error would be at most 7.62% while in the case of 35 mm focal length, it would be at most 13.4%. As the images have been cropped, and nothing is known about this process, this distortion is not reversible but can be used to provide some bounds.

In order to create the ground truth data for the task of resolution estimation, we used the object detector to isolate the calibration cards, writeable areas, and seals. For simplicity's sake, we used FRAT and marked rectangles with a 5 cm diagonal wherever that was possible. When 5 cm was not an option due to image size, we marked 1 cm. In the few cases where there was no visible inch or cm marking, we operated on the assumption that inches are divided by halves, quarters, and eights; while centimeters are divided by halves and millimeters. For all analyses, we used Pixel per Centimeter (PpCm) unit.

3.1   Baseline

For a baseline, we considered the calibration-card object detector described in Sect. 2. Specifically, since almost all calibration cards are variants of the Kodak calibration cards, they have a length of 20.5 cm, even if they differ in other details, while having a height of less than 5 cm. The calibration cards occur always on the sides of the images vertically or horizontally aligned either vertically or horizontally, by choosing the largest side of the bounding box, we can with high certainty obtain an estimate on how many pixels make a length of 20.5 cm on the photographed plane. This heuristic interpretation of the calibration card object detection gave a Spearman's correlation of 89.13% and makes it a reasonably strong baseline. To the knowledge of the authors there is no other baseline available in the literature that is applicable in this experimental context.

3.2   ResResNet

We introduce a network called ResResNet which is a typical ResNet [4] where the final fully connected layer is substituted with a small multi-layer perceptron (MLP) consisting of three fully connected layers with 512, 256, and 1 nodes. While the ResResNet can utilise any ResNet architecture as a backend, we mostly experimented with ResNet18 in order to limit the capacity of the model to suit our relatively small dataset. Due to their average pooling layer, ResResNet can operate on images of arbitrary size. Image resolution can be considered as a texture-base analysis. Thus, during inference, we run ResResNet on a sliding window over the initial image and consider the median prediction as a result for the image. Note: in Fig. 7 this inference method is called stable.

3.3   Training

We used cropped images of the writeable areas as detected with the object detector for inputs and the estimated PpCm as deduced from the annotations on the respective calibration cards. For training, each input image was randomly cropped to a square of 512×512 pixels providing sufficient data augmentation. We employed a Mean Square Error (MSE) loss function, a learning rate of 0.0001 and used an ADAM optimiser. The batch-size was set to 16. The backbone of the ResResNet was initialised with weights pre-trained on Imagenet [2] and would not converge at all when initializing with random weights. In total, we trained the model for 1000 epochs. Figure 6 shows that the model converges to its validation performance after epoch 200 while the training error continues to decrease, it is evident that no problematic overfitting occurs at least for the 1000 epochs.

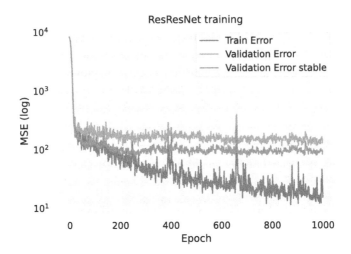

Fig. 6. ResResNet training convergence

3.4   Results and Analysis

Resolution estimation is a regression task. Hence, typical performance metrics such as Accuracy and F-score are not applicable. For our performance analysis, we use MSE and we consider Spearmann's correlation to be more appropriate as the errors do not follow a Gaussian distribution in a very loose sense. MSE quantifies the size of mistakes in total while Spearman's correlation quantifies how often predictions are correct. Figure 7 shows the results. While both valida- tion and train sets were curated such that they would only refer to images that contained a calibration card out of necessity, they fail to demonstrate the base limitation of the baseline system, which is that a calibration card is required for an estimate to be possible at all.

Therefore, we gradually remove samples from the validation set from the most erroneous one (according to the squared error) to the least erroneous one. The results are shown in Fig. 8. We notice that the calibration card baseline performs the best for most samples but overall ResResNet is more reliable as it handles hard samples more gracefully. Furthermore, ResResNet deals with images that do not contain any calibration card at all. The fact that "texture" analysis is enough to learn to predict physical resolution indicates that historical textual data is by nature scale-sensitive and because the trained image analysis models operate on various scales, we can speculate that they implicitly learn multiple scale representations of the same phenomenon. Further experiments that might help validate these hypotheses would be to see whether re-scaling train and test sets for typical textural image analysis tasks such as binarization would improve performance on these tasks. Another approach to better make use of this hypothesis would be to add ResResNet as a Spatial Transformer Network (STN) [5] with PpCm being the single degree of freedom.

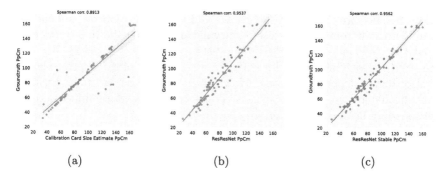

Fig. 7. Scatter plots of the resolution estimates with the ground-truth for the validation set. The resolution estimate from object detection on (a) calibrations cards, (b) the resolution estimate from ResResNet on the cropped charter, and (c) the resolution estimate from ResResNet on the cropped charter after outlier removal.

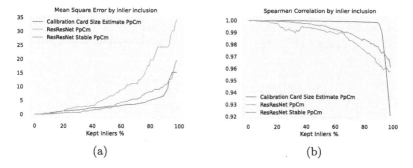

Fig. 8. Outlier contribution to PpCm prediction. (a) MSE growth and (b) Spearmann's correlation growth as more outliers are included.

4   Discussion and Conclusion

4.1   Data Annotation Efficiency

In this paper, we present a work on Diplomatic charters motivated by making optimal use and re-use of domain experts. Our principal research question was how much effort can we economise on annotating and how far will these weak annotations take us in the use of subsequent pipelines. While observing the annotation task, it was evident that marking the seals on a charter would only take a few seconds while waiting for the image to load might take quite a bit longer. To a great extent the proposed class ontology was designed to make better use of the scholar's time by trying opportunistically to ask more questions while they wait. Essentially we tried to increase the objects to be annotated on each image in order to dilute waiting time of annotators to the extent where it would not cause frustration.

The defined classes were chosen not only on the theoretical completeness but also on what is easy to obtain and what might be easier to learn from a machine learning perspective. Modeling the classes and the annotation recommendation benefited greatly from the interdisciplinary nature of the team. It is worth pointing out that the class that made diplomatists sceptical was the class **Wr:Fold**. Since it does not coincide totally with the established "plica" charter element. It is also the worst performing one, while this could also be interpreted with the fact that folds are not well defined in rectangles and therefore object detection might possibly not be well suited for these elements.

4.2   Towards a Full Diplomatics Pipeline

It also appears that in the case of diplomatics, there is no need for a dewarping stage as in other DIA (Document Image Analysis) pipelines [10]. While dewarping is a very simple step on its own, it complicates a DIA pipeline quite a bit because pixel coordinates are not consistent across the whole pipeline and from a user interface perspective it becomes a major challenge to align localised outputs to the original images. The performance trade-off of using simple object detection as the only supervision data for layout analysis in other use cases has yet to be determined. Thus, there is strong evidence that it is sufficient in the diplomatics case.

4.3   Future Work

The work presented in this paper is mostly about presenting a preliminary visual analysis of a large corpus of charters. Several questions for further research are still open. The most interesting finding from a DIA perspective is the fact that, in the case of diplomatic charters, we can predict the PpCm of a charter by looking at patches of 512×512 pixels. An open question is whether resolution can be predicted because of the features of the foreground, the background (parchment), or the acquisition process and how we can exploit that. The other direction in which the annotated data can be used is in high quality synthesis of data for training the subsequent stages of the pipeline such as binarization and HTR.

Acknowledgements. The work presented in this paper has been supported by ERC Advanced Grant project (101019327) "From Digital to Distant Diplomatics" and by the DFG grant No. CH 2080/2-1 "Font Group Recognition for Improved OCR".

References

1. Boroş, E., et al.: A comparison of sequential and combined approaches for named entity recognition in a corpus of handwritten medieval charters. In: 2020 17th International conference on frontiers in handwriting recognition (ICFHR), pp. 79–84. IEEE (2020)
2. Deng, J., Dong, W., Socher, R., Li, L.J., Li, K., Fei-Fei, L.: Imagenet: a large-scale hierarchical image database. In: 2009 IEEE Conference on Computer Vision and Pattern Recognition, pp. 248–255. IEEE (2009)

3. Filatkina, N.: Historische formelhafte sprache. In: Historische formelhafte Sprache. de Gruyter (2018)
4. He, K., Zhang, X., Ren, S., Sun, J.: Deep residual learning for image recognition. In: Proceedings of the IEEE Conference on Computer Vision and Pattern Recognition, pp. 770–778 (2016)
5. Jaderberg, M., Simonyan, K., Zisserman, A., Kavukcuoglu, K.: Spatial transformer networks. In: Cortes, C., Lawrence, N., Lee, D., Sugiyama, M., Garnett, R. (eds.) Advances in Neural Information Processing Systems, vol. 28. Curran Associates, Inc. (2015)
6. Jocher, G.: YOLOv5 by Ultralytics (2020). https://doi.org/10.5281/zenodo.3908559, https://github.com/ultralytics/yolov5
7. Jocher, G., et al.: ultralytics/yolov5: v3.1 - Bug Fixes and Performance Improvements (2020). https://doi.org/10.5281/zenodo.4154370
8. Leipert, M., Vogeler, G., Seuret, M., Maier, A., Christlein, V.: The notary in the haystack – countering class imbalance in document processing with CNNs. In: Bai, X., Karatzas, D., Lopresti, D. (eds.) DAS 2020. LNCS, vol. 12116, pp. 246–261. Springer, Cham (2020). https://doi.org/10.1007/978-3-030-57058-3_18
9. Lin, T.-Y., et al.: Microsoft COCO: common objects in context. In: Fleet, D., Pajdla, T., Schiele, B., Tuytelaars, T. (eds.) ECCV 2014. LNCS, vol. 8693, pp. 740–755. Springer, Cham (2014). https://doi.org/10.1007/978-3-319-10602-1_48
10. Neudecker, C., et al.: Ocr-d: an end-to-end open source ocr framework for historical printed documents. In: Proceedings of the 3rd International Conference on Digital Access to Textual Cultural Heritage, pp. 53–58 (2019)
11. Nicolaou, A., Christlein, V., Riba, E., Shi, J., Vogeler, G., Seuret, M.: Tormentor: deterministic dynamic-path, data augmentations with fractals. In: Proceedings of the IEEE/CVF Conference on Computer Vision and Pattern Recognition, pp. 2707–2711 (2022)
12. Pletschacher, S., Antonacopoulos, A.: The page (page analysis and ground-truth elements) format framework. In: 2010 20th International Conference on Pattern Recognition, pp. 257–260. IEEE (2010)
13. Seuret, M., et al.: Combining ocr models for reading early modern printed books (2023)

Greek Literary Papyri Dating Benchmark

Asimina Paparrigopoulou[1]([✉]), Vasiliki Kougia[2], Maria Konstantinidou[3],
and John Pavlopoulos[1]

[1] Department of Informatics, Athens University of Economics and Business,
Athens, Greece
{paparrigopoul20,annis}@aueb.gr
[2] University of Vienna, Vienna, Austria
vasiliki.kougia@univie.ac.at
[3] Democritus University of Thrace, Komotini, Greece
mkonst@helit.duth.gr

Abstract. Dating papyri accurately is crucial not only to editing their texts but also for our understanding of palaeography and the history of writing, ancient scholarship, material culture, networks in antiquity, etc. Most ancient manuscripts offer little evidence regarding the time of their production, forcing papyrologists to date them on palaeographical grounds, a method often criticized for its subjectivity. In this work, with data obtained from the Collaborative Database of Dateable Greek Bookhands, an online collection of objectively dated Greek papyri, we created a dataset of literary papyri, which can be used for computational papyri dating. We also experimented on this dataset, by fine-tuning four convolutional neural networks pre-trained on generic images.

Keywords: image classification · papyri images · computational dating

1 Introduction

The object of papyrology is reading, studying, interpreting, and exploiting ancient texts preserved on papyrus [1]. In reality, however, we cannot define this discipline based on their writing material [2], considering that a papyrologist also studies texts surviving on parchment, ostraca, wood, bone, stone, and fabric (but not inscriptions, therefore the writing medium must be portable). These texts are exactly the same as the ones surviving on papyrus and they come from the same societies and date to the same periods of time [1]. Therefore, it would be more appropriate to adopt Bagnall's definition [2] that "papyrology is a discipline concerned with the recovery and exploitation of ancient artifacts bearing writing and of the textual material preserved on such artifacts". In terms of content, we can define two main categories of papyri: literary papyri, bearing texts of literary interest, and documentary ones, bearing texts of various topics of daily life, such as contracts, tax receipts, business letters, etc. [1][1].

---

[1] The ones not clearly falling within either category are called subliterary papyri and include texts like commentaries, school exercises, magical charms and curses, scientific treatises, etc. [2].

© The Author(s), under exclusive license to Springer Nature Switzerland AG 2023
M. Coustaty and A. Fornés (Eds.): ICDAR 2023 Workshops, LNCS 14193, pp. 296–306, 2023.
https://doi.org/10.1007/978-3-031-41498-5_21

Dating papyri is considered particularly important for the interpretation and the assessment of their content [1]. Documents are often much easier to date, since they frequently bear a date or some reference to known people, institutions, offices or other evidence helpful to that direction. Nonetheless, chronological attribution is not always straightforward: the writers of private letters for the most part did not record dates, while literary texts remain dateless [3]. So what methods do papyrologists apply in these cases?

1.1 Background

Turner [3] in his work "Greek Manuscripts of the Ancient World" describes some of the methods employed for papyrus dating. In some cases, archaeological evidence may be of assistance, like the papyri from Herculaneum, which we know were written before 79 BCE., when the volcano of Vesuvius erupted. Furthermore, when a document and a literary papyrus are found together in a mummy cartonnage, we can trust the date of the dated documents as a terminus ante quem for the literary text, since both papyri were discarded at the same moment as useless paper. More trustworthy are the dates we can extract when the backside of a papyrus is reused. More specifically, when there is a dated document on the front side (the recto side), then we know that the text on the back (the verso side) was written or copied after the date of the dated document. Conversely, if the dated document is on the back of the papyrus, we know that the text on the front was written or copied before the date of the document. However, in this case we cannot be sure of the time gap between the two. In the event that none of the above evidence is offered for dating, we can take into account the content, such as events that are described or "exploit fashions in 'diplomatic' usage, such as the use of and form taken by abbreviations" [3].

The method used predominantly to get more accurate results, especially when all the other criteria are absent, is based on palaeography, i.e. the study of the script. Dating on palaeographical grounds is based on the assumption that graphic resemblance implies that the two manuscripts are contemporary [4], as literary papyri are written in elaborate and conservative more formal writing styles that remain unchanged for decades or even centuries, whereas documentary papyri are almost always written in cursive scripts that can be dated with relative accuracy [1]. However, this distinction is not absolute, considering that, as stated by Choat [5], "many dated documents, and the scripts of some of these are sufficiently similar to those of literary papyri for them to form useful comparanda to the latter" and, as Mazza [4] adds, frequently documentary papyri are written in literary scripts and vice versa literary papyri are copied in documentary scripts. Therefore, it is obvious that relying solely on palaeography is a great challenge that presents plenty and considerable difficulties. For the chronological attribution of a papyrus, the papyrologist should have "a wide range of potential comparanda and have them available for easy consultation" [5]. This is not an easy task nor can be achieved without the proper training. Besides, as stated above, the fact that literary texts almost never bear a chronological indication, results in a very small number of literary papyri, securely dated, that can form a basis for comparison. On the other hand, to estimate the date of a papyrus one should take into account all the parameters, like the provenance, the context, the content, the language, the dialect, the codicology, the page layout, the general appearance of the script, the specific letter shapes of the papyrus under examination [5]. Lastly but most importantly, we should not overlook the subjectivity of the whole method, a parameter to which, according to Choat [5], is given less regard than should be.

In recent years, there have been efforts to date manuscripts of various languages with the help of computational means [6]. In reality, what these tools and techniques are trying to achieve is the chronological attribution of the manuscripts, based on the palaeographic assumption of the affiliation of scripts, described above, trying, nonetheless, to eliminate the subjective element of this method. However, our study of the literature makes it clear that most of these computational approaches disregard manuscripts written in the Greek language.

1.2   Our Contribution

Greek papyri form a distinct collection of ancient manuscripts. Despite sharing characteristics shared by all ancient (as well as modern for that matter) handwritten artefacts, they also have a number of properties unique to them, which call for new research, specifically dictated by and centred on their specificities. Such properties are the time and geography of their production (which includes the materials involved, e.g. papyrus and ink), format, state of conservation and, most importantly, writing culture in the Graeco-Roman world and the evolution of Greek script and writing. Unlike ancient manuscripts in other languages, collections of Greek papyri are both plenty and scarce, both uniform and diverse. They are scarce compared to medieval Greek manuscripts (particularly in size), but still numerous. They (almost) unfailingly come from Egypt, a small fraction of the Greek-speaking ancient world, but they exhibit sufficient diversity in their content, form, and script, to merit separate and distinct examination.

This research is an initial investigation into the computational dating of Greek papyri. It exploits available data resources to explore machine learning methods that may assist papyrologists by computing a date for papyri of unknown dates. Using data obtained from an online collection of securely dated papyri, a machine-actionable dataset was created, suitable for the task of computational dating of Greek literary papyri. We used this dataset to fine-tune deep learning classifiers, benchmarking their ability to estimate the papyri date. We also experimented with different transformations of the images in order to give them as input to Convolutional Neural Networks (CNNs) pre-trained on ImageNet [26], which contains images from the generic domain. The best results were achieved by a DenseNet operating on images resized to (448, 448). The remainder of this article first summarises the related work and then describes the presented dataset. The methodology used is presented next, followed by the experiments that were undertaken on the dataset, and a discussion. A summary of the findings and suggestions for future work concludes this article.

2   Related Work

Dating of papyri (images) with computational means has been studied for many languages [8–12], but not for Greek. Dating the *text image* is very different from dating *the text of the image*, as has been done for ancient Greek inscriptions [13]. For instance, the latter requires transcription of the text in the image, which is a time-consuming process. Such a technique is also irrelevant in the case of literary papyri, because the texts that they transmit typically date much earlier than the actual manuscripts (e.g. a scribe in Late Antiquity copying on papyrus a Homeric poem composed more than a thousand years earlier). Also, any information regarding the script or clues aside of the text will be disregarded. Given the absence of dating methods for Greek, the following overview focuses on the studied dating methods for other languages.

The studied languages are Latin [8,14,15], Hebrew [9], Dutch [11,16–19], Arabic [12], Swedish [14,15], French [8] and English [10]. A collection of 595 Dead Sea Scrolls written in Hebrew alphabet (derived from Aramaic script) was the dataset with the oldest manuscripts, dated from 250 to 135 BCE [9]. The rest of the datasets comprised more data, ranking from approx. two [12] to more than ten thousand manuscripts [15], while the one with the most recent manuscripts comprises historical English-language documents [10], printed between the 15th and 19th century.

The employed methods usually were standard machine learning methods, such as KNN [12], decision trees [8], random forests [8] and support vector machines [9,11,17–19]. Textural features, such as Gabor filters, Uniform Local Binary Patterns and Histogram of Local Binary Patterns are extracted and then fed to the classifiers [16]. The writing style evolution, however, has also been used as an intermediate step [9,12]. In this case, the periods are first aligned with specific writing styles. Then, any new manuscript is dated based on the detected style.

Pre-trained convolutional neural networks have been used to extract features, which are passed to a classifier or regressor [11,14], or used in combination with text features extracted with optical character recognition methods [10]. Transfer learning has been reported to lead to human performance [14]. This was deemed to be the most promising direction for the present study on Greek manuscripts, and was, hence, employed.

3   Data

To create a dataset that can be used to train and assess computational approaches (machine-actionable dataset) for the purpose of papyrus dating, images from an online collection of manuscripts, the Collaborative Database of Dateable Greek Bookhands [20], were downloaded manually and saved. Each image was stored in a formatted manner: [CENTURY]_[MANUSCRIPT NAME].JPG, to preserve the date (learning target) within the filename. With this process a new machine-actionable dataset for literary papyri (PaLIT) was developed. It should be clarified that the online collection used was compiled on the basis of the script and not the content of each papyrus (there are some documents and subliterary papyri in the CDDGB) and, therefore, the dataset developed maintain this criterion of categorization of manuscripts. In some cases only a limited sample of the papyrus images was used for training, as only a small number of papyri was available, allowing for a slight increase in training data in the future.

3.1   PaLIT

The Collaborative Database of Dateable Greek Bookhands is an online catalogue of ancient Greek manuscripts written in literary script, from the 1st to the 9th century A.D, hosted by Baylor University. The data it contains can be dated based on some kind of objective dating criterion, such as the presence of a document that contains a date on the reverse side, or a datable archaeological context associated with the manuscript. The list of papyri included in this dataset could have been more comprehensive, as extensive bibliographic information is not included and secondary literature has not been consulted. Such tasks have already been undertaken by two ongoing - and highly anticipated - projects (report from the introduction of the CDDGB website). However, for lack of a better alternative and since the collection of objectively dated bookhands goes beyond the scope of this study, this collection is deemed adequately reliable. Moreover, it is unlikely that a complete list of securely dated literary papyri would increase the number of specimens beyond the lower hundreds. Figure 1 below shows in detail the distribution by century of the image data taken from this collection. The total number of images used form PaLIT is 161. Collected images were in JPG or PNG format and their resolution varied. A few images in gif format are also included in the collection but we excluded them owing to their poor quality. Specimens in the collection written in minuscule script were excluded too, due to the fact that minuscule Greek cannot be placed confidently into the script evolution process and appears after the period on which this study focuses. Finally, a challenge we had to deal with was duplicates, that is the multiple images that are in many cases available for a single papyrus, each of which depicts a different part of it. Thus, we chose one representative image for each papyrus to the exclusion of the rest.

4   Methodology

Transfer Learning

We employed four CNN architectures to predict the date of a papyri image including DenseNet [25], VGG [21], EfficientNet [24] and ResNet [23]. These models consist of multiple convolutional layers followed by pooling layers and a classifier, which varies from one dense layer to an MLP with several layers. In contrast with traditional CNNs where the layers are connected subsequently, DenseNet connects each layer with every other layer. VGG uses convolutional layers with small filters (3×3), and max and spatial pooling layers, while

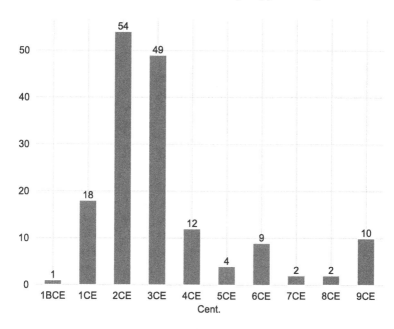

Fig. 1. Distribution of images per century

ResNets are deep CNNs that are trained with residual learning. EfficientNets is a family of models that achieve state-of-the-art results and were created using neural architecture search and compound coefficient scaling. For each model we load a set of weights pre-trained on ImageNet, a dataset of over 14 million images belonging to 1000 classes [26]. Then, we remove the classifier, which is pre-trained on these classes and replace it with a dense layer and softmax in order to obtain a probability distribution over the centuries that serve as labels in our task. The final model is fine-tuned end-to-end on the PaLIT dataset.

5   Experiments

The experimental settings and the selected evaluation measures are discussed in this section, followed by a presentation of the results.

5.1   Experimental Settings

For our experiments, we removed instances of centuries that occurred less than 10 times. We used 5-fold cross-validation to split the remaining 144 instances into 80% for training and 20% for testing. For each fold, 20% of the training set was used for validation and early stopping. Experiments were undertaken with Google Colaboratory, using a 15GB NVIDIA Tesla T4 GPU. The code was implemented in PyTorch and the pre-trained weights of the models were

Table 1. F1 and MAE with 5-fold cross-validation in dating (standard error of the mean) of fine-tuned CNNs operating on three different image transformations.

Model	Resize (224)		Random crop (224)		Resize (448)	
	F1	MAE	F1	MAE	F1	MAE
DenseNet	0.268 (0.076)	1.120 (0.187)	**0.306 (0.030)**	1.124 (0.109)	**0.346 (0.049)**	**0.980 (0.119)**
VGG	0.114 (0.019)	1.145 (0.047)	0.114 (0.012)	1.152 (0.035)	0.133 (0.039)	1.235 (0.053)
EfficientNet	**0.323 (0.063)**	**0.971 (0.112)**	0.215 (0.050)	1.302 (0.131)	–	–
ResNet	0.224 (0.062)	1.056 (0.050)	0.184 (0.033)	**1.116 (0.081)**	0.235 (0.053)	1.077 (0.061)

loaded from Torhvision. We used the following versions of the CNNs architectures: DenseNet-121, VGG-16, EfficientNetV2-L and ResNet-101. These models require a minimum input size of (224, 224). The images in the PaLIT dataset have an average width of 1616.3 and a height of 1774.8. We experimented with three different settings: 1. Resize (224): the images were resized to the minimum required size, 2. Random crop (224): the images were cropped randomly to the minimum required size, so that each image will be randomly transformed every time it is loaded to the model, 3. Resize (448): the images were resized to (448, 448).[2] In addition, the images were transformed to RGB and normalized with the mean and standard deviation of ImageNet. We trained the models using cross-entropy loss and SGD with momentum as the optimizer. Early stopping was used with patience of three epochs. Each model was trained three times with different seed initialization.

5.2   Evaluation Measures

The performance of all models was measured using F1, which is the harmonic mean of Precision and Recall. Macro-averaging was used across the centuries, in order to put an equal interest in the performance of all centuries and not only of the most frequent ones (e.g., 2nd and 3rd; see Fig. 1). Classification evaluation measures do not capture how far a model has failed in its estimates (here, regarding the century) in relation to the ground truth. Hence, we employ also the confusion matrix and the Mean Absolute Error (MAE), defined as the sum of the absolute difference of each predicted value from the respective ground truth of all the test set samples, divided by the total number of samples.

5.3   Experimental Results

Table 1 presents the experimental results. We observe that overall the best F1 score is achieved by DenseNet using resizing to (448, 448), while EfficientNet with 224-resizing has the best MAE score. For DenseNet better scores are achieved when using random cropping and 448-resizing, compared to 224-resizing. An improvement is also shown for VGG and ResNet when resizing to (448, 448)

---

[2] For EfficientNet we could not use a size larger than 224 as a CUDA out-of-memory error would occur.

compared to the other two settings, but only in terms of F1 score. It is reasonable to infer that resizing to smaller sizes makes the task more difficult and important information is missed. Interestingly, however, this is not the case for EfficientNet, which performs better when resizing the images to 224. It even achieves competitive results in the standard resizing setting (224), which can also be seen in the confusion matrices in Fig. 2. Hence, we conclude that resizing to smaller sizes should not be disregarded as an option when fine-tuning pre-trained image classifiers.

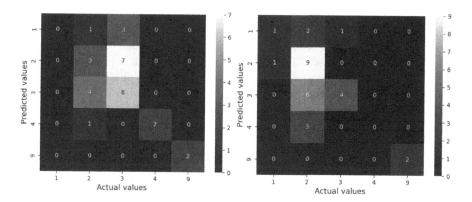

Fig. 2. Confusion matrices for the two best performing models: EfficientNet with 224-resizing (left) and DenseNet with 448 resizing (right).

6   Discussion

6.1   Century-Level Dating

The use of the century as the unit for dating the papyri was dictated by the image labelling in the training set of literary papyri (PaLIT). As broad a range as it may seem to the non-expert, most papyrologists assign specimens to dates spanning a whole century, sometimes even two centuries. Rarely, papyri are assigned to one half of a century, or a vague 'early' or 'late' part of a century. Even in documents, which frequently carry a precise date (often down to a day) the ones that do not, are also assigned dates spanning whole centuries.[3] Apart from constituting standard practice, there are also valid theoretical concerns about assigning narrow ranges when dating on palaeographical grounds [7].

---

[3] See for example the POxy Oxyrhynchus Online Database [22]. Out of the approx. 2,500 non-objectively dated papyri, no more than 10% are assigned a date span of less than a whole century.

6.2 Challenges and Limitations

The lack of data, meaning objectively dated papyri that could be used for model training, is an important and great challenge that we faced. The lack of publicly available machine-actionable data is most probably due to papyri licensing issues. A thorough investigation of these issues has not been performed by the authors.

Class imbalance characterises the presented dataset. The distribution of papyri per century is very heterogeneous (Fig. 1), to the point that some centuries have few to almost no samples, while others have a significant representation. For example, a single sample of literary papyri is included in PaLIT from the 1st BCE and two from the 7th CE. By contrast, the 2nd and 3rd centuries CE are supported with 87 and 71 samples respectively. This imbalance naturally affects the results and introduces a limitation, because it leads to a poor performance of the models when they are called to date manuscripts of the centuries that have minimal representation from samples.

Inaccurate ground truth is a final limitation. The dates that are assigned to the papyri by objective criteria [3] are often not entirely precise or accurate but estimates with an error probability of about 50 years. This means that a manuscript attributed to a certain century may have been written in the previous or the following century. These dates, however, serve as the ground truth for machine learning, and hence noise may be present.

7   Conclusion

This study presented experiments with transfer learning for the challenging task of dating Greek papyri. We used data from two online collections of objectively dated papyri, organised in a machine-actionable form. Experimental analysis showed that a DenseNet CNN, fine-tuned on images resized to (448, 448), achieves the best F1 score (0.346). Given that the ground truth estimates come with an error probability of 50 years, we find that room for improvement exists. Future work will attempt to establish better results, by extending our dataset and by experimenting with augmentation.

References

1. Παπαθωμάς,   Α.:   Εισαγωγή στην παπυρολογία. 3η Επαυξημένη Έκδοση. Αμφιλόχιος Παπαθωμάς, Athens (2016)
2. Bagnall, R.S.: The Oxford Handbook of Papyrology. Oxford University Press, Oxford (2012)
3. Turner, E.G.: Greek manuscripts of the ancient world. In: Parsons, P.J. (ed.) Revised and Enlarged edn. Institute of Classical Studies, London (1987)
4. Mazza, R.: Dating early Christian papyri: old and new methods - introduction. J. Study New Testament **42**(1), 46–57 (2019)
5. Choat, M.: Dating papyri: familiarity, instinct and guesswork. J. Study New Testament **42**(1), 58–83 (2019)

6. Omayio, E.O., Indu, S., Panda, J.: Historical manuscript dating: traditional and current trends. Multimed. Tools Appl. 1–30 (2022)

7. Nongbri, B.: Palaeographic analysis of codices from the early Christian period: a point of method. J. Study New Testament **42**(1), 84–97 (2019)

8. Baledent, A., Hiebel, N., Lejeune, G.: Dating ancient texts: an approach for noisy French documents. In: Proceedings of LT4HALA 2020 - 1st Workshop on Language Technologies for Historical and Ancient Languages, pp. 17–21. European Language Resources Association (ELRA), Marseille (2020)

9. Dhali, M., Jansen, C.N., de Wit, J.W., Schomaker, L.: Feature-extraction methods for historical manuscript dating based on writing style development. Pattern Recogn. Lett. **131**(1), 413–420 (2020)

10. Li, Y., Genzel, D., Fujii, Y., Popat, A.C.: Publication date estimation for printed historical documents using convolutional neural networks. In: Proceedings of the 3rd International Workshop on Historical Document Imaging and Processing (HIP 2015), pp. 99–106. Association for Computing Machinery, New York (2015)

11. Hamid, A., Bibi, M., Moetesum, M., Siddiqi, I.: Deep learning based approach for historical manuscript dating. In: International Conference on Document Analysis and Recognition (ICDAR), pp. 967–972. Institute of Electrical and Electronics Engineers (IEEE), Sydney (2019)

12. Adam, K., Baig, A., Al-Maadeed, S., Bouridane, A., El-Menshawy, S.: KERTAS: dataset for automatic dating of ancient Arabic manuscripts. Int. J. Doc. Anal. Recognit. (IJDAR) **21**(4), 283–290 (2018). https://doi.org/10.1007/s10032-018-0312-3

13. Assael, Y., et al.: Restoring and attributing ancient texts using deep neural networks. Nature **603**(7900), 280–283 (2022)

14. Wahlberg, F., Wilkinson, T., Brun, A.: Historical manuscript production date estimation using deep convolutional neural networks. In: 15th International Conference on Frontiers in Handwriting Recognition (ICFHR), pp. 205–210. Institute of Electrical and Electronics Engineers (IEEE), Shenzhen (2016)

15. Wahlberg, F., Mårtensson, L., Brun, A.: Large scale style based dating of medieval manuscripts. In: Proceedings of the 3rd International Workshop on Historical Document Imaging and Processing (HIP 2015), pp. 107–114. Association for Computing Machinery, New York (2015)

16. Hamid, A., Bibi, M., Siddiqi, I., Moetesum, M.: Historical manuscript dating using textural measures. In: International Conference on Frontiers of Information Technology (FIT), pp. 235–240. Institute of Electrical and Electronics Engineers (IEEE), Islamabad (2018)

17. He, S., Samara, P., Burgers, J., Schomaker, L.: Towards style-based dating of historical documents. In: 14th International Conference on Frontiers in Handwriting Recognition, pp. 265–270. Institute of Electrical and Electronics Engineers (IEEE) Hersonissos, Greece (2014)

18. He, S., Samara, P., Burgers, J., Schomaker, L.: Image-based historical manuscript dating using contour and stroke fragments. Pattern Recogn. **58**(1), 159–171 (2016)

19. He, S., Samara, P., Burgers, J., Schomaker, L.: Historical manuscript dating based on temporal pattern codebook. Comput. Vis. Image Underst. **152**(1), 167–175 (2016)

20. Edwards, G.: The collaborative database of dateable Greek bookhands (CDDGB). Baylor University. https://www.baylor.edu/classics/index.php?id=958430. Accessed 21 Dec 2021

21. Simonyan, K., Zisserman, A.: Very deep convolutional networks for large-scale image recognition. In: International Conference on Learning Representations (ICLR), pp. 1–14 (2014). arXiv:1409.1556

22. EES (Egypt Exploration Society): POxy: Oxyrhynchus Online. http://www.papyrology.ox.ac.uk/POxy/. Accessed 21 Dec 2021

23. He, K., Zhang, X., Ren, S., Sun, J.: Deep residual learning for image recognition. In: Proceedings of the IEEE Conference on Computer Vision and Pattern Recognition, 770–778 (2016)

24. Tan, M., Le, Q.: EfficientNetv2: smaller models and faster training. In: International Conference on Machine Learning, pp. 10096–10106 (2021)

25. Huang, G., Liu, Z., Van Der Maaten, L., Weinberger, K. Q.: Densely connected convolutional networks. In: Proceedings of the IEEE Conference on Computer Vision and Pattern Recognition, pp. 4700–4708 (2017)

26. Russakovsky, O., et al.: ImageNet large scale visual recognition challenge. Int. J. Comput. Vision **115**, 211–252 (2015)

Stylistic Similarities in Greek Papyri Based on Letter Shapes: A Deep Learning Approach

Isabelle Marthot-Santaniello[1]($\boxtimes$), Manh Tu Vu[2], Olga Serbaeva[1],
and Marie Beurton-Aimar[2]

[1] University of Basel, Basel, Switzerland
{i.marthot-santaniello,olga.serbaevasaraogi}@unibas.ch
[2] LaBRI, Bordeaux University, Bordeaux, France
{manh.vu,marie.beurton}@labri.fr

Abstract. This paper addresses the issue of clustering historical handwritings according to similarity in absence of metadata on date or style. While releasing a new dataset called AlphEpMu, it proposes to use SimSiam deep neural network model to evaluate similarity between images of individual characters from three Greek letter categories (alpha, epsilon and mu). Two specimens of a given letter category are defined as similar if they come from the same piece of manuscript (penned by a single writer). Similarity is then computed between pairs of images of the same letter category from different manuscripts. Last, scores of the three letter categories are merged to express similarity between the manuscripts. Applied to Greek Literary papyri, this approach is proved useful to paleographers since it allows organizing a complex group of 72 manuscripts (AlphEpMu-72) into a meaningful network but also spotting micro phenomena of similarity which can help explaining the general evolution of handwritings.

Keywords: Computational Paleography · Character similarity · Greek papyri

1 Introduction

Papyri preserved thanks to the dry climate of the Egyptian desert are a key source for ancient Greek literature, providing many masterpieces that would have otherwise been lost to our knowledge. These ancient manuscripts have however usually reached us as damaged fragments scattered in various collections without much information on their context of production like the precise date of their copy or the identity and location of their writer. Being able to use whatever is left of an ancient manuscript to recover its context of production would allow specialists to make major advances in the interpretation of these literary works and shed new light on ancient culture.

M. Coustaty and A. Fornés (Eds.): ICDAR 2023 Workshops, LNCS 14193, pp. 307–323, 2023.
https://doi.org/10.1007/978-3-031-41498-5_22

Fig. 1. Samples of full images from the dataset: 61240 (Sorbonne Université, Institut de Papyrologie), 60471 (University of Hamburg), 61210 (University of Michigan Library, Papyrology Collection)

Evaluating similarities among handwritings is a fundamental task in paleography as the field of study devoted to ancient handwritings: the highest degree of similarity will point to two fragments belonging to the same manuscript. A high but lower degree of similarity would be found between two manuscripts penned by the same writer (at a different time, with a different writing implement, etc.). At an even lower degree of similarity would stand manuscripts written in the same "style", by copyists having been trained at the same school, in the same place, at the same period or sharing the same cultural influences.

Information regarding the identity of the copyist, the date and place of production is, however, not explicitly recorded in Greek papyri as they can be later in "colophons" of medieval manuscripts. Assigning a common writer to several manuscripts or a date and a style to a given piece relies on paleographical analyses. Consensus among experts is hard to reach because similarity is apprehended differently, often based on personal expertise whose criteria are difficult to communicate. Measuring and visualising similarity among handwritings with objective, reproducible methods can thus offer solid grounds to improve paleographical interpretations.

The goal of this present study is to tackle the question of defining, expressing and interpreting similarity among historical handwritings. To be as solid as possible, it uses only criteria that reach consensus among experts. Its methodological choices are to focus on the character (letter) level and to define as "similar" specimens of letters coming from the same ancient document. The applied method to classify automatically characters belongs to Deep Learning models. The used convolutional network is a specific siamese network, called SimSiam, which has the benefit of not requiring ground truth for the training step. The obtained results are expressed as a distance matrix between each pair of character images. As usual the trained network has no knowledge nor apriori about definition of similarity between two characters. In order to facilitate the interpretation of results, graphs are presented and interpreted in terms of stylistic similarities.

2   State of the Art

2.1   Papyrology and Digital Paleography

Because papyri witness the continuous phenomenon of evolution of Greek hand-writings over a millennium (3rd century BCE - 7th century CE), they provide extremely diverse writing specimens that do not easily fit into strictly defined categories. There is no consensus on a typology that would allow assigning a predefined "style" (and thus a date) to any writing sample. Reference works [3,17,27] and studies dedicated to a particular style (Severe Style in [7], Biblical Majuscule in [16]) allow defining some "stylistic classes" but their transitional (i.e. initial and final) phases are still matter of interpretation.

Papyri have proved to be challenging for computational methods because of their complex background, their degree of degradation, and their scattering across various collections who have their own digitisation processes. Although the papyrological documentation is considered to be rich for classical Antiquity (ca. 80,000 texts of all nature already published), the quantity of homogeneous material is limited because the texts are short (as opposed to hundreds of pages of a medieval manuscript). The elaboration of securely labeled datasets that can be used as ground truth is specifically challenging. There have however been some attempts these recent years: papyri were included in ICDAR 2019 Competition on Document Image Binarization (DIBCO 2019) [22], several investigations on Writer Identifications have followed the publication of the GRK-Papyri dataset [14], the most recent being Christlein et al. [6]. Last, Hebrew papyri from the Dead See Scroll have been the object of computational paleography investigations [21].

The difficulties mentioned above are not proper to Greek papyri: "objective" definitions of styles are also difficult to produce for instance for medieval Latin scripts [26]. The Archetype framework was elaborated to address the challenge of efficiently describing handwritings: by "tagging" individual letters and sharing the annotations, it aimed at "showing" them instead of "describing" them [2]. This is also one of the functionalities of the Research Environment for Ancient Documents (READ)[1] that is used in the present work and has proved its efficiency for comparative paleography analysis, for instance with Sanskrit [25].

2.2   Deep Learning State of the Art

In computer vision, early research focused on extracting writer-specific features from handwritten patterns using techniques such as SIFT, Contour, Hinge, and Path Signature [13,28]. These features were then classified using traditional classifiers such as distance-based, Support Vector Machines (SVM), Hidden Markov Model (HMM), and Fuzzy-based classifiers. Recently, deep learning has been emerging as a powerful tool to analyse complex data such as ancient historical documents. Several approaches have been proposed that combine the mentioned

---

[1] https://github.com/readsoftware/read.

handcrafted techniques with deep neural network, such as [6,15], or completely use end-to-end deep neural network [9]. However, most of these approaches rely on supervised learning, which requires a large amount of annotated data that is not available in some cases.

Other research directions, such as unsupervised [5], semi-supervised [24], or self-supervised [18] learning, require less annotated data. These methods aim to extract underlying patterns or structures from data using a small amount of labelling or no labelling at all. For example, Pirrone et al. [20] proposed a self-supervised, Siamese-based deep neural network [1] that can identify papyrus fragments that look similar to each other, while Zhang et al. [29] proposed a self-supervised method based on SimSiam network [4] to recognise handwritten text. Compared to Siamese networks, the SimSiam network is simpler and easier to train since it only requires positive pairs, i.e., pairs of images that look similar to each other. For our problem, experts have checked that all the individual letters from the same manuscript have been penned by the same writer (no manuscript of our dataset present a change of hand), but we are unsure whether several manuscripts could have been from the same writer or not. A naive approach for generating negative pairs could be picking randomly an image from other manuscripts and consider that they are written by different hands. But this is a false assumption because they could also be written by the same writer. Hence, the SimSiam network which requires only positive pairs, seems to be the best choice for this particular problem.

3   Description of the Datasets

3.1   The Preliminary Annotation Work

There was no existing dataset of images of Literary papyri. We thus collected images from online catalogues or requested them to their owning institutions to gradually reach an initial dataset of 144 literary manuscripts from the 3rd century BCE to the 7th century CE coming from Egypt (see Fig. 1). The manuscripts are referred to according to their unique identifier (a five- or more digit "TM" number) from trismegistos.org.

Each of the images was uploaded to and processed in READ[2] namely, we have manually drawn a box around each letter present on the papyri and linked the boxes to the transcription. It should be kept in mind that the focus on individual letter shapes has its limits since stylistic characteristics are also in other handwriting aspects: the layout (writing restrained within two lines like modern upper case or within four lines like lower case), the contrast among letters (opposition small/tall, large/thin letters) and the width variations of the strokes (opposition thick/thin strokes within a letter). It has however the advantage of reducing the complexity of the problem since everyone would agree that an alpha is an alpha, all the most on known literary texts like the *Iliad*, leaving the

---

[2] https://github.com/readsoftware/read/tree/maindev2/model; https://prezi.com/vi ew/VLUMNCTWZ4TamZJU9Vcr/.

possibility open for further holistic investigations that would take into account other aspects of handwritings.

This annotation work of all the letters on the images of the 144 papyri generated a total of 51024 small images of individual letters, called "cliplets". It can be seen on the PalEx viewer[3]. The cliplets were then exported from the "paleography report" function in READ, and from those only the best quality ones (manually assigned a specific tag "bt1") were accepted for the further analysis (see Fig. 2).

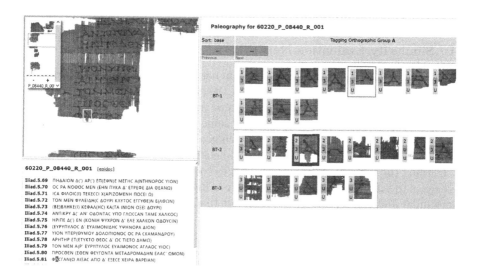

Fig. 2. Papyrus 60220 annotated at the character level using READ interface. On the right, the paleographic report allows assigning tags, here related to preservation state. Papyrussammlung der Staatlichen Museen zu Berlin

3.2 The AlphEpMu Dataset

The Greek alphabet counts 24 letters (or "graphemes" which we call in what follows "letter categories"). Because of time constraint, we chose to focus on three letter categories. Preliminary analysis has shown that all the letter categories were not equally significant to spot palaeographical stylistic similarities among manuscripts. For instance iota and omicron which are among the most frequent letters in ancient Greek (each represents app. 10 percent of the letters of a random Greek text) do not vary much in shape since they roughly always look like respectively a vertical bar and a round. We have thus chosen alpha and epsilon which are the two most frequent letter categories: each has a frequency of app. 11 percent, both are at least represented once in 98 percent of our texts. Besides, their shapes evolve significantly according to periods and styles. Mu was

---

[3] https://showcase.d-scribes.philhist.unibas.ch/viewer.

the third letter category chosen because although it is a medium frequent letter (ca. 3 percent), it is one of those that vary the most in terms of shape and one that is traditionally used by paleographers to discriminate handwriting styles (see [23] for Coptic Manuscripts). Annotating 216 papyrus images (sometimes papyri are digitized on several images), we formed the AlphEpMu dataset that counts a total of 6598 cliplets: 1575 alphas, 4194 epsilons, and 829 mus from 144 TMs. This dataset is made public on a dedicated webpage along with other material mentioned below[4].

4   The Deep Learning Approach

4.1   Similarity Evaluation Network

In this work, we have employed the SimSiam network architecture with ResNet18 network [11] pretrained on ImageNet [8] as the encoder of the network. Although the original idea of the SimSiam network is to train the network using two differ-ent views of same image (augmenting the same image twice), we can still apply this network to our specific user case, which is about measuring the similarity of the two input cliplets containing two specimens of a given Greek letter category. By training the network with pairs of images coming from the same manuscript, we expect the network to be able to assign a high similarity score to pairs that look similar and low similarity score to pairs that look different. The architecture overview of the network can be seen in Fig. 3.

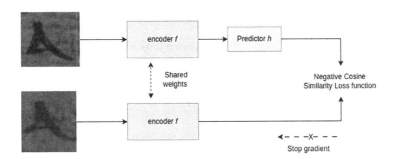

Fig. 3. Overview of the network architecture. Two images of the same letter cate-gory and same TM are processed by the same encoder network f (a backbone plus a projection multilayer-perceptron MPL).

4.2   Dataset Preprocessing and AlphEpMu* Dataset

From the AlphEpMu dataset, we discarded the cliplets that are extremely small (width or height less than 32 pixels). TMs that have less than two cliplets of each of the three letter categories (alpha, epsilon and mu) are also discarded.

---

[4] https://d-scribes.philhist.unibas.ch/en/case-studies/iliad-208/alphepmu-dataset/.

Finally, the cleaned-up version of AlphEpMu dataset, called AlphEpMu* counts $5,698$ cliplets, including $1,375$ alphas, $3,591$ epsilon and 732 mus. The precise number of cliplets of each letter category used for each TM is given in Appendix A and on the webpage.

From AlphEpMu*, cliplets are resized to 64×64 pixels and padded with white background for preserving their aspect ratio. We also apply some augmentation techniques to ease the training process while preserving the letter shapes, which include Random Image Translation, Colour Jitter, and Random Grayscale.

The whole network system is implemented using PyTorch framework [19]. During the training phase, Adam optimiser [12] was employed with the initial learning rate of $lr \times BatchSize/256$ (Linear scaling [10] with base $lr = 4e - 4$). The batch size is set to 196 images per batch.

4.3   Cliplets Similarity Evaluating

In training, for each cliplet x_1^t in the AlphEpMu* dataset, we randomly pick another x_2^t cliplet which comes from the same letter category and same TM t. We train the network using this pair as input. The output of the encoder f (ResNet18 network) includes two vector feature representations of the two input images (cliplets). Then a prediction multilayer-perceptron (MLP) h is applied on one side, and a stop-gradient operation is applied on the other side (see [4]). This network architecture setup allows the model to maximise the similarity between the two input images, while avoiding collapsing solutions (i.e. all model outputs collapse to a constant) by using stop gradient mechanism. Once we have a fully trained SimSiam network, we can use this network to compute the similarities between different cliplets of a given TM in our dataset (within the same letter category alpha, epsilon and mu). Specifically, let p_1^t and p_2^t be the two output vectors of our network, such as $p_1^t = h(f(x_1^t))$ and $p_2^t = h(f(x_2^t))$. The similarity score between these two cliplets is computed as follows:

$$S(p_1^t, p_2^t) = \frac{p_1^t}{\|p_1^t\|_2} \cdot \frac{p_2^t}{\|p_2^t\|_2} \tag{1}$$

which is basically the cosine similarity between the two vectors p_1^t and p_2^t. Since the output value of S is ranging between $[-1, 1]$, we shift this output value to the range of $[0, 1]$ by replacing S by $S' = \frac{S+1}{2}$. With the function S', we can also measure the similarity of cliplets that come from different TMs, just by replacing the parameter p_1^t or p_2^t by a feature vector from another cliplet. To measure the similarity between two different TM t_1 and TM t_2 for each of the three letter categories, we calculate the similarity score between each pair formed by a cliplet from t_1 and a cliplet from t_2 and we average these scores as follows:

$$SS(t_1, t_2) = \frac{1}{m \times n} \sum_{i=0}^{m} \sum_{j=0}^{n} S'(p_i^{t_1}, p_j^{t_2}) \tag{2}$$

where m and n are the total numbers of cliplets that belong to TM t_1 and TM t_2, respectively. This computing has been done first for the three letter categories separately and then the scores were averaged to obtain a global similarity score per TM pairs.

4.4 Results

A usual way to visualize global results of similarity matrix is heatmap. We generated a heatmap for each of the three letter category, available on the dedicated webpage. They show, as expected, that the similarity scores among cliplets coming from the same TM are extremely high, thus that the intra-writer variability at the document level appears in our experiment as a minor phenomenon compared to the inter-writer variation. A few other high scores can be spotted but due to the large number of pairs, thus of line and columns, heat maps are difficult to use.

To validate the approach, we have a test case, two fragments of the same manuscript TM 59170 divided in two items: 59170Oxy and 59170Yale (further described below). Into the similarity matrix this pair reaches the best similarity score (smallest distance) for each letter category ($\alpha = 0.987, \epsilon = 0.98, \mu = 0.994$).

5   Stylistic Similarities Among Manuscripts

5.1   The AlphEpMu-72 Subset

As said above, we made the deliberate choice to train with all the specimens of the three chosen letter categories available to us to reflect real research scenarios. However, in order to analyse stylistic similarities between the manuscripts, we had to make the data coherent. Only 72 TMs preserved at least 2 good quality specimens of the 3 letter categories. They form the AlphEpMu-72 subset whose list, along with metadata, is in Appendix B and on the webpage. Metadata come from trismegistos.org (except a few additions by the present authors marked with an *). All have a date assigned by experts upon paleographical criteria and 33 could be assigned a "style" label because they belong to clearly identified types (Biblical Majuscule, Severe style, etc.). Four are written on parchment, the others on papyrus (to evaluate the impact of the background). Some are formed by a single fragment, others are composed of many pieces that can be preserved inside several frames (different images) or can even be located in different institutions (different digitization processes). All but one bear passages of Homer's *Iliad*. As already mentioned, one manuscript has been introduced as a test case: TM 59170 preserves excerpts of Apollonius Rhodius' *Argonautica*. It counts two fragments kept in two collections (Yale and Oxford) which were digitized independently one from the other. This manuscript written in Biblical Majuscule has been chosen to test the hypothesis of an expert that it was penned by the same writer as two other items included in the dataset (TM 60842 and 60764). TM 60891 was also found extremely similar to this group by one of the present authors, while two others labeled Biblical Majuscule on trismegistos.org (60457 and 60965) are also in the dataset.

5.2   General Repartition

We provide the detailed similarity scores of the AlphEpMu-72 dataset on our webpage. Since 72 line/column heatmaps are difficult to handle, another possibility is to turn to clustering computing. A dendrogram has been created from the clustering method WARD.D2 for the whole AlphEpMu-72 dataset. Figure 4 shows the obtained visualisation with average scores of the three letter categories.

It is useful to spot the most similar pairs (here clearly the two 59170 Oxy and Yale are the most closely connected, suggesting that we have not missed two other TMs that could in fact be from the same manuscript). The algorithm designed six main clusters (with 6 colors) which are not obvious to interpret in paleographical terms. Although the two 59170 are then connected with a piece suspected to be from the same writer (60764) in the green branch, the other Biblical Majuscule items (60965, 60457, 60891 and 60842) are found in the orange branch, close but not clearly connected to each other. Some other remarks can be made here and there: the brown branch groups three texts (61165, 60214 and 60492) that use an original combination of similar shapes of alpha, mu and epsilon, while the most closely connected texts in the red branch (61239, 61117 and 61240) are among the most ancient of the dataset. The general meaning of the texts distribution is however not easy to grasp, most likely because this presentation does not allow to visualise that a text can be similar to many others.

This is why we turn to another visualization type based on network graph. This view shows only a link between two nodes (TMs) if the similarity value is equal or greater than a defined threshold. We experimentally chose to comment on the network obtained with a minimum threshold value 0.78 because it showed isolated items and clear clusters.

Looking at the network graph at 0.78, coherent groups are more discernible (see Fig. 5):

- The most ancient items of the dataset ("Early Ptolemaic" marked in orange and "Late Ptolemaic" in yellow)
- Alexandrian Majuscule among the most recent (in pink)
- The Biblical Majuscule group (in dark blue)
- The Severe style (in green)
- The Formal round (in light blue)
- The group combining similar shapes of alpha, mu and epsilon (in purple) that constituted the brown branch in Fig. 4.

Besides, seven items are isolated that have indeed unusual features. Parchment items have not grouped together because of the common background but are linked to similar writing shapes.

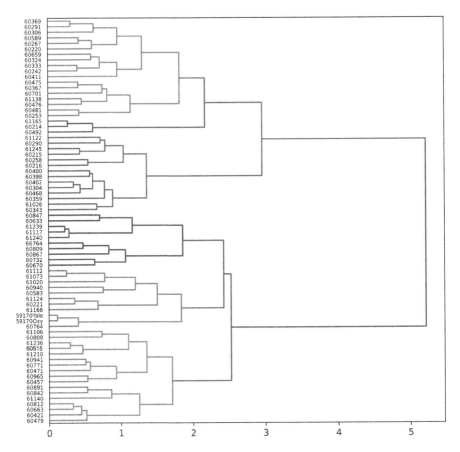

Fig. 4. Dendrogram resulting of the Ward clustering algorithm from the distance matrix).

5.3   Interpretations Within Letter Categories

Highest similarity scores are expected from fragments of the same manuscript (confirmed by the test case of TM 59170 mentioned above), then of the same writer and last of the same style. Our hypothesis was that TM 59170 would be most similar to TM 60842, 60764 and 60891 (possibly the same writer) while very close to the two other Biblical Majuscule on trismegistos.org (60457 and 60965). For the alphas, at the second and third rank were two Biblical Majuscules 60965 (0.952) and 60764 (0.926) before the score drops to 0.77 for a non Biblical item (61020), see Fig. 6. For mus, four out of the five most similar are Biblical Majuscule (60764, 60457, 60891, 60842) but surprisingly a Severe style mu comes in the way at the third rank with a high similarity score (0.947 for 60771). Similarity scores of epsilons do not show stylistic coherence, the most likely explanation is that Biblical Majuscule epsilons do not have a specific shape that would be characteristic of this style, it can be found in several other styles.

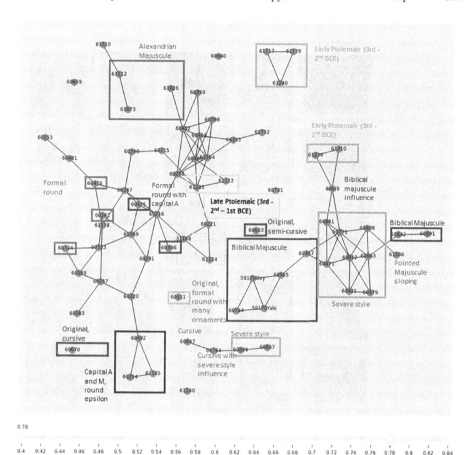

Fig. 5. Network graph at 0.78 showing stylistics similarities grouped together

Looking at the detail of similarity scores for each letter category can be relevant to spot partial similarities among items that would have otherwise been unnoticed. For instance 60306 and 61168 have an extremely high similarity score among alphas, (see Fig. 6), which indeed look like they have been penned by the same writer, while the mus are totally different (see Fig. 7). Comparing the rest of the letter categories, one can notice many other common features, like the nus that are very similar but in mirror view in one manuscript compared to the other (see the details in PalEx viewer mentioned above). These observations suggest a strong cultural link between these two manuscripts that deserves further investigations.

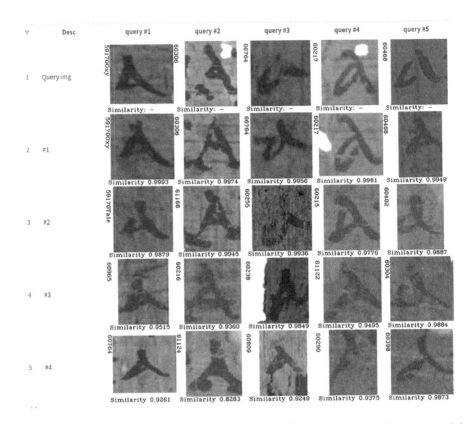

Fig. 6. The query TMs and their corresponding ordered similarities that our model has predicted.

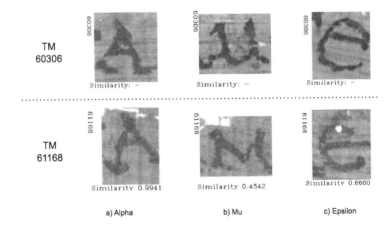

Fig. 7. TM 60306 and 61168

6   Conclusion

The goal of this paper was to introduce a new dataset called AlphEpMu and to present a method to analyse similarities among historical handwritings at the character level without introducing prior expert knowledge. A test on two fragments of the same manuscript has been successful, texts written in well identified style (Biblical Majuscule, Severe style, Alexandrian Majuscule) were grouped together while original pieces were isolated. Observations of similarity scores for each letter category allows spotting cultural influences and nuances that have to be taken into account to apprehend the complexity of the writing phenomenon, for instance the transitional phases that cause difficulties in paleography. Results can still be improved (for instance not all the Biblical Majuscule items achieved the highest scores as they would have expected) but also the influence of parameters needs to be further investigated (state of preservation, digitization process, number of cliplets per TM to have a meaningful result). More comparisons with different neural network architectures would be done in future works in order to improve parameter value setting.

Machine Learning methods like the one presented in this paper are expandable to large datasets: if all the 80,000 papyri of all nature are not suited for letter segmentation because of the cursivity of their handwriting, at least 8,000 literary papyri (mostly in book-hand or semi-cursive scripts) are already published that could potentially be organized in coherent stylistic clusters in the future. The present study shows that it is not necessary to have precise metadata (paleographical descriptions of the handwriting features) on each text: they require time and expertise without necessarily reaching consensus and would not be feasible for a large dataset. What are needed are a few solid milestones (clear canonical types) whose distributions can be studied and around which interpretations of the remaining material can be built.

Since this experiment with only 72 manuscripts and three letter categories gives promising results, we can expect that an extension to a bigger dataset and a larger list of letter categories will fine-tune the clustering. Expanding the list of items would require automatic detection and recognition of the Greek characters, a topic at the center of a competition organized as part of ICDAR 2023[5].

Acknowledgement. This work was partially supported by the Swiss National Science Foundation as part of the project no. PZ00P1-174149 "Reuniting fragments, identifying scribes and characterizing scripts: the Digital paleography of Greek and Coptic papyri (d-scribes)".

---

[5] https://lme.tf.fau.de/competitions/2023-competition-on-detection-and-recognition-of-greek-letters-on-papyri/.

Appendix A   AlphEpMu* Dataset

dataset description(2)

TM num.	A	E	M	AEM	Ciiplets
60214	2	36	6	44	72
60215	27	98	16	141	543
60216	2	8	2	12	49
60217	4		2		53
60219		4			49
60220	12	32	10	54	570
60221	8	20	7	35	271
60238	2	2			53
60242	12	24	4	40	141
60246	7	16			103
60248	4	10			43
60251	2	12			110
60253	5	12	2	19	200
60255	2	2			57
60258	8	16	4	28	97
60267	12	16	8	36	162
60276	5	8			69
60281		22			596
60287		3	3		167
60290	4	6	2	12	90
60291	4	12	6	22	141
60304	13	26	12	51	270
60306	89	244	58	391	2821
60312		12			60
60324	24	42	8	74	247
60326aFrag		8			93
60326a		47			609
60326b		36			550
60329		9			281
60333	17	32	12	61	306
60337	5		2		58
60343	14	26	2	42	157
60347		4			81
60358		15			258
60359	38	72	14	124	869
60364		8	5		55
60367	4	10	2	16	77
60368		11			318
60369	4	8	4	16	85
60396		6			208
60398	24	52	8	84	416
60400	2	10	2	14	191
60402	8	38	6	52	199
60411	34	68	22	124	422
60421	32	4	6	42	432
60457	15	32	2	49	381
60462	4				746
60468	7	22	2	31	187
60471	20	26	4	50	337
60472		4			72
60475	4	8	2	14	54
60476	24	72	12	108	552
60479	8	24	2	34	112
60481	2	10	6	18	79
60485		2			103
60487		8			171
60489		12			305
60492	5	4	2	11	66
60494		5			78
60515	6	4	2	12	91
60558		9			180
60560		24			455
60571		474			10020
60583	32	112	56	200	1266
60589	5	8	4	17	79
60600	12	22			314
60633	26	72	8	106	431
60650		25			515
60658		15			209
60659	10	20	10	40	250
60662		4			77
60683	12	4	4	20	78
60666		10			216
60670	5	20	6	31	255
60682		6			131
60684		15			310
60701	12	40	12	64	326
60732	27	72	32	131	720
60734		25			768
60740	36		16		401
60761		2			178
60764a		4			140
60764b	28	64	24	116	471
60771	6	14	4	24	93
60808	4	2	2	8	50
60809	10	8	2	20	81
60810	2	4			28
60812	3	6	4	13	92
60842	8	8	2	18	149
60845		6			101
60847	24	48	8	80	464
60855		9			177
60867	50	24	4	78	318
60885		26			404
60891	140	176	42	358	1768
60896a		5			123
60896b	6	6			111
60901	3	14			116
60910	2	4			78
60934		4			82
60935	3	3			136
60940	5	10	10	25	185
60941	14	46	12	72	287
60965	12	16	2	30	126
60968	2	8			81
60997		4			169
61020	17	18	15	50	906
61022		3			107
61026	5	2	4	11	286
61051		9			393
61073	30	90	34	154	1173
61106	7	12	2	21	361
61112	5	6	4	15	64
61117	14	35	26	75	564
61122	32	54	12	98	505
61124	6	28	6	40	349
61138	7	8	6	21	81
61140	30	34	4	68	198
61141		6			439
61165	44	66	29	139	697
61167		16			504
61168	8	8	3	19	204
61175		11			281
61210	22	60	20	102	467
61212	7	16			187
61213	3	6			99
61223	2				1062
61228	14	8			221
61236	13	23	2	38	255
61239	7	8	2	17	288
61240	23	16	8	47	344
61244	8	8			80
61245	6	17	6	29	143
61246		14	2		136
61256		5			125
62623		10			289
62892		6			206
65858		6			28
66784	13	54	6	73	286
66757		4			67
827766		7			203
827767		5			201
59170-Oxy	8	8	2	18	74
59170-Yale	20	26	6	52	107

60847* see 60515

Taken for analysis (white lines): A = 1240, E = 2457, M = 702; Number of ciiplets in 72 chosen TMs in READ = 25530
Total annotated in 144 TMs (white and grey lines): A = 1375, E = 3589, M = 732; Number of ciiplets in 144 TMs in READ = 51024

Appendix B   AlphEpMu-72 Dataset

TM num.	Edition	Mat.	Date	Prov.	Iliad sect. range	Handwriting st.
			dataset description(1)			
60214	Pintaudi (ed.), Miscellanea papyrologica (Pap. Flor. 7):281(1)	pap.	AD01	Eg.	05.486-495	none
60215	Mus. Helveticum 24 (1967):62(2); Fs.150 Berlin:373f(10); BKT 9 1	pap.	AD01	Eg., Arsin.	09.181-210	none
60216	Fs. 150 Berlin:381(17)	pap.	AD01	Eg.	012.459-471	none
60220	BKT 5.1:4 (8440) descr.	pap.	AD01	Eg., Arsin.	05.68-103	none
60221	BKT 5.1:4 (9584) descr.	pap.	AD01	Eg.	01.449-461	none
60242	ZPE 18 (1975):311-312	pap.	AD01	Eg.	05.159-168	Formal round
60253	P. Köln Gr. 1 34	pap.	AD01	Eg.	014.311-326 (gl.)	none
60258	Pap. Congr. XII (Ann Arbor 1968):20-22	pap.	AD01	Eg.	01.522-546	none
60267	ZPE 46 (1982):76-77	pap.	AD01	Eg.	06.211-221	none
60290	Wiener Studien 76 (1963):159-160	pap.	AD01	Eg.	02.137-144	none
60291	Archiv für Bibliographie 1 (1926):90(11)	pap.	AD01	Eg.	08.436-461	none
60304	FuB 8 (1967):101-102	pap.	AD01-02	Eg.	011.237-254	none
60306	BKT 5.1:3 (6869 e.a.), CdE 92 (2017):105-110	pap.	AD01	Eg.	01.44-611	Formal round
60324	PSI 14 1376	pap.	AD01-02	Eg.	07.50-66	Formal round
60333	P. Hamb. 2 158	pap.	AD01-02	Eg.	07.141-158	none
60343	BASP 12 (1975):19-20(1)	pap.	AD01	Eg.	02.89-110	none
60359	P. Hamb. 2 157	pap.	AD01-02	Eg.	03.384-461	none
60367	Wiener Studien 76 (1963):161	pap.	AD01-02	Eg.	020.86-100	none
60369	Archiv für Bibliographie 1 (1926):90(12)	pap.	AD01-02	Eg.	09.152-161	none
60396	Pintaudi (ed.), Miscellanea papyrologica (Pap. Flor. 7):282f(2)	pap.	AD02	Eg., Herm.	06.75-127	none
60400	AfP 44 (1998):11(6)	pap.	AD02	Eg.	023.400-455	none
60402	AfP 39 (1993):7-8	pap.	AD02	Eg.	03.23-36	none
60411	Lameere, Aperçus de paléographie homérique:433-447	pap.	AD01-02	Eg.	08.433-447	Formal round
60421	PSI 12 1275 Ro [1]	pap.	AD02	Eg., Oxy.	023.877-897	Severe st. sloping
60457	PSI 14 1377	pap.	AD03	Eg.	09.682-709	Bibl. Maj.
60468	P. Hamb. 2 156	pap.	AD02	Eg.	02.12-26	none
60471	P. Hamb. 3 198 Ro	pap.	AD02	Eg., Arsin.	015.399-456	Severe st. sloping*
60475	CdE 88 (2013):101-104	pap.	AD02	Eg., Oxy.	017.80-94	Formal maj.
60476	ZPE 186 (2013):18-21	pap.	AD01-02	Eg., Oxy.	022.1-57	Formal round
60479	P. Köln Gr. 1 23	pap.	AD02	Eg.	01.560-580	Severe st. sloping*
60481	P. Köln Gr. 1 38	pap.	AD02	Eg.	024.303-311	none
60482	P. Köln Gr. 5 207	pap.	AD02	Eg.	01.150-163	none
60515	Schwendner, Literary and non-literary papyri (Mich. Diss.) 3	pap.	AD02	Eg.	02.745-754	none
60583	AfP 59 (2013):17-28	pap.	BC01-01	Eg., Oxy.	014.227-283	Cap. round (*Bibl. Maj. μ)
60589	Études de papyrologie 3 (1936):49(2)	pap.	AD02	Eg.	013.297-302	none
60633	CdE 26 (1938):387	pap.	AD02-03	Eg.	012.265-292	Semi-cursive
60659	ZPE 18 (1975):309-310	pap.	AD02-03	Eg.	05.114-132	none
60663	ZPE 186 (2013):15-16	pap.	AD02-03	Eg., Oxy.	020.241-250	Severe st.
60670	P. Köln Gr. 7 300	pap.	AD02-03	Eg.	010.7-28	Cursive hand
60701	BASP 41 (2004):64-65	pap.	AD02-03	Eg., Oxy.	011.39-52	none
60732	Stud. Pal. 5 74 [A]	pap.	AD02-03	Eg., Arsin.	05.541-896	none
60764	BKT 5.1:4 (7499 e.a.) descr.	pap.	AD02-03	Eg., Arsin.	08.169-324	Bibl. Maj.
60771	P. Oxy. 4 759 descr.	pap.	AD03	Eg., Oxy.	05.662-682	Severe st.
60808	ZPE 186 (2013):14-15	pap.	AD03	Eg., Oxy.	019.417-421	Severe st.*
60809	ZPE 186 (2013):16-17	pap.	AD03	Eg., Oxy.	021.372-382	Severe st.
60812	AfP 59 (2013):12-14	pap.	AD03	Eg., Oxy.	08.109-123	Severe st. sloping
60842	P. Ryl. Gr. 3 542; ZPE 46 (1982):123f(1)	pap.	AD03	Eg.	05.473-495	Bibl. Maj.
60847	ZPE 46 (1982):71-72(8)	pap.	AD03	Eg.	03.1-25	Cursive*
60867	P. Oxy. 3 540 descr.	pap.	AD03	Eg., Oxy.	02.672-684	Severe st.
60891	P. Oxy. 52 3663	pap.	AD02-03	Eg., Oxy.	018.33-408	Bibl. Maj.
60940	PSI 14 1375	parch.	AD03-04	Eg.	04.476-517	none
60941	ZPE 18 (1975):312-313	pap.	AD03	Eg.	02.57-71	Severe st. sloping
60965	BKT 5.1:5 (10574) descr.	pap.	AD04	Eg.	01.406-419	Bibl. Maj.
61020	P. Oxy. 56 3826	parch.	AD04-05	Eg., Oxy.	04.517-522, 05.1-75	Bibl. + pointed maj.
61026	Archiv für Bibliographie 1 (1926):90f(14)	parch.	AD05-06	Eg., Arsin.	011.449-485	Alexandrian maj.
61106	PSI 13 1298; Bastianini/Casanova, I papiri Homerici (SeTP N.S. 14):289-290, Laur. 4 130	pap.	AD06	Eg.	013.232-831	Alexandrian maj. (irreg.)
61106	ZPE 42 (1981):39-41	pap.	AD06-07	Eg.	013. 425-469	Pointed maj. sloping
61112	SCO 22 (1973):34-36	parch.	AD06-07	Eg.	01.476-518	Alexandrian maj.
61117	BIFAO 65 (1967):63(1)	pap.	BC03-02	Eg.	017.566-578b	none
61122	Fs. 150 Berlin:382-383(20)	pap.	BC01	Eg.	015.5-31	none
61124	BKT 5.1:18-20(I 3)	pap.	BC01	Eg.	018.585-608d	none
61138	P. Oxy. 4 761 descr.	pap.	BC01	Eg., Oxy.	06.147-149	none
61140	ZPE 14 (1974):31-32	pap.	BC01	Eg.	015.625-657	epsilon theta st.
61165	AfP 44 (1998):8-10(4)	pap.	BC01-01	Eg., Herm.	017.51-98	none
61168	BKT 5.1:4 (9949) descr.	pap.	BC01-01	Eg.	023.718-732	none
61210	ZPE 56 (1984):11-15	pap.	BC02	Eg.	010.421-460	none
61236	P. Hibeh 1 22	pap.	BC03	Eg.	021.302-611*	none
61239	P. Sorb. 1 4	pap.	BC03	Eg.	012.228-265	none
61240	BIFAO 65 (1967):67(2)	pap.	BC03	Eg.	06.280-292	none
61245	Fs. 150 Berlin:371(8)	pap.	BC03-02	Eg.	07.183-195	none
66764	Biblos 49 (2000):131-134 (2000)	pap.	AD03	Eg., Oxy.	04.83-4.104	Cursive, severe st. inf.*
59170Oxy	P. Oxy. 75 5027	pap.	AD03	Eg., Oxy.	Ap. Rhodius, Argon.: 02.101-150	Bibl. Maj.

Notes: 59170Yale see Oxy; 60847* see 60515

References

1. Bromley, J., Guyon, I., LeCun, Y., Säckinger, E., Shah, R.: Signature verification using a "Siamese" time delay neural network. In: Advances in Neural Information Processing Systems, vol. 6 (1993)
2. Brookes, S., Stokes, P.A., Watson, M., De Matos, D.M.: The digipal project for European scripts and decorations. Essays Stud. **68**, 25–59 (2015)
3. Cavallo, G.: La scrittura greca e latina dei papiri: una introduzione. Studia erudita, F. Serra (2008)
4. Chen, X., He, K.: Exploring simple Siamese representation learning (2020)
5. Christlein, V., Gropp, M., Fiel, S., Maier, A.: Unsupervised feature learning for writer identification and writer retrieval. In: 2017 14th IAPR International Conference on Document Analysis and Recognition (ICDAR), vol. 1, pp. 991–997. IEEE (2017)
6. Christlein, V., Marthot-Santaniello, I., Mayr, M., Nicolaou, A., Seuret, M.: Writer retrieval and writer identification in Greek papyri. In: Carmona-Duarte, C., Diaz, M., Ferrer, M.A., Morales, A. (eds.) IGS 2022. LNCS, vol. 13424, pp. 76–89. Springer, Cham (2022). https://doi.org/10.1007/978-3-031-19745-1_6
7. Corso, L.D.: Lo 'stile severo' nei p.oxy.: una lista. Aegyptus **86**, 81–106 (2006)
8. Deng, J., Dong, W., Socher, R., Li, L.J., Li, K., Fei-Fei, L.: ImageNet: a large-scale hierarchical image database. In: 2009 IEEE Conference on Computer Vision and Pattern Recognition, pp. 248–255. IEEE (2009)
9. Fiel, S., Sablatnig, R.: Writer identification and retrieval using a convolutional neural network. In: Azzopardi, G., Petkov, N. (eds.) CAIP 2015. LNCS, vol. 9257, pp. 26–37. Springer, Cham (2015). https://doi.org/10.1007/978-3-319-23117-4_3
10. Goyal, P., et al.: Accurate, large minibatch SGD: training ImageNet in 1 hour (2018)
11. He, K., Zhang, X., Ren, S., Sun, J.: Deep residual learning for image recognition (2015)
12. Kingma, D.P., Ba, J.: Adam: a method for stochastic optimization (2017)
13. Lai, S., Jin, L.: Offline writer identification based on the path signature feature. In: 2019 International Conference on Document Analysis and Recognition (ICDAR), pp. 1137–1142. IEEE (2019)
14. Mohammed, H., Marthot-Santaniello, I., Märgner, V.: GRK-papyri: a dataset of Greek handwriting on papyri for the task of writer identification. In: 2019 International Conference on Document Analysis and Recognition (ICDAR), pp. 726–731. IEEE (2019)
15. Nguyen, H.T., Nguyen, C.T., Ino, T., Indurkhya, B., Nakagawa, M.: Text-independent writer identification using convolutional neural network. Pattern Recogn. Lett. **121**, 104–112 (2019)
16. Orsini, P.: Studies on Greek and Coptic Majuscule Scripts and Books. De Gruyter, Berlin, Boston (2019). https://doi.org/10.1515/9783110575446
17. Orsini, P., Clarysse, W.: Early new testament manuscripts and their dates: a critique of theological palaeography. Ephemer. Theol. Lovan. **88**(4), 443–474 (2012)
18. Paixão, T.M., et al.: Self-supervised deep reconstruction of mixed strip-shredded text documents. Pattern Recogn. **107**, 107535 (2020)
19. Paszke, A., et al.: PyTorch: an imperative style, high-performance deep learning library. In: Advances in Neural Information Processing Systems, vol. 32, pp. 8024–8035. Curran Associates, Inc. (2019). http://papers.neurips.cc/paper/9015-pytorch-an-imperative-style-high-performance-deep-learning-library.pdf

20. Pirrone, A., Beurton-Aimar, M., Journet, N.: Self-supervised deep metric learning for ancient papyrus fragments retrieval. Int. J. Doc. Anal. Recognit. (IJDAR) **24**(3), 219–234 (2021). https://doi.org/10.1007/s10032-021-00369-1

21. Popović, M., Dhali, M.A., Schomaker, L.: Artificial intelligence based writer identification generates new evidence for the unknown scribes of the dead sea scrolls exemplified by the great Isaiah scroll (1qisaa). PLoS ONE **16**(4), e0249769 (2021)

22. Pratikakis, I., Zagoris, K., Karagiannis, X., Tsochatzidis, L., Mondal, T., Marthot-Santaniello, I.: ICDAR 2019 competition on document image binarization (DIBCO 2019). In: 2019 International Conference on Document Analysis and Recognition (ICDAR), pp. 1547–1556 (2019). https://doi.org/10.1109/ICDAR.2019.00249

23. Richter, S.G., Schröder, K.D.: Digitale Werkzeuge zur Systematisierung koptischer Handschriften, pp. 439–448. De Gruyter, Berlin, Boston (2020)., https://doi.org/10.1515/9783110591682-030

24. Romain, K., Abdel, B.: Semi-supervised learning through adversary networks for baseline detection. In: 2019 International Conference on Document Analysis and Recognition Workshops (ICDARW), vol. 5, pp. 128–133. IEEE (2019)

25. Serbaeva, O., White, S.: READ for solving manuscript riddles: a preliminary study of the manuscripts of the 3rd ṣaṭka of the *Jayadrathayāmala*. In: Barney Smith, E.H., Pal, U. (eds.) ICDAR 2021, Part II. LNCS, vol. 12917, pp. 339–348. Springer, Cham (2021). https://doi.org/10.1007/978-3-030-86159-9_24

26. Stutzmann, D.: Clustering of medieval scripts through computer image analysis: towards an evaluation protocol. Digit. Medievalist **10** (2016)

27. Turner, E.G., Parsons, P.J.: Greek manuscripts of the ancient world. Bulletin supplement, University of London. Institute of classical studies, London, 2, edn. revised and enlarged edn. (1987)

28. Xiong, Y.J., Wen, Y., Wang, P.S., Lu, Y.: Text-independent writer identification using sift descriptor and contour-directional feature. In: 2015 13th International Conference on Document Analysis and Recognition (ICDAR), pp. 91–95. IEEE (2015)

29. Zhang, X., Wang, T., Wang, J., Jin, L., Luo, C., Xue, Y.: Chaco: character contrastive learning for handwritten text recognition. In: Porwal, U., Fornés, A., Shafait, F. (eds.) ICFHR 2022. LNCS, vol. 13639, pp. 345–359. Springer, Cham (2022). https://doi.org/10.1007/978-3-031-21648-0_24

Author Index

Printed in the United States
by Baker & Taylor Publisher Services